FOUNDATIONS OF FUZZY CONTROL

FOUNDATIONS OF FUZZY CONTROL

A PRACTICAL APPROACH

Second Edition

Jan Jantzen
University of the Aegean at Chios, Greece

This edition published in 2013

First Edition published in 2007

Registered office
John Wiley & Sons Ltd, The Atrium, Southern Gate, Chichester, West Sussex, PO19 8SQ, United Kingdom

For details of our global editorial offices, for customer services and for information about how to apply for permission to reuse the copyright material in this book please see our website at www.wiley.com.

Library of Congress Cataloging-in-Publication Data

Jantzen, Jan.
Foundations of fuzzy control : a practical approach / Jan Jantzen.
1 online resource.
Includes bibliographical references and index.
Description based on print version record and CIP data provided by publisher; resource not viewed.
ISBN 978-1-118-53557-8 (MobiPocket) – ISBN 978-1-118-53558-5 (Adobe PDF) –
ISBN 978-1-118-53559-2 (ePub) – ISBN 978-1-118-50622-6 (hardback) 1. Automatic control.
2. Fuzzy systems. 3. Fuzzy automata. I. Title.
TJ213
629.8′312–dc23

2013023628

A catalogue record for this book is available from the British Library.

ISBN: 978-1-118-50622-6

Typeset in 10/12pt Times by Aptara Inc., New Delhi, India

1 2013

In memory of Ebrahim (Abe) Mamdani (1 Jun 1942–22 Jan 2010) and Lauritz Peter Holmblad (23 Aug 1944–30 Mar 2005)

Figure 1 EH Mamdani (1942–2010)

Contents

The * in the heading denotes that the section can be skipped on the first reading as it contains background material catered for advanced readers of this book.

Foreword

Since the objective of Foundations of Fuzzy Control is to explain why fuzzy controllers behave the way they do, I would like to contribute a historical perspective.

Before the 1960s, a cement kiln operator controlled a cement kiln by looking into its hot end, the burning zone, and watching the smoke leaving the chimney. The operator used a blue glass to protect his eyes. He controlled the fuel/air ratio in order to achieve steady operation of the kiln.

Central control was introduced in the cement industry in the 1960s. PID controllers were installed, mainly for uniform feeding of the raw materials and the fuel. Computers for process supervision and control were introduced in the cement industry in the late 1960s.

During experimental work in the 1970s, the fuel control strategy was programmed as a two-dimensional decision table with an error signal and the change in error as inputs.

The first time we heard about fuzzy logic was at the fourth IFAC/IFIP International Conference on Digital Computer Applications to Process Control, held in Zürich, Switzerland, in 1974. As a postscript to a paper on learning controllers, Seto Assilian and Abe Mamdani proposed fuzzy logic as an alternative approach to human-like controllers.

Experimental work was carried out at the Technical University of Denmark. The theoretical understanding and inspiration in relation to process control was gained mainly from papers written by Lotfi Zadeh and Abe Mamdani, and control experiments were performed in laboratory-scale processes such as, for example, a small heat exchanger. The rule based approach that underlies the decision tables was also inspired by the instructions that we found in a textbook for cement kiln operators, which contained 27 basic rules for manual operation of a cement kiln.

The first experiments using a real cement kiln were carried out at the beginning of 1978 at an FL Smidth cement plant in Denmark. At this stage of the development work, the attitude of the management was sceptical, partly because of the strange name, 'fuzzy'. Other names were suggested, but eventually, with an increasing understanding by the management of the concept, it was decided to stay with the word fuzzy, a decision that has never been regretted since.

In 1980, FL Smidth launched the first commercial computer system for automatic kiln control based on fuzzy logic. To date, hundreds of kilns, mills and other processes have been equipped with high-level fuzzy control by FL Smidth and other suppliers of similar systems.

Jens-Jørgen Østergaard
FL-Soft, Copenhagen

Foreword

Preface to the Second Edition

This second edition of Foundations of Fuzzy Control includes new chapters on gain scheduling, fuzzy modelling and demonstration examples. Fuzzy gain scheduling is a straightforward extension of the usual PID type fuzzy controllers in the sense that fuzzy rules can interpolate naturally between PID controllers. Broadly speaking, the concept of local fuzzy models is dual to fuzzy gain scheduling. The demonstration chapter includes five larger examples that can be used as teaching modules. Furthermore, the chapter on stability has been extended to include performance. The intent has been to reach farther than mere analysis, that is, to devise a design method that starts from specifications of performance. The book adopts a practical approach, which is reflected in the new subtitle, *A Practical Approach*.

The guiding principle has been to try to reach the bottom of the matter by means of geometry. Thus, the PID controller can be seen as an inner product. Together with viewpoints from adaptive control and the self-organizing controller, this has led to a set of tuning recommendations, where the starting point is a performance specification, namely, the desired settling time (Chapter 7). The tuning recommendations are applied to an unstable chemical reactor tank and for the control of the idle speed in a car engine, in order to test and demonstrate how it works (Chapter 10). Hopefully, the reader will find the second edition of the book even more fundamental and coherent than the first edition owing to the geometric approach.

My students requested more examples and illustrations, and this second edition tries to fulfil that wish. A simulator (Autopilot) was developed to illustrate concepts in nonlinear control, such as equilibria, and the tool can be used as a stand-alone teaching tool. The book contents have been reorganized, and each chapter consists now of two parts, clearly separated by a summary: the first part is intended for an introductory course, and the part after the summary is for an advanced course. The advanced part is also a research guideline for students who wish to write their thesis within fuzzy control.

I teach an introductory course on the Internet using one of the demonstration examples. Access to the course is through the companion website www.wiley.com/go/jantzen, which is devoted to this book. The website also contains downloadable material, such as the MATLAB® programs that produced the figures, lecture slides and error corrections.

Finally, I wish to acknowledge the inspiration and help I have received from Abe Mamdani, especially in connection with the idle speed project (Chapter 10). He died, much too early, in 2010, and he is sadly missed. This second edition is dedicated to him, as well as to Peter Holmblad – two giants in the history of fuzzy control.

Jan Jantzen
University of the Aegean at Chios, Greece

Preface to the First Edition

In summary, this textbook aims to explain the behaviour of fuzzy logic controllers. Under certain conditions a fuzzy controller is equivalent to a proportional-integral-derivative (PID) controller. The equivalence enables the use of analysis methods from linear and nonlinear control theory. In the linear domain, PID tuning methods and stability criteria can be transferred to linear fuzzy controllers. The Nyquist plot shows the robustness of different settings of the fuzzy gain parameters. As a result, a fuzzy controller can be guaranteed to perform as well as any PID controller. In the nonlinear domain, the stability of four standard control surfaces can be analysed by means of describing functions and Nyquist plots. The self-organizing controller (SOC) is shown to be a model reference adaptive controller. There is the possibility that a nonlinear fuzzy PID controller performs better than a linear PID controller, but there is no guarantee. Even though a fuzzy controller is nonlinear in general, and commonly built in a trial and error fashion, we can conclude that control theory does provide tools for explaining the behaviour of fuzzy control systems. Further studies are required, however, to find a design method such that a fuzzy control system exhibits a particular behaviour in accordance with a set of performance specifications.

Fuzzy control is an attempt to make computers understand natural language and behave like a human operator. The first laboratory application (mid-1970s) was a two-input-two-output steam engine controller by Ebrahim (Abe) Mamdani and Seto Assilian, UK, and the first industrial application was a controller for a cement kiln by Holmblad and Østergaard, FL Smidth, Denmark. Today there is a tendency to combine the technology with other techniques. Fuzzy control together with artificial neural networks provide both the natural language interface from fuzzy logic and the learning capabilities of neural networks. Lately hybrid systems, including machine learning and artificial intelligence methods, have increased the potential for intelligent systems.

As a follow-up to the pioneering work by Holmblad and Østergaard, which started at the Technical University of Denmark in the 1970s, I have taught fuzzy control over the Internet to students in more than 20 different countries since 1996. The course is primarily for graduate students, but senior undergraduates and PhD students also take the course. The material, a collection of downloadable lecture notes at 10–30 pages each, formed the basis for this textbook.

A fuzzy controller is in general nonlinear, therefore the design approach is commonly trial and error. The objective of this book is to explain the behaviour of fuzzy logic controllers, in order to reduce the amount of trial and error at the design phase.

Much material has been developed by applied mathematicians, especially with regard to stability analysis. Sophisticated mathematics is often required which unfortunately makes the material inaccessible to most of the students on the Internet course. On the other hand, application-oriented textbooks exist, easily accessible, and with a wide coverage of the area. The design approach is nevertheless still trial and error. The present book is positioned between mathematics and heuristics; it is a blend of control theory and trial and error methods. The key features of the book are summarized in the following four items.

- *Fundamental.* The chapter on fuzzy reasoning presents not only fuzzy logic, but also classical set theory, two-valued logic and two-valued rules of inference. The chapters concerning nonlinear fuzzy control rely on phase plane analysis, describing functions and model reference adaptive control. Thus, the book presents the parts of control theory that are the most likely candidates for a theoretical foundation for fuzzy control, it links fuzzy control concepts back to the established control theory and it presents new views of fuzzy control as a result.
- *Coherent.* The analogy with PID control is the starting point for the analytical treatment of fuzzy control, and it pervades the whole book. Fuzzy controllers can be designed, equivalent to a P controller, a PD controller, a PID controller or a PI controller. The PD control table is equivalent to a phase plane, and the stability of the nonlinear fuzzy controllers can be compared mutually, with their linear approximation acting as a reference. The self-organizing controller is an adaptive PD controller or PI controller. In fact, the title of the book could also have been *Fuzzy PID Control.*
- *Companion web site.*[1] Many figures in the book are programmed in MATLAB® (trademark of The MathWorks, Inc.), and the programs are available on the companion web site. For each such figure, the name of the program that produced the figure is appended in parentheses to the caption of the figure. They can be recognized by the syntax *.m, where the asterisk stands for the name of the program. The list of figures provides a key and an overview of the programs.
- *Companion Internet course.* The course concerns the control of an inverted pendulum problem or, more specifically, rule based control by means of fuzzy logic. The inverted pendulum is rich in content, and is therefore a good didactic vehicle for use in courses around the world. In this course, students design and tune a controller that balances a ball on top of a moving cart. The course is based on a simulator, which runs in the MATLAB® environment, and the case is used throughout the whole course. The course objectives are: to teach the basics of fuzzy control, to show how fuzzy logic is applied and to teach fuzzy controller design. The core means of communication is email, and the didactic method is email tutoring. An introductory course in automatic control is a prerequisite.

The introductory chapter of the book shows the design approach by means of an example. The book then presents set theory and logic as a basis for fuzzy logic and fuzzy reasoning, especially the so-called generalized modus ponens. A block diagram of controller components and a list of design choices lead to the conditions for obtaining a linear fuzzy controller, the prerequisite for the fuzzy PID controller.

[1] www.wiley.com/go/jantzen

The following step is into the nonlinear domain, where everything gets more difficult, but also more interesting. The methods of phase plane analysis, model reference adaptive control and describing functions provide a foundation for the design and fine-tuning of a nonlinear fuzzy PID controller.

The methods are demonstrated in a simulation of the inverted pendulum problem, the case study in the above-mentioned course on the Internet. Finally, a short chapter presents ideas for supervisory control based on experience in the process industry.

The book aims at an audience of senior undergraduates, first-year graduate students and practising control engineers. The book and the course assume that the student has an elementary background in linear differential equations and control theory, corresponding to an introductory course in automatic control. Chapters 1, 2, 3 and 9 can be read with few prerequisites, however. Chapter 4 requires knowledge of PID control and Laplace transforms and Chapters 5, 6 and 7 require more and more background knowledge. Even the simulation study in chapter 8 requires some knowledge of state-space modelling to be fully appreciated. Mathematical shortcuts have been taken to preserve simplicity and avoid formalism.

Sections marked by an asterisk (*) may be skipped on a first reading; they are either very mathematical or very practically oriented, and thus off the main track of the book.

It is of course impossible to cover in one volume the entire spectrum of topic areas. I have drawn the line between fuzzy control and neuro-fuzzy control. The latter encompasses topics such as neural networks, learning and model identification that could be included in a future edition.

Acknowledgements. I am pleased to acknowledge the many helpful suggestions I received from the late Lauritz Peter Holmblad, who acted as external supervisor on Masters projects at the Technical University of Denmark, and Jens-Jørgen Østergaard. They have contributed process knowledge, sound engineering solutions and a historical continuity. Thanks to Peer Martin Larsen, I inherited all the reports from the early days of fuzzy control at the university. I also had the opportunity to browse the archives of Abe Mamdani, then at Queen Mary College, London. I am also pleased to acknowledge the many helpful suggestions from Derek Atherton and Frank Evans, both in the UK, concerning nonlinear control, and in particular state-space analysis and describing functions. Last but not least, former and present students at the university and on the Internet have contributed collectively with ideas and suggestions.

Jan Jantzen
University of the Aegean at Chios

1

Introduction

Fuzzy control uses sentences, in the form of rules, to control a process. The controller can take many inputs, and the advantage of fuzzy control is the ability to include expert knowledge. The interface to the controller is more or less natural language, and that is what distinguishes fuzzy control from other control methods. It is generally a nonlinear controller. There are, however, very few design procedures in the nonlinear domain compared to the linear domain. This book proposes to stay as long as possible in the linear domain, on the solid foundations of linear control theory, before moving into the nonlinear domain with the design. The design method consists accordingly of four steps: design a PID controller, replace it with a linear fuzzy controller, make it nonlinear, and fine-tune the resulting controller. A nonlinear process may have several equilibrium points, and the local behaviour can be different from the behaviour far from an equilibrium, which makes it difficult to control. In order to demonstrate various aspects of nonlinear control, the book uses a simulator of a train car on a hilly track.

Fuzzy controllers appear in consumer products such as washing machines, video cameras, and cars. Industrial applications include cement kilns, underground trains, and robots. A *fuzzy controller* is an *automatic controller*, that is, a self-acting or self-regulating mechanism that controls an object in accordance with a desired behaviour. The object can be, for instance, a robot set to follow a certain path. A fuzzy controller acts or regulates by means of rules in a more or less natural language, based on the distinguishing feature: fuzzy logic. The rules are invented by plant operators or design engineers, and fuzzy control is thus a branch of artificial intelligence.

1.1 What Is Fuzzy Control?

Conventionally, computer programs make rigid *yes* or *no* decisions by means of decision rules based on two-valued logic: true/false, yes/no, or one/zero. An example is an air conditioner with a thermostatic controller that recognizes just two states: above the desired temperature or below the desired temperature. *Fuzzy logic*, on the other hand, allows intermediate truth-values between true and false.

Foundations of Fuzzy Control: A Practical Approach, Second Edition. Jan Jantzen.

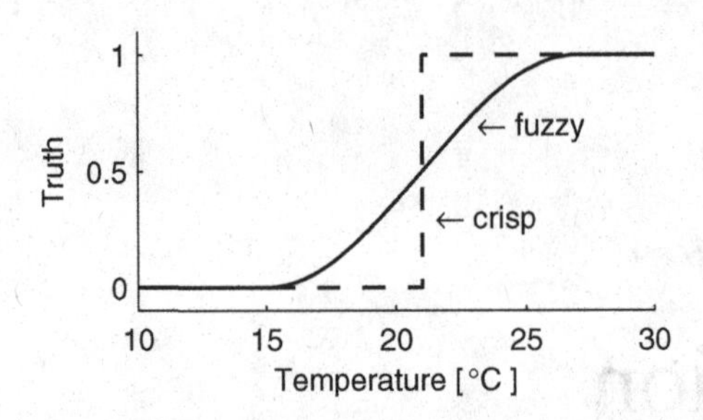

Figure 1.1 A warm room. The crisp air conditioner considers any temperature above 21°C warm. The fuzzy air conditioner considers gradually warmer temperatures. (figwarm.m)

A fuzzy air conditioner may thus recognize 'warm' and 'cold' room temperatures. The rules behind are less precise, for instance:

- *Rule*. If the room temperature is warm and slightly increasing, then increase the cooling.

Many classes or *sets* have *fuzzy* rather than sharp boundaries, and this is the mathematical basis of fuzzy logic: the set of 'warm' temperature measurements is one example of a fuzzy set.

The core of a fuzzy controller is a collection of *linguistic* (*verbal*) rules of the *if–then* form. Several variables may appear in each rule, both on the *if* side and on the *then* side. The rules can bring the reasoning used by computers closer to that of human beings.

In the example of the fuzzy air conditioner, the controller works on the basis of a temperature measurement. The room temperature is just a number, and more information is necessary to decide whether the room is warm. Therefore, the designer must incorporate a human being's perception of warm room temperatures. A straightforward approach is to evaluate beforehand all possible temperature measurements. For example, on a scale from 0 to 1, truly *warm* corresponds to 1 and definitely *not warm* corresponds to 0,

Grades of *warm*	0.0	0.0	0.4	0.9	1.0
Temperature (°C)	10	15	20	25	30

This example uses *discrete* temperature measurements, whereas Figure 1.1 shows the same idea graphically, in the form of a *continuous* mapping of temperature measurements to truth-values. The mapping is arbitrary, that is, based on preference, not mathematical reason.

1.2 Why Fuzzy Control?

If PID control (proportional-integral-derivative control) is inadequate – for example, in the case of higher-order processes, systems with a long deadtime, or systems with oscillatory modes (Åström and Hägglund 2006) – fuzzy control is an option. But first, let us consider why one would not use a fuzzy controller:

- The PID controller is well understood, easy to implement – both in its digital and analogue forms – and it is widely used. By contrast, the fuzzy controller requires some knowledge of fuzzy logic. It also involves building arbitrary membership functions.

- The fuzzy controller is generally nonlinear. It does not have a simple equation like the PID, and it is more difficult to analyse mathematically; approximations are required, and it follows that stability is more difficult to guarantee.
- The fuzzy controller has more tuning parameters than the PID controller. Furthermore, it is difficult to trace the data flow during execution, which makes error correction more difficult.

On the other hand, fuzzy controllers are used in industry with success. There are several possible reasons:

- Since the control strategy consists of *if–then* rules, it is easy for a process operator to read. The rules can be built from a vocabulary containing everyday words such as 'high', 'low', and 'increasing'. Process operators can embed their experience directly.
- The fuzzy controller can accommodate many inputs and many outputs. Variables can be combined in an *if–then* rule with the connectives *and* and *or*. Rules are executed in parallel, implying a recommended action from each. The recommendations may be in conflict, but the controller resolves conflicts.

Fuzzy logic enables non-specialists to design control systems, and this may be the main reason for its success.

1.3 Controller Design

Established design methods such as pole placement, optimal control, and frequency response shaping only apply to linear systems, whereas fuzzy control is generally nonlinear. Since our knowledge of the behaviour of nonlinear systems is limited, compared with the situation in the linear domain, this book is based on a design procedure founded on linear control:

1. Design a PID controller.
2. Replace it with a linear fuzzy controller.
3. Make it nonlinear.
4. Fine-tune it.

The idea is to exploit the design methods within PID control and carry them forward to fuzzy control. The design procedure is feasible because it is possible to build a linear fuzzy controller that functions exactly as any PID controller does. The following example introduces the design procedure.

1.4 Introductory Example: Stopping a Car

Assume that we are to design a controller that automatically stops a car in front of a red stop light, as a part of future safety equipment. Figure 1.2 illustrates the situation, and it defines the symbols for the brake force (F), the mass of the car (m), and its position (y). Assume also that we can only apply the brakes, not the accelerator pedal, in order to keep the example simple. Even though the example is simple, it is representative; think of parking a robot in a charging dock, parking a ferry at the quay, or stopping a driver-less metro train at a station.

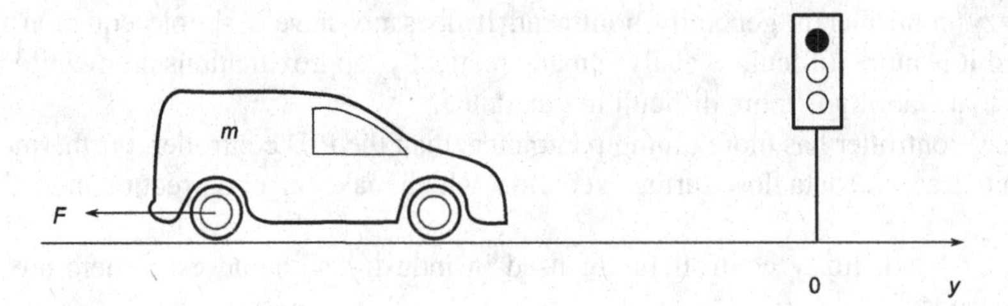

Figure 1.2 Stopping a car. The position y is positive towards the right, with zero at the stop light. The brakes act with a negative force F on the mass m.

Figure 1.3 shows a simulation model in Simulink (trademark of The MathWorks, Inc.). The block diagram includes a limiter block on the brake force, and the model is therefore nonlinear.

Step 1: Design a PID controller

Our first attempt is to try a proportional (P) controller,

$$F = K_p e \tag{1.1}$$

where K_p is the *proportional gain*, which can be adjusted to achieve the best response. The error $e \geq 0$ is the position error measured from the reference point *Ref* to the current position $y \leq 0$, that is,

$$e = Ref - y \tag{1.2}$$

Since y is negative and $Ref = 0$, then e is positive. But K_p is also positive, and the P controller in Equation (1.1) would demand a positive force F – in other words, acceleration by means of the accelerator pedal. The problem definition above ruled out the accelerator pedal, however, and we can conclude that a proportional controller is inadequate.

Our second attempt is to apply a proportional-derivative (PD) controller, since it includes a prediction. The controller is

$$F = K_p(e + T_d\dot{e}) \tag{1.3}$$

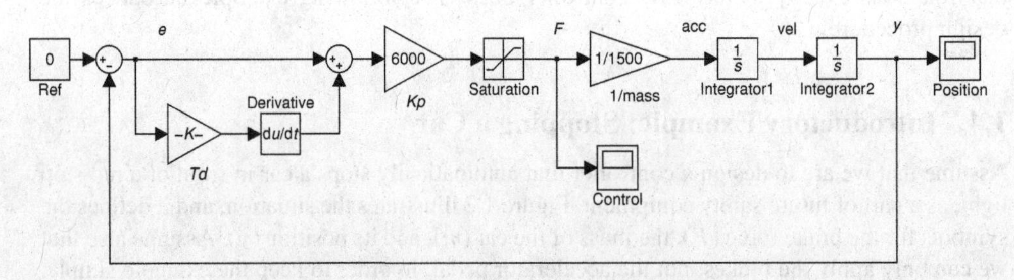

Figure 1.3 Simulink block diagram. A PD controller brakes the car from initial conditions $y(0) = -15$ and $\dot{y}(0) = 10$. (figcarpd.mdl)

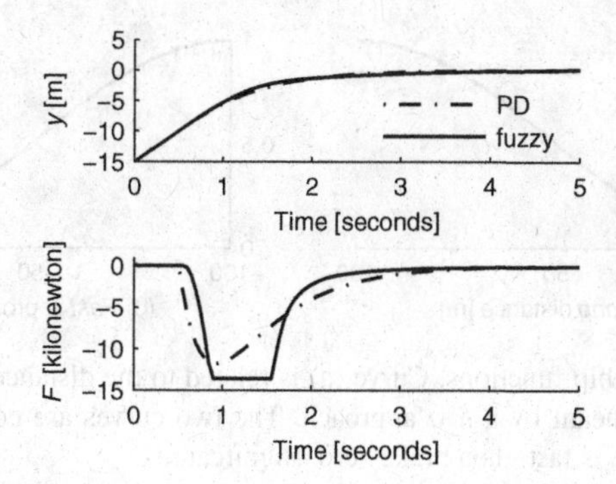

Figure 1.4 Stopping a car. Comparison between a PD controller and a fuzzy controller. (figstopcar.m)

where T_d is the *derivative gain*, which can be adjusted. Now the control signal is proportional to the term $e + T_d\dot{e}$ which is the predicted error T_d seconds ahead of the current error e. Compared to the P controller, the PD controller calls for extra brake force when the velocity is high. Figure 1.4 shows the response and the brake force with

$$K_p = 6000$$
$$T_d = 1$$

During the first 0.5 s, the control signal is zero. Thereafter the derivative action takes over and starts to brake the car. In other words, the controller waits 0.5 s until it kicks in, it quickly increases the braking force, and after about 1 s it relaxes the brake gently. It takes about 5 s to stop the car.

We *tuned* the gains K_p and T_d in order to achieve a good closed loop performance. Hand tuning is possible, but it generally requires patience and a good sense of how the system responds. It is easier to use rules, for example the *Ziegler–Nichols tuning rules*. Although the rules often result in less than optimal settings, they are a good starting point for a manual fine tuning.

Step 2: Replace it with a linear fuzzy controller

A fuzzy controller consists of *if–then* rules describing the action to be taken in various situations. We will consider the situations where the distance to the stop light is long or short, and situations where the car is approaching fast or slowly. The linguistic terms must be specified precisely for a computer to execute the rules.

The following chapters will show how to design a *linear* fuzzy controller, with a performance that is exactly the same as the PD controller in the previous step. It is a design aid, because the PD controller, with its tuning, settles many design choices for the fuzzy controller. One requirement is that the membership functions should be linear.

At the end of this step, we have a fuzzy controller, with a response (not shown) exactly as the PD response in Figure 1.4.

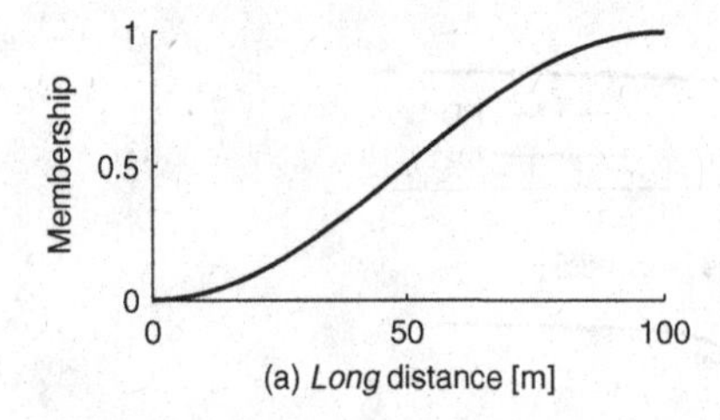

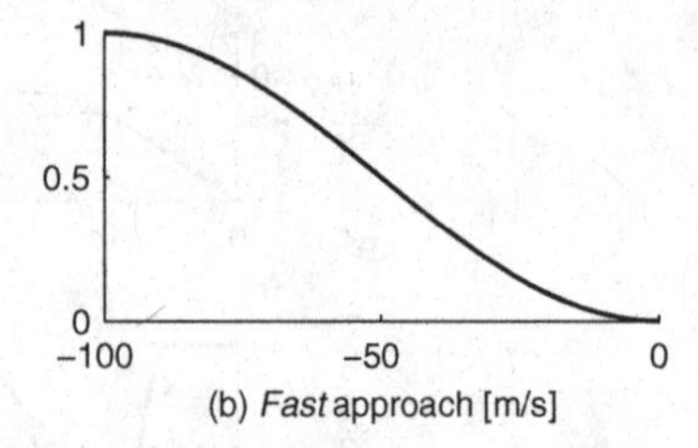

Figure 1.5 Fuzzy membership functions. Curve (a) is related to the distance from the stop light, and curve (b) specifies what is meant by a *fast* approach. The two curves are components of the rule: If distance is long and approach is fast, then brake zero. (figmfcar.m)

Step 3: Make it nonlinear

A complete rule base of all possible input combinations contains four rules:

If distance is long and approach is fast, then brake zero (1.4)

If distance is long and approach is slow, then brake zero (1.5)

If distance is short and approach is fast, then brake hard (1.6)

If distance is short and approach is slow, then brake zero (1.7)

The linguistic terms must be specified precisely for a computer to execute the rules. Figure 1.5 shows how to implement 'long', as in 'distance is long'. It is a fuzzy *membership function*, shaped like the letter *s*. The horizontal axis is the *universe*, which is the interval [0, 100]% of the full range of 15 m. The vertical axis is the *membership grade*, that is, how compatible a distance measurement is with our perception of the term 'long'. For instance, a distance of 15 m (100%) has membership 1, because the distance is definitely long, while half that distance is long to a degree of just 0.5. Note that the horizontal axis corresponds to the previously defined error e, scaled onto a standard range relative to the maximum distance.

The term 'fast', as in 'approach is fast', is another membership function. The horizontal axis is again percentages of full range (10 m/s), but the numbers are negative to emphasize that the distance is decreasing rather than increasing. The horizontal axis corresponds to the previously defined time derivative $\dot{e}$ scaled onto the universe. The -100% corresponds to the maximum speed of 10 m/s. Similarly, the membership function for 'short' is just a mirror image of the membership function 'long', and the membership function 'slow' is just a mirror image of 'fast'.

Turning to the *then*-side of the rules, the term 'zero' means to apply the brake force $F = 0$. The term 'hard' is the full brake force of -100%.

The nonlinear domain is poorly understood in general, and it usually calls for a trial and error design procedure. Nevertheless, the following chapters provide methods such that at least some *analysis* is possible.

Step 4: Fine-tune it

Figure 1.4 shows the response with the nonlinear controller, together with the initial PD response, after adjusting one tuning factor (input gain on the error, GE). The response is close to the initial PD response, but a little faster. The lower plot with the control signals shows the difference: the fuzzy controller waits longer before it kicks in, then it uses all the available brake force, and thereafter it releases the brake quicker than the PD controller.

The behaviour is not necessarily better than PD control. But since the fuzzy controller in step 2 is guaranteed to perform the same way, it is safe to say that the fuzzy controller is at least as good. Whether it performs better after steps 3 and 4 is an open question, but at least the fuzzy controller provides extra options to shape the control signal. This could be important if passenger comfort has a high priority.

Example 1.1 *Tuning by means of process knowledge*

Is it possible to use a mathematical model to find optimal settings for the PD controller?

► *Solution*

Disregarding engine dynamics, skidding, slip, and friction – other than the frictional forces in the brake pads – the force F causes an acceleration a according to Newton's second law of motion $F = ma$. Acceleration is the derivative of the velocity which in turn is the derivative of the position. Thus $a = \ddot{y}$, where the dots are Newton's dot notation for the differentiation operator $\mathrm{d}/\mathrm{d}t$. *We can rewrite the differential equation that governs the motion of the car as*

$$F = m\ddot{y} \Leftrightarrow \ddot{y} = \frac{F}{m} \tag{1.8}$$

For a Volkswagen Caddy Van (diesel, 2-L engine) the mass, without load and including the driver, is approximately 1500 kg. *Assume that the stop light changes to red when the car is* 15 m (49 ft) *away at a speed of* 10 m/s (36 km/h *or* 23 mph). *We have thus identified the following constants:*

$$m = 1500$$
$$y(0) = -15$$
$$\dot{y}(0) = 10$$

Here $y(0)$ means the initial position, that is $y(t)$ at time $t = 0$, and $\dot{y}(0)$ is the initial speed. The force F arises not from the engine, but from an opposite friction force in the brakes, and it is directed in the negative direction. Since the brake is our only means of control, the control signal F is constrained to the interval

$$-13\,600 \le F \le 0 \tag{1.9}$$

This can be seen as follows. According to its data sheet, the car requires at least 27.3 m *to stop when driving at a speed of* 80 km/h. *As all the kinetic energy is converted to work, we have, on the average,*

$$\frac{1}{2}m\,(\dot{y})^2 = Fy$$

and thus

$$|F| = \frac{1}{y}\frac{1}{2}m(\dot{y})^2$$
$$= \frac{1}{27.3}\frac{1}{2}1500\left(\frac{80\,000}{3600}\right)^2$$
$$\approx 13\,600$$

We therefore assume that the anti-lock braking system limits the magnitude of the brake force to 13 600 N (newton).

The closed loop characteristic equation *is obtained by inserting Equation (1.3) into Equation (1.8):*

$$\ddot{y} = \frac{K_p(e + T_d\dot{e})}{m} = -\frac{K_pT_d}{m}\dot{y} - \frac{K_p}{m}y \tag{1.10}$$

There will be a steady state solution, since insertion of $\ddot{y} = \dot{y} = 0$ yields the solution $y = 0$; this is just a check that a solution in accordance with the problem definition is feasible.

Disregarding the nonlinearity, the transfer function in the Laplace domain is the forward path gain in the block diagram divided by 1 minus the loop gain (Mason's rule),

$$\frac{y(s)}{Ref} = \frac{K_p(1 + T_ds)\frac{1}{m}\frac{1}{s^2}}{1 + K_p(1 + T_ds)\frac{1}{m}\frac{1}{s^2}}$$
$$= \frac{\frac{K_p}{m}T_ds + \frac{K_p}{m}}{s^2 + \frac{K_p}{m}T_ds + \frac{K_p}{m}} \tag{1.11}$$

The denominator is the closed loop characteristic polynomial, *compare Equation (1.10), and it is a second-order polynomial in s. The general transfer function of a second-order system is*

$$T = \frac{\omega_n^2}{s^2 + 2\zeta\omega_ns + \omega_n^2} \tag{1.12}$$

Here ω_n is the natural frequency *– the frequency of oscillation without damping – and ζ is the* damping ratio. *It is very useful here, because we are looking for the response without overshoot, which is as fast as possible. This is the case when $\zeta = 1$, which yields a* critically damped *response. Comparing with Equation (1.11) our damping ratio is*

$$\zeta = \frac{1}{2}\sqrt{\frac{K_p}{m}}T_d$$

and taking $\zeta = 1$ *gives us an optimal tuning relationship between* T_d *and* K_p,

$$T_d = \frac{2}{\sqrt{\frac{K_p}{m}}} \tag{1.13}$$

Keeping this relationship ensures that the response has no overshoot, and consequently the velocity will be zero when the car arrives at the stop light. We used this relationship to choose the previously mentioned settings $T_d = 1$ *and* $K_p = 6000$.

So far we have turned a blind eye to the numerator in Equation (1.11): the first term contains a differentiation s which is not present in the general transfer function. Our Ref is constant, however, with a zero derivative, in which case we can ignore the first term.

The example illustrates that good knowledge of the process to be controlled is beneficial, and it is in some cases crucial. A mathematical model is even necessary in order to analyse stability. But more importantly, even though the fuzzy controller is based on expert rules, it must still be tuned, and the PD controller gave us a tuning (K_p and T_d) that we could carry forward to the fuzzy controller. The PD controller furthermore provides us with a reference for the performance of the fuzzy controller, which is very useful in practice.

1.5 Nonlinear Control Systems

The designer is faced with many choices in the design of a nonlinear fuzzy controller, and some choices are only suitable for some processes. We therefore have to take into account typical kinds of nonlinear processes. Even though nonlinear systems can exhibit a local behaviour that is different from the overall behaviour, there are a few general aspects to consider such as local equilibria and their stability.

Example 1.2 *Chemical tank reactor*

Given a nonlinear chemical tank reactor that requires cooling and heating by means of a surrounding jacket of water, describe the control task.

► *Solution*

A reactor can be in three different temperature states for a given flow of cooling liquid, but only one of them is desirable.

Hot water ramps the temperature of the tank up from the ambient temperature (near 20° C) to a temperature where the reaction begins to accelerate. If the reaction is exothermic *(it develops heat by itself), the heat released through the reaction must be removed by circulating cool water through the jacket.*

The controller must then be able to change from heating to cooling at a specified point that takes account of a certain time delay.

A simple simulation tool, *Autopilot*, will be used for testing nonlinear controllers. Autopilot simulates a driver-less train car on a track as in Figure 1.6. We model the train car using Newton's second law as previously, but this time the forces acting on the train car are nonlinear.

At points where the track is horizontal, such as A, B and C, the normal force counterbalances the gravitational force when the train is standing still, and the sum of the external forces is thus

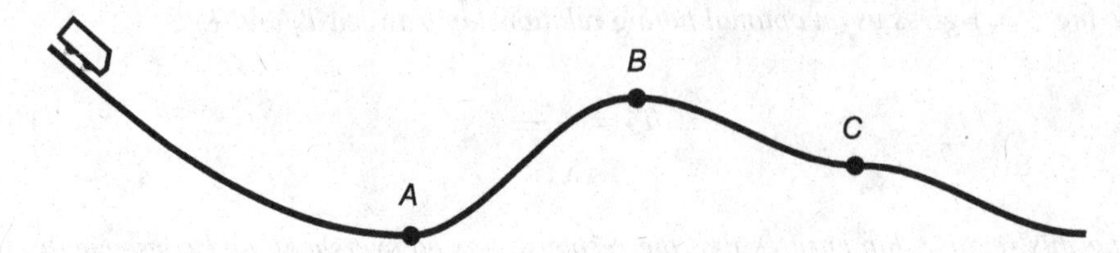

Figure 1.6 Autopilot. A driver-less train car moves on a hilly track. The drawing shows three kinds of equilibrium at stations A, B and C.

zero, that is, they are equilibrium points. Point A is a *stable equilibrium* because, when the train is disturbed slightly from the equilibrium, the forces tend to restore it to the equilibrium point. Point B is an *unstable equilibrium* since a slight disturbance away from the point results in a force that tends to move the train even farther away. Point C is a *saddle point* since a slight disturbance towards the left results in a force that tends to move the train back, while a slight disturbance towards the right results in a force that tends to move the train even farther away.

These are the only three types of equilibrium that a designer can meet. The example is therefore representative of many control problems. Furthermore, we can change the shape of the track as we wish, in order to emphasize other realistic situations in nonlinear control.

A controller affects the motion of the train according to our design. Assume that the task is to drive the train to station A as fast as possible, but without overshoot, then we have to craft a controller that applies a brake force in the correct amount. Assume instead that the task is to balance the train at station B, then we have to devise a rather tight controller that is able to take it there, and keep it in position, even if it is disturbed. The two equilibrium types call for two different control strategies. If the task is to stabilize the train in station C the controller must react in opposite ways to the left and to the right of the equilibrium point. On the left side, the curve helps the train to get back, while on the right side, the controller must work against the curve and use a larger effort.

The control strategy is non-symmetric in this respect, and it is possible to accommodate this in a fuzzy controller. If the task is to stabilize the train at a saddle point or on a slope of the curve, a linear PID controller may well turn out to be inadequate. In that case, a nonlinear fuzzy controller could be considered.

The Autopilot simulator could also model a chemical reactor, at least in principle, because the reactor behaves much like a saddle point. We shall use Autopilot throughout the book as a means to test various design choices and controller configurations.

Example 1.3 *Stop the car on a curve.*

Use Autopilot to simulate a car, but this time on a road shaped like a parabola. The mass of the car is $m = 1500$ kg *as previously, and the initial conditions are the same. The road is curved vertically according to the equation* $(x, y) = (x(t), 0.02x^2(t))$. *How does the previously designed PD controller perform when trying to stop the car at the bottom at* $x = 0$?

► *Solution*

The car is now affected by a downward slope, which affects speed. We therefore expect to use more brake force in order to stop the car at the horizontal position $x = 0$.

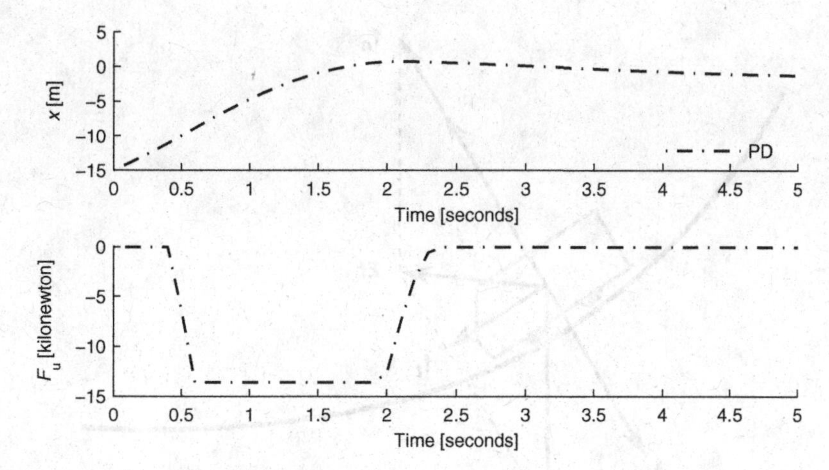

Figure 1.7 Stop the car at the bottom of a parabolic valley. The car starts at $x = -15$ m, and the controller tries to stop it at $x = 0$. (figrunautopilot1.m)

With the previous tuning ($K_p = 6000$ and $T_d = 1$) we get the response in Figure 1.7. Comparing this with Figure 1.4 the controller uses more control force. It kicks in early, it saturates quickly in the limit of the anti-lock braking system, it uses full brake force for more than a second, and then it quickly releases the brake power. But the car did not stop at the stop signal (the reference point); the upper plot shows that there is overshoot, before the car returns to the stop light. It is the upward curvature of the road that makes the car return, because the controller is only allowed to use the brake.

It is apparently difficult for the controller in the previous example to avoid overshoot. This could probably be fixed by tuning the controller a little tighter. The example demonstrates that one setting may not fit all situations. In that case, a solution might be to tune several controllers and interpolate between them, depending on a scheduling variable, such as the position. The method is called *gain scheduling*, and Chapter 8 will discuss how to interpolate between PID controllers by means of fuzzy rules.

1.6 Summary

It is relatively difficult to design a fuzzy controller, because it is in general nonlinear, and nonlinear systems are more or less unpredictable. Instead we propose to stay as long as possible in the linear domain, in accordance with the proposed design procedure. The idea is to start from a PID controller, and then to design a linear fuzzy controller that is equivalent to the initial PID controller. At this point, all the results from linear control theory can be applied, including tuning methods and stability calculations. In the next phase, the fuzzy controller is made nonlinear.

The design procedure has limited scope in the sense that it requires a PID controller. The reward is that the design procedure provides reliability: it guarantees that the fuzzy controller

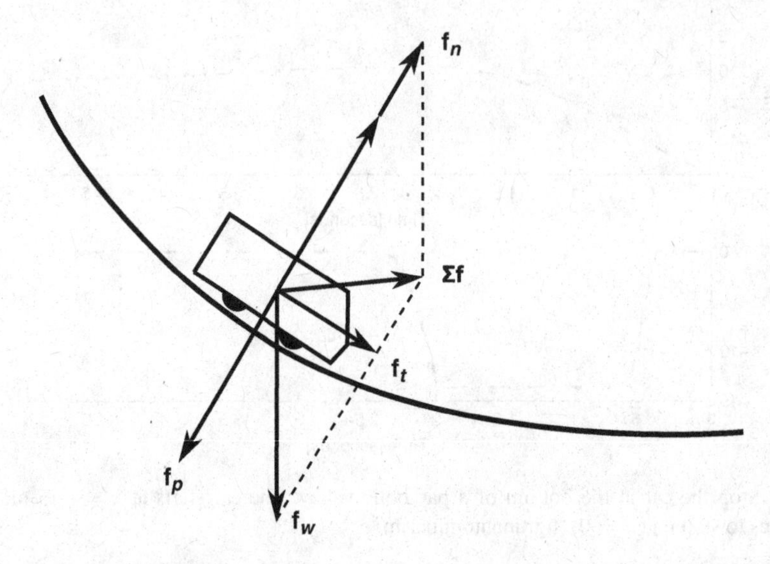

Figure 1.8 External forces affecting the train car.

performs at least as well as its initial PID controller. There is the possibility, but no guarantee, that it will perform better.

Some of the following chapters provide tools for analysing the nonlinear fuzzy controller, in particular, phase plane analysis and describing functions. Still, trial and error is a characteristic of fuzzy control.

1.7 The Autopilot Simulator*

This section is the first *-marked section in the book. The * in the heading signals that the section can be skipped on a first reading, because it contains background material which is unnecessary for the overall understanding. For the advanced student, however, the *-marked sections provide deeper insight and topics for further research. In this case, the section documents the inner workings of the Autopilot simulator written in Matlab (a programming environment for algorithm development, trademark of The MathWorks, Inc.[1]).

Autopilot simulates a driver-less train car on a track, which lies in a vertical plane to keep it simple. The train is affected by a downward gravitation force, a normal force from the track, and a control force from its motor/brakes.

According to Newton's second law, the equation of motion is $\sum \mathbf{f} = m\mathbf{a}$ where $\sum \mathbf{f}$ is the sum of forces acting on the train car, m is its mass, and $\mathbf{a}$ is the acceleration. The equation is valid for vectors, indicated by lower-case letters and set in a bold typeface. At a given horizontal position x, Figure 1.8 defines the external forces: the gravitational force $\mathbf{f}_w$ and the normal force $\mathbf{f}_n$. The former is resolved into its tangential component $\mathbf{f}_t$ and its perpendicular component $\mathbf{f}_p$.

[1] www.mathworks.com

Example 1.4 *Parabolic track*

Given the track curve $(x, y) = (x(t), ax^2(t))$ $(a > 0)$, characterize the external forces affecting the train car and characterize the equilibrium point.

► *Solution*

Notice first that the track curve has the shape of an upward bowl (parabola) possessing a minimum.

The normal force is the track's reaction. It balances $\mathbf{f}_p$, and it also contains an extra velocity-dependent centripetal force because of the track's curvature. Since the track curves upwards, the magnitude of $\mathbf{f}_n$ is larger than the magnitude of $\mathbf{f}_p$. The resultant force $\sum \mathbf{f} = \mathbf{f}_w + \mathbf{f}_n$ consists of a tangential component, which accelerates the car, and a centripetal force, which is responsible for the curvature of the motion.

The resultant force $\sum \mathbf{f}$ points to the right, when x is negative, and it points to the left, when x is positive. In other words, the force always points towards the y-axis of the coordinate system. By intuition, this could result in an oscillatory motion around $x = 0$, the character of which depends on the initial position.

At an equilibrium point, the sum of the forces is zero. This happens at $x = 0$ when the train is standing still. The upward bowl shape of the curve indicates that the equilibrium is a stable equilibrium.

1.8 Notes and References*

In the mid-1960s, Lotfi A Zadeh (born in 1921) of the University of California at Berkeley, USA, invented the theory of fuzzy sets. He argued that, more often than not, the classes of objects encountered in the real physical world have imprecisely defined criteria for membership (Zadeh 1965). For example, the 'class of numbers that are much greater than 1' or the 'class of tall human beings' have ill-defined boundaries. Yet such imprecisely defined classes play an important role in human reasoning and communication.

Ebrahim (Abe) H Mamdani, a control engineer at Queen Mary College in London, was attempting to develop an adaptive system that could learn to control an industrial process. He used a steam engine as a laboratory model, and with his colleagues set up a program that would teach the computer to control the steam engine by monitoring a human operator. At this point Mamdani's research student, Seto Assilian, tried to apply fuzzy logic. He created a set of simple rules in fuzzy terms, and Mamdani and Assilian then studied ways to use fuzzy rules of thumb directly in automating process controls. A few years later, Mamdani and Procyk managed to develop a linguistic self-organizing controller (Procyk and Mamdani 1979). It was an adaptive controller that was able to learn how to control a wide variety of processes, nonlinear and multi-variable, in a relatively short time. It was called *self-organizing* because at that time the meaning of the words 'adaptive' and 'learning' had not yet been agreed upon. The work of the pioneers led to a growing literature in fuzzy control and wide-ranging applications, as Table 1.1 illustrates.

In Japan, Michio Sugeno developed a self-learning fuzzy controller (Sugeno, Murofushi, Mori, Tatematsu, and Tanaka 1989). Twenty control rules determined the motion of a model car. Each rule recommends a specific change in direction, based on the car's distance from the walls of a corridor. The controller drives the car through angled corridors, after a learning session where a 'driving instructor' pulls it through the route a few times. Self-learning

Table 1.1 Milestones in early fuzzy history.

Year	Event	Reference
1965	First article on fuzzy sets	Zadeh 1965
1972	A rationale for fuzzy control	Zadeh 1972
1973	Linguistic approach	Zadeh 1973
1974	Fuzzy logic controller	Assilian and Mamdani 1974
1976	Warm water process	Kickert and van Nauta Lemke 1976
1977	Table based controller	Mamdani 1977
1977	Heat exchanger	Østergaard 1977
1977	Self-organizing controller	Procyk and Mamdani 1979
1980	Fuzzy conditional inference	Fukami *et al.* 1980
1980	Cement kiln controller	Holmblad and Østergaard 1982
1983	Train operation	Yasunobu *et al.* 1983
1984	Parking control of a model car	Sugeno *et al.* 1989
1985	Fuzzy chip	Togai and Watanabe 1985
1986	Fuzzy controller hardware system	Yamakawa and Miki 1986
1987	Sendai subway in operation	Yasunobu *et al.* 1983
1989	Fuzzy home appliances sold in Japan	
1989	The LIFE project is started in Japan	
1990	Rule learning by neural nets	Kosko 1992
1990	Hierarchical controller	Østergaard 1990, 1996

controllers that derive their own rules automatically are interesting because they could reduce the effort needed for translating human expertise into a rule base.

The first industrial application was in 1978 where a fuzzy controller was operating in closed loop on a rotary cement kiln in Denmark. Fuzzy control then became a commercial product of the Danish cement company FL Smidth & Co (now FLSmidth). The fuzzy control research program in Denmark was initiated in 1974 (Larsen 1981).

The Laboratory for International Fuzzy Engineering (LIFE), Yokohama, was set up by the Japanese Ministry of International Trade and Industry in 1989. It had a six-year budget and a research staff of around 30. LIFE conducted basic research with universities and member Japanese companies and subsidiaries of US and European companies, including Matsushita, Hitachi, Omron, and VW – about 50 companies in all. The research program was trimmed to five major projects: image understanding, fuzzy associative memory, fuzzy computing, intelligent interface, and the intelligent robot. They were all carried out with two themes in view: a navigation system for the blind and home computing.

A European network of excellence called ERUDIT was initiated in 1995 with support from the European Commission. ERUDIT, which lasted six years, was an open network for uncertainty modelling and fuzzy technology, aimed at putting European industry at the leading edge. The network was followed by another network EUNITE[2] with a broader scope: smart adaptive systems. That network was in turn followed by a coordinated action NISIS[3] with an even broader scope: nature-inspired systems.

[2] www.eunite.org
[3] www.nisis.de

For further information

Beginners may start with two articles in Institute of Electrical and Electronics Engineers (IEEE) Spectrum (Zadeh 1984, Self 1990) and then move on to the more advanced textbook by Timothy Ross (2010); it is oriented towards applications in engineering and technology with many calculated examples. The most efficient way to learn about fuzzy logic and fuzzy control is to study the Fuzzy Logic Toolbox for Matlab, especially the Fuzzy Inference System (MathWorks 2012). The toolbox covers other related techniques such as clustering, neuro-fuzzy systems, and genetic algorithms; for an in-depth description, see the related book by toolbox author Roger Jang and his two coauthors (Jang, Sun and Mizutani 1997).

The terms rule base and inference engine are loans from the field of expert systems, and Lee (1990) uses these to give a wide survey of the whole area of fuzzy control. The article lists 150 references.

A major reference on fuzzy control is the book by Driankov, Hellendoorn and Reinfrank (1996). It is explicitly targeted at the control engineering community, engineers in industry, and university students. The more recent book by Michels, Klawonn, Kruse and Nürnberger (2006) is a comprehensive reference book that includes the current state of the art at the time of writing. It views fuzzy control from the viewpoint of classical control theory, just as the present book does. Chapter 3 gives more specific references related to fuzzy control.

Industrial applications are described in a special issue of the journal *Fuzzy Sets and Systems*, for instance, the fuzzy car by Sugeno *et al.* (1989) and an arc welding robot by Murakami *et al.* (1989). There are more early applications in the classical book by Sugeno (1985). Ten years later, Constantin von Altrock (1995) described more than 30 case studies from companies that employed fuzzy and neuro-fuzzy methods. The FLSmidth controller is described in detail by Holmblad and Østergaard (1982).

There are more than ten journals related to fuzzy sets. Two of the major journals are *Fuzzy Sets and Systems* and *International Journal of Approximate Reasoning*, both published by Elsevier, and a third one is *Journal of Intelligent and Fuzzy Systems*, published by IOS Press, Netherlands. The Institute of Electrical and Electronics Engineers, IEEE, started a journal in 1992 called *IEEE Transactions on Fuzzy Systems*. The *Int. J. of Uncertainty, Fuzziness and Knowledge-Based Systems* is published four times per year by World Scientific Publishing Co. It is a forum for research on imprecise, vague, uncertain and incomplete knowledge. Other journals that occasionally have fuzzy control articles are *Automatica*, the control section of *IEE Proceedings*, *IEEE Transactions on Systems Man and Cybernetics*, *IEEE Transactions on Computers, Control Engineering Practice,* and the *International Journal of Man-Machine Studies*.

There is an active newsgroup called `comp.ai.fuzzy`.[4] It supplies useful news, conference announcements, and discussions.

There are two major professional organizations. The International Fuzzy Systems Association (IFSA) is a worldwide organization dedicated to fuzzy sets. IFSA publishes the *International Journal of Fuzzy Sets and Systems* (24 issues per year), holds international conferences, establishes chapters and sponsors activities. The other organization is the North American Fuzzy Information Processing Society, NAFIPS,[5] with roughly the same purpose.

---

[4] groups.google.com/group/comp.ai.fuzzy

[5] nafips.ece.ualberta.ca/

2

Fuzzy Reasoning

Fuzzy reasoning is based on fuzzy set theory, which is a generalization of classical set theory. An object is a member of a fuzzy set to a degree, which is one, zero, or something in between. Fuzzy sets thus have a fuzzy boundary with a gradual transition from membership to non-membership. Fuzzy logic is built on fuzzy set theory, and in fuzzy logic a proposition can be true, false, or partially true. Fuzzy reasoning is then the ability to draw conclusions from fuzzy propositions using if–then rules and rules of inference. Theoretical fuzzy logic extends the concepts of tautology, implication, and inference by means of modus ponens.

A washing machine equipped with a fuzzy reasoning ability can interpret a linguistic statement, such as 'if the washing machine is only half full, then use less water'. We say that fuzzy logic adds intelligence to the machine since the internal computer infers an action from a set of if–then rules that are specified by a human being. Fuzzy logic is 'computing with words', to quote the creator of fuzzy logic, Lotfi A Zadeh.

In classical logic, an assertion is either true or false – not something in between. In fuzzy logic, however, an assertion can be more or less true. Fuzzy logic is based on the theory of fuzzy sets, where an object's membership of a set is gradual rather than just member/not member. Fuzzy logic uses truth-values from the whole interval of real numbers between zero (*False*) and one (*True*).

This chapter explains a subset of the theory – enough to enable engineering students to write programs that make computers reason and act.

2.1 Fuzzy Sets

Fuzzy sets are a further development of mathematical set theory, first studied formally by the German mathematician Georg Cantor (1845–1918). It is possible to express most of mathematics in the language of set theory, and researchers are today looking at the consequences of 'fuzzifying' set theory, resulting in, for example, fuzzy logic, fuzzy numbers, fuzzy intervals, fuzzy arithmetic, and fuzzy functions. Fuzzy logic is based on fuzzy sets, and with

Foundations of Fuzzy Control: A Practical Approach, Second Edition. Jan Jantzen.

fuzzy logic a computer can process words from natural language, such as 'small', 'large', and 'approximately equal'.

Although elementary, the following sections include definitions from classical set theory in order to shed light on the underlying ideas. Only the definitions that are necessary and sufficient will be presented; students interested in delving deeper into classical set theory and logic can, for example, read the comprehensive treatment by Stoll (1979).

2.1.1 Classical Sets

According to Cantor a *set* $\mathcal{X}$ is a collection of definite, distinguishable objects of our intuition that can be treated as a whole. The objects are the *members* of $\mathcal{X}$. The concept 'objects of our intuition' gives us great freedom of choice, even with respect to sets with infinitely many members. Objects must be definite: given an object and a set, it must be possible to determine whether the object is, or is not, a member of the set. Objects must also be distinguishable: given a set and its members, it must be possible to determine whether any two members are different or the same.

The members completely define a set. To determine membership, it is necessary that the sentence 'x is a member of $\mathcal{X}$', where x is replaced by an object and $\mathcal{X}$ by the name of a set, is either true or false. We use the symbol $\in$ and write $x \in \mathcal{X}$ if object x is a member of the set $\mathcal{X}$. The assumption that the members determine a set is equivalent to saying, 'two sets $\mathcal{X}$ and $\mathcal{Y}$ are equal, $\mathcal{X} = \mathcal{Y}$, if and only if they have the same members'. The set whose members are the objects $x_1, x_2, \ldots, x_n$ is written as

$$\{x_1, x_2, \ldots, x_n\}$$

In particular, the set with no members is the *empty set* symbolized by $\emptyset$. The set $\mathcal{X}$ is included in $\mathcal{Y}$,

$$\mathcal{X} \subseteq \mathcal{Y}$$

if and only if each member of $\mathcal{X}$ is a member of $\mathcal{Y}$. We also say that $\mathcal{X}$ is a *subset* of $\mathcal{Y}$, and it means that, for all x,

$$\text{if } x \in \mathcal{X}, \text{ then } x \in \mathcal{Y}$$

The construct *if–then* is fundamental for fuzzy rules, and the sentence above shows that it is related to the concept of subsets. The empty set is a subset of every set.
Almost anything called a set in ordinary conversation is acceptable as a mathematical set, as the next example indicates.

Example 2.1 *Classical sets*

Give examples of collections of definite and distinguishable objects that are sets in the classical mathematical sense.

► *Solution*

(a) The set of non-negative integers less than 3. This is a finite set with three members $\{0, 1, 2\}$.

(b) The set of live dinosaurs in the basement of the British Museum. This set has no members and is the empty set $\emptyset$.

(c) The set of measurements greater than 10 volts. Even though this set is infinite, it is possible to determine whether a given measurement is a member.

(d) The set $\{0, 1, 2\}$ *is the set from (a). Since* $\{0, 1, 2\}$ *and* $\{2, 1, 0\}$ *have the same members, they are equal sets. Moreover,* $\{0, 1, 2\} = \{0, 1, 1, 2\}$ *for the same reason.*

(e) The members of a set may themselves be sets. The set

$$\mathcal{X} = \{\{1, 3\}, \{2, 4\}, \{5, 6\}\}$$

is a set with three members, namely, $\{1, 3\}$, $\{2, 4\}$, *and* $\{5, 6\}$. *Matlab supports sets of sets, or* nested sets, *in* cell arrays.

(f) It is possible in Matlab to assign an empty set, for instance: x = [ ].

Although the brace notation $\{\cdot\}$ is practical for listing sets of a few elements, it is impractical for large sets and impossible for infinite sets. How do we then define a set with a large number of members?

For an answer we require a few more concepts to be defined. A *proposition* is an assertion (declarative statement) which can be classified as either true or false. By a *predicate* in x we understand an assertion formed using a formula in x. For instance, '$0 < x \leq 3$', or '$x > 10$ volts' are predicates. They are not propositions, however, since they are not necessarily true or false. Only if we assign a value to the variable x does each predicate become a proposition. A predicate $P(x)$ in x defines a set $\mathcal{X}$ by the convention that the members of $\mathcal{X}$ are exactly those objects a such that $P(a)$ is true. In mathematical notation,

$$\{x \mid P(x)\}$$

is 'the set of all x such that $P(x)$ is true'. Thus $a \in \{x \mid P(x)\}$ if and only if $P(a)$ is a true proposition.

2.1.2 Fuzzy Sets

A system in which propositions must be either true or false, but not both, uses a two-valued logic. As a consequence, what is not true is false and vice versa; this is the *law of the excluded middle*. But a two-valued logic is only an approximation to human reasoning, as Zadeh observed (1965):

> Clearly, the 'class of all real numbers which are much greater than 1,' or 'the class of beautiful women,' or 'the class of tall men,' do not constitute classes or sets in the usual mathematical sense of these terms.

We might call it *Zadeh's challenge*, because he focuses on the elasticity in the meaning of terms such as 'much', 'beautiful', and 'tall'. To define the set of tall men as a classical set, one would use a predicate $P(x)$, for instance $x \geq 176$, where x is the height of a person, and the

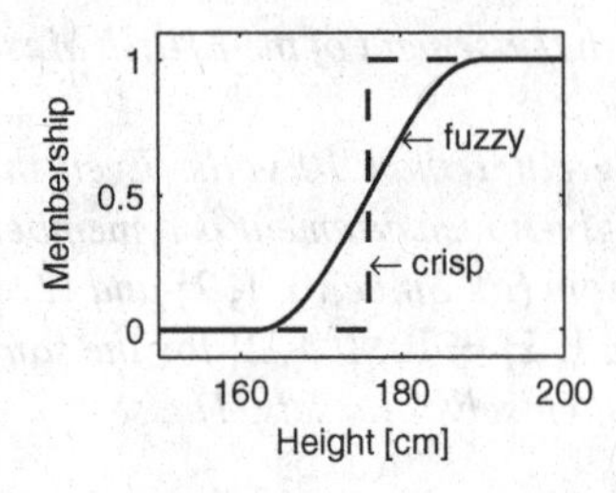

Figure 2.1 Two definitions of the set of tall men, a crisp set and a fuzzy set. (figtall.m)

right-hand side of the inequality is a threshold value in centimetres (176 cm $\simeq$ 5 ft 9 in). This is an abrupt approximation to the meaning of 'tall' (Figure 2.1).

Following Zadeh, a *membership grade* allows finer detail, such that the transition from membership to non-membership is gradual rather than abrupt. A membership function, defined on a universe of discourse, associates with each object x in the universe, the degree to which x is a member of the set. The membership grade for all members defines the fuzzy set.

Definition 2.1 *Fuzzy set. Given a collection of objects $\mathcal{U}$, a fuzzy set $\mathcal{A}$ in $\mathcal{U}$ is defined as a set of ordered pairs*

$$\mathcal{A} \equiv \{\langle x, \mu_{\mathcal{A}}(x)\rangle \mid x \in \mathcal{U}\} \tag{2.1}$$

where $\mu_{\mathcal{A}}(x)$ is called the membership function *for the set of all objects x in $\mathcal{U}$.*

The symbol '$\equiv$' stands for 'defined as'. The membership function relates to each x a membership grade $\mu_{\mathcal{A}}(x)$, which is a real number in the closed interval [0, 1]. Notice that it is now necessary to work with pairs $\langle x, \mu_{\mathcal{A}}(x)\rangle$, whereas for classical sets a list of objects suffices, since their membership is understood. An *ordered pair* $\langle x, y\rangle$ is a list of two objects, in which the object x is considered as the first and y as the second (note: in the set $\{x, y\}$ the order is insignificant).

The term *fuzzy* (synonymous with indistinct) suggests a boundary zone, rather than a sharp boundary, as in Figure 2.2. Indeed, fuzzy logicians speak of classical sets being *crisp sets,*

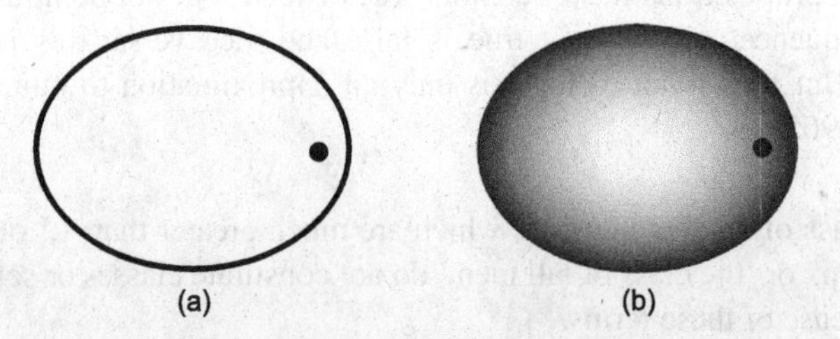

Figure 2.2 Diagrams of a crisp set (a) and a fuzzy set (b). The boundary of a fuzzy set is gradual, and an object (the dot) can be a partial member of the set illustrated by a gradual transition from black to white.

to distinguish them from fuzzy sets. As with crisp sets, we are only guided by intuition in deciding which objects are members and which are not; a formal basis for how to determine the membership grade of a fuzzy set is absent. The membership grade is a precise, but arbitrary, measure: that is, it rests on personal preference, and not on reason.

The definition of a fuzzy set extends the definition of a classical set, because membership values μ are permitted in the interval $0 \leq \mu \leq 1$, and the higher the value, the higher the membership. A classical set is consequently a special case of a fuzzy set, with membership values restricted to the end points or $\mu \in \{0, 1\}$.

A single pair $\langle x, \mu(x) \rangle$ is a fuzzy *singleton*; thus the whole set can be viewed as the union of its constituent singletons.

Example 2.2 *Fuzzy sets*

Give examples of typical fuzzy sets.

► *Solution*

(a) The set of real numbers x much greater than one ($x \gg 1$).

(b) The set of high temperatures, the set of strong winds, or the set of nice days are fuzzy sets in weather reports.

(c) The set of young people. A one-year-old baby will clearly be a member of the set of young people, and a 100-year-old person will not be a member of this set. A person aged 30 might be young to the degree 0.5.

(d) The set of adults. The Danish railways allow children under the age of 15 to travel at half price. An adult is thus defined by the set of passengers aged 15 or older. By this definition, the set of adults is a crisp set.

(e) A predicate may be crisp but it may be perceived as fuzzy: a speed limit of 60 km/h is taken to be an elastic range of more or less acceptable speeds within, say, 60–70 km/h (37–44 mph) by some drivers. Notice how, on the one hand, the traffic law is crisp while, on the other hand, those drivers' understanding of the law is fuzzy.

2.1.3 Universe

Members of a fuzzy set are taken from a *universe of discourse,* or *universe* for short. The universe consists of all objects that can come into consideration; refer the set $\mathcal{U}$ in Equation (2.1). The universe depends on the context, as the next example shows.

Example 2.3 *Universes*

Give examples of typical universes.

► *Solution*

(a) The set of real numbers x much greater than one ($x \gg 1$) *could have as a universe all real numbers, alternatively all positive integers.*

(b) The set of young people could have all human beings in the world as its universe. Alternatively, it could have the numbers between 0 and 100; these would then represent the age in years.

(c) The universe depends on the measuring unit; a duration in time depends on whether it is measured in hours, days, or weeks.

(d) A non-numerical quantity, for instance, taste, *must be defined on a* psychological continuum*; an example of such a universe is* $\mathcal{U} = \{bitter, sweet, sour, salt, hot\}$.

A programmer can exploit the universe to suppress faulty measurement data, for instance, negative values for a duration of time, by making the program consult the universe.

2.1.4 Membership Function

There are two alternative ways to represent a membership function: continuous or discrete. A continuous fuzzy set $\mathcal{A}$ is defined by a continuous membership function $\mu_{\mathcal{A}}(x)$. A *trapezoidal* membership function is a piecewise linear, continuous function, controlled by four parameters $\{a, b, c, d\}$ (Jang *et al.* 1997)

$$\mu_{Trapezoid}(x; a, b, c, d) = \left\{ \begin{array}{ll} 0 & , x \leq a \\ \dfrac{x-a}{b-a} & , a \leq x \leq b \\ 1 & , b \leq x \leq c \\ \dfrac{d-x}{d-c} & , c \leq x \leq d \\ 0 & , d \leq x \end{array} \right\}, x \in \mathbb{R} \tag{2.2}$$

The parameters $a \leq b \leq c \leq d$ define four breakpoints. For convenience, we name them: left foot-point (a), left shoulder-point (b), right shoulder-point (c), and right foot-point (d). Figure 2.3 (a) illustrates a trapezoidal membership function.

A *triangular* membership function is piecewise linear, and derived from the trapezoidal membership function by merging the two shoulder-points into one, that is, setting $b = c$, Figure 2.3 (b and d).

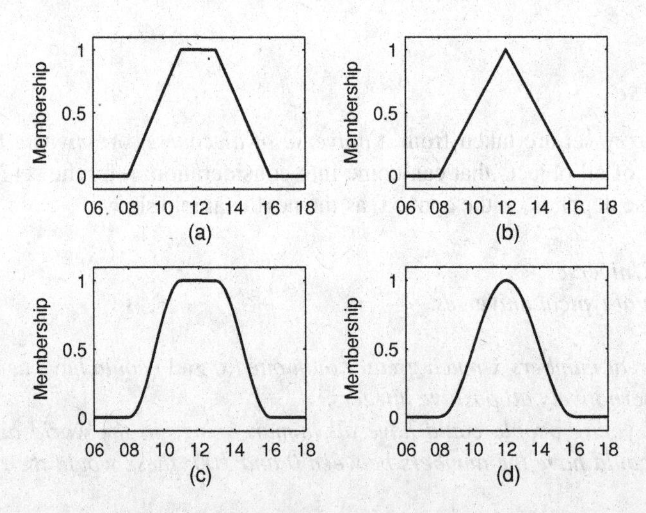

Figure 2.3 Around noon. Four possible membership functions representing the time 'around noon': a) trapezoidal, b) triangular, c) smooth trapezoid, and d) smooth triangular. The universe is the hours of the day in 24-hour format. (figmf0.m)

Smooth, differentiable versions of the trapezoidal and triangular membership functions can be obtained by replacing the linear segments corresponding to the intervals $a \leq x \leq b$ and $c \leq x \leq d$ by a nonlinear function, for instance a half period of a cosine function,

$$\mu_{STrapezoid}(x; a, b, c, d) = \begin{Bmatrix} 0 & , x \leq a \\ \frac{1}{2} + \frac{1}{2}\cos\left(\frac{b-x}{b-a}\pi\right) & , a \leq x \leq b \\ 1 & , b \leq x \leq c \\ \frac{1}{2} + \frac{1}{2}\cos\left(\frac{x-c}{d-c}\pi\right) & , c \leq x \leq d \\ 0 & , d \leq x \end{Bmatrix}, x \in \mathbb{R}$$

We name it *STrapezoid*, which stands for '*smooth trapezoid*' or '*soft trapezoid*'. Figures 2.3 (c–d) illustrate the smooth membership functions. Other possibilities exist for generating smooth trapezoidal functions, for example, Gaussian, generalized bell, and sigmoidal membership functions (Jang *et al.* 1997).

Discrete fuzzy sets are defined by means of a discrete variable x_i $(i = 1, 2, \ldots)$. A discrete fuzzy set $\mathcal{A}$ is defined by ordered pairs,

$$\mathcal{A} = \{\langle x_1, \mu(x_1)\rangle, \langle x_2, \mu(x_2)\rangle, \ldots \mid x_i \in \mathcal{U}, i = 1, 2, \ldots\}$$

Each membership value $\mu(x_i)$ is an evaluation of the membership function μ at a discrete point x_i in the universe $\mathcal{U}$, and the whole set is a collection, usually finite, of pairs $\langle x_i, \mu(x_i)\rangle$.

Example 2.4 *Discrete membership function*

It may be more convenient to work with vectors in a program rather than a mathematical definition of a membership function. How can we do that?

► *Solution*

To achieve a discrete triangular membership function, assume the universe is a vector **u** *of seven elements,*

$$\mathbf{u} = \begin{bmatrix} 9 & 10 & 11 & 12 & 13 & 14 & 15 \end{bmatrix}$$

Assume that the defining parameters are $a = 10, b = 12, c = 12,$ *and* $d = 14$ *then, by Equation (2.2), the corresponding membership values are,*

$$\mu = \begin{bmatrix} 0 & 0 & 0.5 & 1 & 0.5 & 0 & 0 \end{bmatrix}$$

Each membership value belongs to one element of the universe, more clearly displayed as a table with the universe in the bottom row, and the membership values in the top row, as follows:

0	0	0.5	1	0.5	0	0
9	10	11	12	13	14	15

When the table format is impractical, the universe and the membership values can be kept in separate vectors.

As a rule of thumb, the continuous form is computationally intensive but requires less storage compared to the discrete form.

2.1.5 Possibility

A fuzzy set induces a *possibility distribution* on the universe, meaning that, one can interpret the membership values as possibilities. How then are possibilities related to probabilities?

First of all, probabilities must add up to one, or the area under a density curve must be one. Memberships may add up to anything (discrete case), or the area under the membership function may be anything (continuous case). Secondly, a probability distribution concerns the likelihood of an event occurring, based on observations, whereas a possibility distribution (membership function) is subjective.

The word 'probably' is synonymous with presumably, doubtless, likely, presumptively. The word 'possible' is synonymous with doable, feasible, practicable, viable, workable. Their relationship is best described in the sentence: what is probable is always possible, but not vice versa. This is illustrated next.

Example 2.5 *Probability versus possibility*

Show the difference between probability and possibility.

▶ *Solution*

(a) Consider the statement 'Hans ate x eggs for breakfast', where $x \in \mathcal{U} = \langle 1, 2, ..., 8 \rangle$ *(Zadeh in Zimmermann 1993). We may associate a probability distribution p by observing Hans eating breakfast for 100 days,*

$$p = \begin{bmatrix} 0.1 & 0.8 & 0.1 & 0 & 0 & 0 & 0 & 0 \\ 1 & 2 & 3 & 4 & 5 & 6 & 7 & 8 \end{bmatrix}$$

A fuzzy set expressing the degree of ease with which Hans can eat x eggs may be the following possibility distribution,

$$\pi = \begin{bmatrix} 1 & 1 & 1 & 1 & 0.8 & 0.6 & 0.4 & 0.2 \\ 1 & 2 & 3 & 4 & 5 & 6 & 7 & 8 \end{bmatrix}$$

The possibility is $\pi(3) = 1$, *and the probability is* $p(3) = 0.1$.

(b) Consider a universe of four car models

$$\mathcal{U} = \{Trabant, Fiat\ Uno, BMW, Ferrari\}.$$

We may associate a probability $p(x)$ *of each car model driving* 100 *mph* (161 *km/h) on a motorway, by observing cars for* 100 *days,*

$$p(Trabant) = 0, \; p(Fiat\ Uno) = 0.1, \; p(BMW) = 0.4, \; p(Ferrari) = 0.5$$

The possibilities may be

$$\pi(Trabant) = 0,\ \pi(Fiat\ Uno) = 0.5,\ \pi(BMW) = 1,\ \pi(Ferrari) = 1$$

Notice that each possibility is at least as high as the corresponding probability.

2.2 Fuzzy Set Operations

Operations on fuzzy sets are defined by means of the membership functions. For example, in order to compare two fuzzy sets, equality and inclusion are defined as follows.

Definition 2.2 *Equality and inclusion (subset) of fuzzy sets.*

Let $\mathcal{A}$ and $\mathcal{B}$ be two fuzzy sets defined on a mutual universe $\mathcal{U}$. The two fuzzy sets $\mathcal{A}$ and $\mathcal{B}$ are equal, if and only if they have the same membership function,

$$\mathcal{A} = \mathcal{B} \equiv \mu_{\mathcal{A}}(x) = \mu_{\mathcal{B}}(x)$$

for all x. A fuzzy set $\mathcal{A}$ is a subset of (included in) a fuzzy set $\mathcal{B}$, if and only if the membership of $\mathcal{A}$ is less than or equal to that of $\mathcal{B}$,

$$\mathcal{A} \subseteq \mathcal{B} \equiv \mu_{\mathcal{A}}(x) \leq \mu_{\mathcal{B}}(x) \tag{2.3}$$

for all x.

2.2.1 Union, Intersection, and Complement

In order to generate new sets from existing sets, we define two operations, which are in certain respects analogous to addition and multiplication. The (classical) union of the sets $\mathcal{X}$ and $\mathcal{Y}$, symbolized by $\mathcal{X} \cup \mathcal{Y}$ and read '$\mathcal{X}$ union $\mathcal{Y}$', is the set of all objects that are members of $\mathcal{X}$ or $\mathcal{Y}$, or both, that is,

$$\mathcal{X} \cup \mathcal{Y} \equiv \{x \mid x \in \mathcal{X} \text{ or } x \in \mathcal{Y}\}$$

Thus, by definition, $x \in \mathcal{X} \cup \mathcal{Y}$ if and only if x is a member of at least one of $\mathcal{X}$ and $\mathcal{Y}$. For example,

$$\{1, 2, 3\} \cup \{1, 3, 4\} = \{1, 2, 3, 4\}$$

The (classical) intersection of the sets $\mathcal{X}$ and $\mathcal{Y}$, symbolized by $\mathcal{X} \cap \mathcal{Y}$ and read '$\mathcal{X}$ intersection $\mathcal{Y}$', is the set of all objects which are members of both $\mathcal{X}$ and $\mathcal{Y}$. That is,

$$\mathcal{X} \cap \mathcal{Y} \equiv \{x \mid x \in \mathcal{X} \text{ and } y \in \mathcal{Y}\}$$

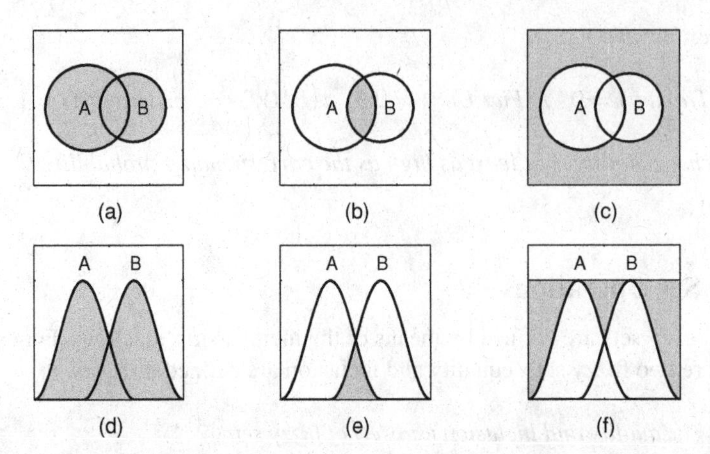

Figure 2.4 Set operations. The top row shows classic Venn diagrams; the universe is represented by the points within the rectangle, and sets by the interior of the circles. The bottom row shows their fuzzy equivalents; the universal set is represented by a horizontal line at membership $\mu = 1$, and sets by membership functions. The shaded areas are: union $A \cup B$ (a, d), intersection $A \cap B$ (b, e), and complement $\overline{A \cup B}$ (c, f). (figvenn2.m)

For example,

$$\{1, 2, 3\} \cap \{1, 3, 4\} = \{1, 3\}$$

The (classical) complement of a set $\mathcal{X}$, symbolized by $\overline{\mathcal{X}}$ and read 'the complement of $\mathcal{X}$', is

$$\overline{\mathcal{X}} \equiv \{x \mid x \notin \mathcal{X}\}$$

That is, the set of those members of the universe that are not members ($\notin$) of $\mathcal{X}$. *Venn diagrams*, Figure 2.4, clearly illustrate the set operations. A Venn diagram is a picture showing sets as circles or closed curves, to illustrate relations between sets and objects.

When dealing with fuzzy sets, we must consider gradual membership. Figure 2.4 (d–f) shows an intuitively acceptable modification of the Venn diagrams. The fuzzy set operations are defined accordingly. The following is a formal definition of fuzzy union,

$$\mathcal{A} \cup \mathcal{B} \equiv \{\langle x, \mu_{\mathcal{A}\cup\mathcal{B}}(x)\rangle \mid x \in \mathcal{U} \text{ and } \mu_{\mathcal{A}\cup\mathcal{B}}(x) = \max(\mu_{\mathcal{A}}(x), \mu_{\mathcal{B}}(x))\}$$

That is, the union is defined as the max operation on the membership functions as x traverses the universe. The notation is a little cumbersome, and we shall use a shorthand notation when there is no ambiguity. Thus, we just define fuzzy union as

$$\mu_{\mathcal{A}\cup\mathcal{B}}(x) \equiv \max(\mu_{\mathcal{A}}(x), \mu_{\mathcal{B}}(x))$$

Definition 2.3 *Fuzzy union, intersection, and complement. Let $\mathcal{A}$ and $\mathcal{B}$ be fuzzy sets defined on a mutual universe $\mathcal{U}$.*

The fuzzy union $\mathcal{A} \cup \mathcal{B}$ *is*

$$\mu_{\mathcal{A}\cup\mathcal{B}}(x) \equiv \max(\mu_{\mathcal{A}}(x), \mu_{\mathcal{B}}(x))$$

The fuzzy intersection $\mathcal{A} \cap \mathcal{B}$ *is*

$$\mu_{\mathcal{A}\cap\mathcal{B}}(x) \equiv \min(\mu_{\mathcal{A}}(x), \mu_{\mathcal{B}}(x)) \quad .$$

The fuzzy complement $\overline{\mathcal{A}}$ *of* $\mathcal{A}$ *is*

$$\mu_{\overline{\mathcal{A}}}(x) \equiv 1 - \mu_{\mathcal{A}}(x)$$

It is, in practice, easy to apply the fuzzy set operations max, min, and $1 - \mu$. These definitions also work on classical sets described by two-valued (0/1) membership functions.

Example 2.6 *Buying a house (after Zimmermann 1993)*
Give an example of how fuzzy set operations can be applied in practice.
► *Solution*
A four-person family wishes to buy a house. An indication of its level of comfort is the number of bedrooms in the house. But the family also wishes to have a large house. The universe $\mathcal{U} = \langle 1, 2, 3, 4, 5, 6, 7, 8, 9, 10\rangle$ *is the set of houses to be considered by the number of bedrooms. The fuzzy set Comfortable may be described as a vector*

$$\mu_c = \begin{bmatrix} 0.2 & 0.5 & 0.8 & 1 & 0.7 & 0.3 & 0 & 0 & 0 & 0 \end{bmatrix}$$

Let μ_l *describe the fuzzy set Large, defined as*

$$\mu_l = \begin{bmatrix} 0 & 0 & 0.2 & 0.4 & 0.6 & 0.8 & 1 & 1 & 1 & 1 \end{bmatrix}$$

The intersection of Comfortable and Large is then,

$$\min(\mu_c, \mu_l) = \begin{bmatrix} 0 & 0 & 0.2 & 0.4 & 0.6 & 0.3 & 0 & 0 & 0 & 0 \end{bmatrix}$$

To interpret, five bedrooms is optimal, having the largest membership value 0.6. *It is, however, not fully satisfactory, since the membership is less than* 1. *The second best solution is four bedrooms, with membership* 0.4. *If the market is a buyer's market, the family may wish to wait until a better offer comes up, hoping to obtain full satisfaction (membership 1).*
The union, Comfortable or Large, is

$$\max(\mu_c, \mu_l) = \begin{bmatrix} 0.2 & 0.5 & 0.8 & 1 & 0.7 & 0.8 & 1 & 1 & 1 & 1 \end{bmatrix}$$

Here four bedrooms is fully satisfactory (membership 1) because it is comfortable. Also 7–10 bedrooms are satisfactory, because the house is large. If the market is a seller's market, the family may wish to buy the house, being content that at least one of the criteria is fulfilled.

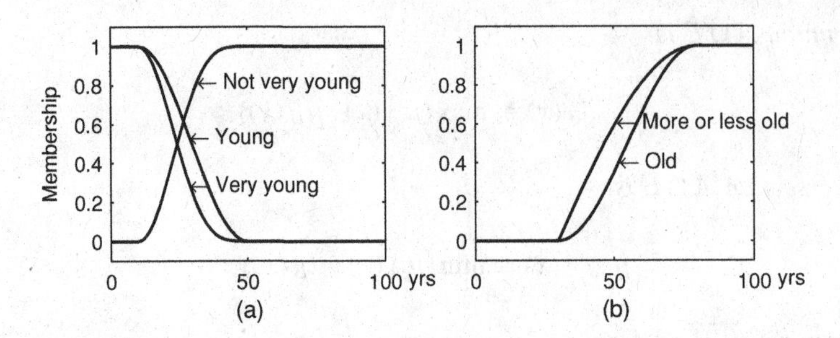

Figure 2.5 The membership functions for *very young* and *not very young* are derived from *young* (a), and the membership function for *more or less old* from *old* (b). (figage.m)

If the children are about to move away from the family within the next couple of years, the parents may wish to buy a house that is Comfortable and Not Large, or

$$\min(\mu_c, 1 - \mu_l) = \begin{bmatrix} 0.2 & 0.5 & 0.8 & 0.6 & 0.4 & 0.2 & 0 & 0 & 0 & 0 \end{bmatrix}$$

In this case, three bedrooms is satisfactory to the degree 0.8.

The example indicates how computer-aided decision support systems can apply fuzzy sets. Another application area is information search, for instance, the World Wide Web search engines.

2.2.2 Linguistic Variables

Whereas an algebraic variable takes numbers as values, a *linguistic variable* takes words or sentences as values. The name of such a linguistic variable is its *label*. The set of values that it can take is called its *term set*. Each value in the term set is a *linguistic value* or *term* defined over the universe. In short, a linguistic variable takes a linguistic value, which is a fuzzy set defined on the universe.

Example 2.7 *Term set Age*

Let 'Age' be a linguistic variable defined in Figure 2.5. Illustrate a term set.

▶ *Solution*

Its term set $\mathcal{T}$ could be defined as

$$\mathcal{T}(age) = \{young,\ very\ young,\ not\ very\ young,\ old,\ more\ or\ less\ old\}$$

Each term is defined on the universe, in this case the numbers from 0 *to* 100 *years.*

A *hedge* is a word that acts on a term and modifies its meaning. For example, in the sentence 'very near zero', the word 'very' modifies the term 'near zero'. Examples of other hedges are 'a little', 'more or less', 'possibly', and 'definitely'. In fuzzy reasoning, a hedge operates on a membership function, and the result is a membership function.

Even though it is difficult to precisely say what effect the hedge 'very' has, it does have an intensifying effect. The hedge 'more or less' (or 'morl' for short) has the opposite effect. Given a fuzzy term with the label $\mathcal{A}$ and membership function $\mu_{\mathcal{A}}(x)$ defined on the universe $\mathcal{X}$, the hedges 'very' and 'morl' are defined as

$$very\ \mathcal{A}: \ \mu_{very\ \mathcal{A}}(x) \equiv \mu_{\mathcal{A}}^{2}(x)$$

$$morl\ \mathcal{A}: \ \mu_{morl\ \mathcal{A}}(x) \equiv \mu_{\mathcal{A}}^{\frac{1}{2}}(x)$$

We have applied squaring and square root, but a whole family of hedges is generated by $\mu_{\mathcal{A}}^{k}$ or $\mu_{\mathcal{A}}^{\frac{1}{k}}$. For example

$$extremely\ \mathcal{A}: \ \mu_{very\ \mathcal{A}}(x) \equiv \mu_{\mathcal{A}}^{3}(x)$$

$$slightly\ \mathcal{A}: \ \mu_{very\ \mathcal{A}}(x) \equiv \mu_{\mathcal{A}}^{\frac{1}{3}}(x)$$

A derived hedge is for example *somewhat* $\mathcal{A}$, defined as *morl* $\mathcal{A}$ *and not slightly* $\mathcal{A}$. With $k = \infty$ the hedge $\mu_{\mathcal{A}}^{k}$ could be named *exactly*, because it would suppress all memberships lower than 1. When $k > 1$ the hedge is a *concentration*, and with $k < 1$ it is a *dilation*.

Example 2.8 *The hedge 'very'*

Assuming a discrete universe $\mathcal{U} = \langle 0, 20, 40, 60, 80 \rangle$ *of ages, show how to apply the hedge 'very' on a discrete membership function.*

▶ *Solution*

Assume the following discrete membership function (implemented in Matlab, for instance):

$$young = \begin{bmatrix} 1 & 0.6 & 0.1 & 0 & 0 \end{bmatrix}$$

The membership vector for the set 'very young' is

$$very\ young = young^{2} = \begin{bmatrix} 1 & 0.36 & 0.01 & 0 & 0 \end{bmatrix}$$

The set 'very very young' is, by repeated application,

$$very\ very\ young = young^{4} = \begin{bmatrix} 1 & 0.13 & 0 & 0 & 0 \end{bmatrix}$$

The derived sets inherit the five-element universe of the primary set.

A *primary term* is a term that must be defined a priori, for example 'young' and 'old' in Figure 2.5, whereas the sets 'very young' and 'not very young' are modified sets. The primary terms can be modified by negation ('not') or hedges ('very', 'more or less'), and the resulting sets can be connected using connectives ('and', 'or', 'implies', 'equals'). Long sentences can be built using this vocabulary, and the result is still a membership function.

2.2.3 Relations

In mathematics, the word *relation* is used in the sense of relationship, for example, the predicates: x is less than y, or y is a function of x. A *binary relation* $\mathcal{R}$ is a set of ordered pairs. We write $x\mathcal{R}y$ for 'x is related to y'. There are established symbols for various relations, for example $x = y$, $x < y$. One simple relation is the set of all pairs $\langle x, y \rangle$, such that x is a member of a set $\mathcal{X}$, and y is a member of a set $\mathcal{Y}$. This is the (classical) *Cartesian product* of $\mathcal{X}$ and $\mathcal{Y}$,

$$\mathcal{X} \times \mathcal{Y} \equiv \{\langle x, y \rangle \mid x \in \mathcal{X}, y \in \mathcal{Y}\}$$

In fact, any binary relation $x\mathcal{R}y$ is a subset of the Cartesian product $\mathcal{X} \times \mathcal{Y}$, and we can think of those instances of $\mathcal{X} \times \mathcal{Y}$ that are members of $\mathcal{R}$ as having membership 1. By analogy, a binary *fuzzy relation* consists of pairs $\langle x, y \rangle$ with an associated fuzzy membership value.

Definition 2.4 *Fuzzy binary relation. Let $\mathcal{A}$ and $\mathcal{B}$ be fuzzy sets defined on $\mathcal{X}$ and $\mathcal{Y}$ respectively, then the fuzzy set in $\mathcal{X} \times \mathcal{Y}$ with the membership function*

$$\mathcal{R} \equiv \{\langle \langle x, y \rangle, \mu_{\mathcal{R}}(x, y) \rangle \mid (x, y) \in \mathcal{X} \times \mathcal{Y}\}$$

is a binary fuzzy relation $\mathcal{R} \subseteq \mathcal{X} \times \mathcal{Y}$.

It is straightforward to generalize the relations from binary to n-ary relations ($n > 2$).

Example 2.9 *Approximately equal*

Given $\mathcal{X} = \mathcal{Y} = \{1, 2, 3\}$ set up a relation 'approximately equal' between all pairs of the three numbers.

▶ *Solution*

Such a relation is most clearly displayed in a table,

		$\mathcal{Y}$		
		1	2	3
	1	1	0.8	0.3
$\mathcal{X}$	2	0.8	1	0.8
	3	0.3	0.8	1

The membership values inside the table express our personal evaluation of 'approximately equal' when comparing a row label with a column label.

In the fuzzy Cartesian product each object is associated with a membership value. The object is a pair in a binary relation. Figure 2.6 illustrates the Cartesian product of two fuzzy sets $\mathcal{A}$ and $\mathcal{B}$. Each set is defined by a membership function on its own universe, and the result is a two-dimensional fuzzy set. Formally, the fuzzy Cartesian product is defined as follows.

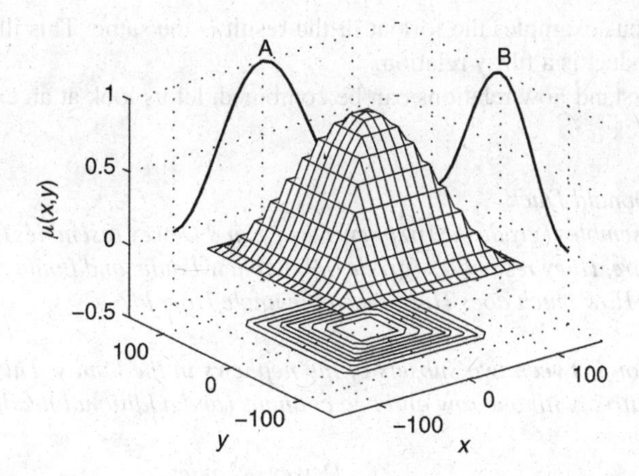

Figure 2.6 Cartesian product of two fuzzy sets $\mathcal{A}$ and $\mathcal{B}$. Each object in the xy-plane is associated with a combined membership value $\mu(x, y) = \min(\mu_{\mathcal{A}}(x), \mu_{\mathcal{B}}(y))$. The membership values form a surface above the xy-plane. The plot under the surface is a contour plot of the surface levels.

Definition 2.5 *Fuzzy Cartesian product. Let $\mathcal{A}$ and $\mathcal{B}$ be fuzzy sets defined on $\mathcal{X}$ and $\mathcal{Y}$ respectively, then the fuzzy set in $\mathcal{X} \times \mathcal{Y}$ with the membership function*

$$\mu_{\mathcal{A}\times\mathcal{B}}(x, y) \equiv \mu_{\mathcal{A}}(x) \cap \mu_{\mathcal{B}}(y)$$

is the Cartesian product $\mathcal{A} \times \mathcal{B}$.

Example 2.10 *Fuzzy Cartesian product*

Assume $\mathcal{X}$ and $\mathcal{Y}$ are given as in Example 2.7. Assume also a fuzzy set $\mathcal{A}$ defined on $\mathcal{X}$ by the membership function $\mu_{\mathcal{A}}(x_i) = \langle 0, 0.5, 1\rangle$ and another fuzzy set $\mathcal{B}$ defined on $\mathcal{Y}$ by the membership function $\mu_{\mathcal{B}}(y_j) = \langle 1, 0.5, 0\rangle$. What is the fuzzy Cartesian product $\mathcal{A} \times \mathcal{B}$?

▶ *Solution*

$\mathcal{A} \times \mathcal{B}$ can be arranged as a two-dimensional fuzzy set,

		$\mathcal{B}$		
		1	*0.5*	*0*
	0	*0*	*0*	*0*
$\mathcal{A}$	*0.5*	*0.5*	*0.5*	*0*
	1	*1*	*0.5*	*0*

The element at row i and column j is the intersection (minimum) of $\mu_{\mathcal{A}}(x_i)$ and $\mu_{\mathcal{B}}(y_j)$. This is an example of an outer product.

Notice that a membership $\mu_{\mathcal{A}\times\mathcal{B}}(x_i, y_j)$ is associated with each object $\langle x_i, y_j\rangle$, whereas the classical Cartesian product consists of objects $\langle x_i, y_j\rangle$ only.

In the two previous examples the format of the result is the same. This illustrates that the fuzzy Cartesian product is a fuzzy relation.

In order to understand how relations can be combined, let us look at an example from the cartoon Donald Duck.

Example 2.11 *Donald Duck*

Nephew Huey resembles (grade 0.8) nephew Dewey, and Dewey resembles (grade 0.5) uncle Donald. Furthermore, Huey resembles (grade 0.9) nephew Louie, and Louie resembles (grade 0.6) uncle Donald. How much does Huey resemble uncle Donald?

► *Solution*

We have a relation between two subsets of the nephews in the family. This is conveniently represented in a matrix, with one row and two columns (and additional headings),

$\mathbf{R}_1 =$

	Dewey	*Louie*
Huey	*0.8*	*0.9*

We have another relation between nephews Dewey and Louie on the one side, and uncle Donald on the other, a matrix with two rows and one column,

$\mathbf{R}_2 =$

	Donald
Dewey	*0.5*
Louie	*0.6*

We wish to find out how much Huey resembles Donald by combining the information in the two matrices:

(i) *Huey resembles Dewey* ($\mathbf{R}_1(1,1) = 0.8$), *and Dewey in turn resembles Donald* ($\mathbf{R}_2(1,1) = 0.5$); *or*
(ii) *Huey resembles Louie* ($\mathbf{R}_1(1,2) = 0.9$), *and Louie in turn resembles Donald* ($\mathbf{R}_2(2,1) = 0.6$).

Assertion (i) contains two relationships combined by 'and'; it seems reasonable to take the intersection. With our previous definition, this corresponds to choosing the smallest membership value for the (transitive) Huey–Donald relationship, or $\min(0.8, 0.5) = 0.5$. *Similarly with statement (ii),* $\min(0.9, 0.6) = 0.6$. *Thus from (i) and (ii), respectively, we deduce:*

(iii) *Huey resembles Donald to the degree* 0.5; *or*
(iv) *Huey resembles Donald to the degree* 0.6.

Although the results in (iii) and (iv) differ, we are equally confident of either result; we have to choose either one or the other, so it seems reasonable to take the one with the strongest result, that is, the union. With our previous definition, this corresponds to $\max(0.5, 0.6) = 0.6$.

Thus, the answer is that Huey resembles Donald to the degree 0.6.

Mathematically speaking, this was an example of *composition* of relations, which is formally defined as follows.

Definition 2.6 *Fuzzy relational composition. Let $\mathcal{R}$ and $\mathcal{S}$ be two fuzzy relations defined on $\mathcal{X} \times \mathcal{Y}$ and $\mathcal{Y} \times \mathcal{Z}$ respectively, then the fuzzy set in $\mathcal{X} \times \mathcal{Z}$ with the membership function,*

$$\mathcal{R} \circ \mathcal{S} \equiv \left\langle \langle x, z \rangle, \bigcup_{y} \mu_{\mathcal{R}}(x, y) \cap \mu_{\mathcal{S}}(y, z) \right\rangle$$

is the composition of $\mathcal{R}$ with $\mathcal{S}$.

When $\mathcal{R}$ and $\mathcal{S}$ are expressed as matrices $\boldsymbol{R}$ and $\boldsymbol{S}$, the composition is equivalent to an *inner product*. The inner product is similar to an ordinary matrix (dot) product, except that multiplication and summation can be replaced by any binary operations. Suppose $\boldsymbol{R}$ is an m-by-p matrix and $\boldsymbol{S}$ is a p-by-n matrix. Then the inner $\cup - \cap$ product (read 'cup-cap product') is an m-by-n matrix $\boldsymbol{T} = (t_{ij})$ whose ijth entry is obtained by combining the ith row of $\boldsymbol{R}$ with the jth column of $\boldsymbol{S}$, such that

$$t_{ij} = \left(r_{i1} \cap s_{1j}\right) \cup \left(r_{i2} \cap s_{2j}\right) \cup \ldots \cup \left(r_{ip} \cap s_{pj}\right) = \bigcup_{k=1}^{p} r_{ik} \cap s_{kj} \tag{2.4}$$

The notation looks unwieldy, but the operation is essentially a matter of combining rows with columns successively, as the next example shows.

Example 2.12 *Inner product*

Use the inner product on the data in the Donald Duck example (Example 2.11).

► *Solution*

For the matrices $\boldsymbol{R}_1$ and $\boldsymbol{R}_2$, the inner product yields,

$$\boldsymbol{R}_1 \circ \boldsymbol{R}_2 = \begin{bmatrix} 0.8 & 0.9 \end{bmatrix} \circ \begin{bmatrix} 0.5 \\ 0.6 \end{bmatrix} = (0.8 \cap 0.5) \cup (0.9 \cap 0.6) = 0.5 \cup 0.6 = 0.6$$

which agrees with the previous result.

With our previous definitions of the set operations, composition is specifically called the *max–min composition* (Ross 2010). If the min operation is replaced by * for multiplication, it is called the *max–star composition*.

2.3 Fuzzy If–Then Rules

Fuzzy controllers are built from if–then rules:

$$\text{If } x \text{ is } \mathcal{A} \text{ then } y \text{ is } \mathcal{B}$$

Here $\mathcal{A}$ and $\mathcal{B}$ are fuzzy sets – defined on universes $\mathcal{X}$ and $\mathcal{Y}$, respectively – carrying labels such as small, medium, large. This is an implication, where the left-hand side 'x is $\mathcal{A}$' is the *antecedent*, and the right-hand side 'y is $\mathcal{B}$' is the *consequent*. The following list gives some examples of such rules in everyday conversation:

1. If it is dark, then drive slowly.
2. If the tomato is red, then it is ripe.
3. If it is early, then John can study.
4. If the room is cold, then increase the heat.
5. If the washing machine is half full, then wash for a shorter time.

Examples 4 and 5 are typical for a heating unit or a washing machine with computer control where the rules are embedded in the computer.

Other forms can be transcribed into the if–then form, for example 'when in Rome, do as the Romans' becomes 'if in Rome, then do as the Romans'.

Example 2.13 *Student John*

Consider rule 3: If it is early, then John can study. Assume that 'early' is a fuzzy set defined on a universe representing the time of the day, in steps of four hours in a 24-hour format to numerically distinguish night from day,

$$\mathcal{U} = \langle 4, 8, 12, 16, 20, 24 \rangle .$$

Define 'early' as a fuzzy set on $\mathcal{U}$,

$$early = \{\langle 4, 0 \rangle , \langle 8, 1 \rangle , \langle 12, 0.9 \rangle , \langle 16, 0.7 \rangle , \langle 20, 0.5 \rangle , \langle 24, 0.2 \rangle\}$$

Define 'can study' as a singleton fuzzy set $\mu_{study} = 1$. What can be concluded from the rule?

► *Solution*

If the hour is truly early, for instance 8 o'clock in the morning, then $\mu_{early}(8) = 1$, and thus John can study to the fullest degree, that is $\mu_{study} = 1$. However, if it is 20.00 hours (8 p.m.), then $\mu_{early}(20) = 0.5$, and accordingly John can study to the degree 0.5.

More precisely, the procedure is as follows: given a particular time of day t_0, the resulting truth-value is computed as $\min(\mu_{early}(t_0), \mu_{study})$.

The example shows that the degree of fulfilment of the antecedent (the *if* side) weights the degree of fulfilment of the conclusion; this is a useful mechanism, which enables one rule to cover a whole range of the universe.

In a fuzzy controller x is a measurement, which is an input to the rule base, and y is the output of the rule base, which is used as a control action. The rule is thus a *condition-action* type of rule, where the action depends on the fulfilment of the condition. A crisp program would execute the action only if the condition is fulfilled. A fuzzy program executes the action always, although in accordance with the degree of fulfilment of the condition.

The rule combines the fuzzy set $\mathcal{A}$ with the fuzzy set $\mathcal{B}$, and formally it is a Cartesian product, that is

$$\text{If } x \text{ is } \mathcal{A} \text{ then } y \text{ is } \mathcal{B} \equiv \mathcal{A} \times \mathcal{B}$$

Figure 2.6, shown earlier, illustrates its implementation.

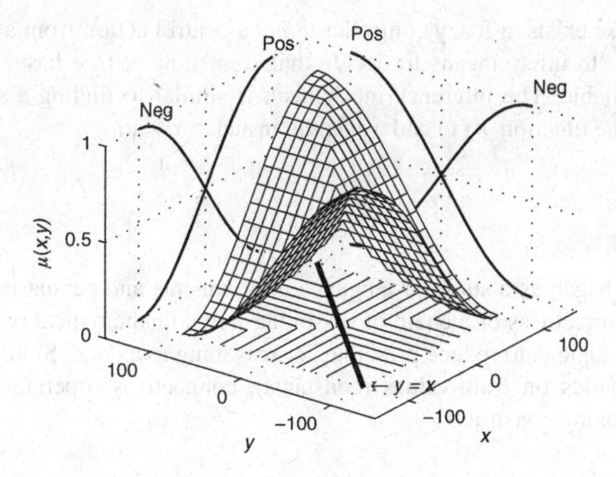

Figure 2.7 Two overlapping granules summarize the function $y = x$ within the universes shown on the x- and y-axes. Each granule corresponds to one rule. (figgranule.m)

2.3.1 Several Rules

Fuzzy controllers execute several rules at a time in parallel. Below is an example of a collection of rules, a *rule base*, consisting of two rules:

1. if x is *Neg* then y is *Neg*
2. if x is *Pos* then y is *Pos*

The term *Neg* is a label of a fuzzy set referring to negative values, and *Pos* is a label referring to positive values. The rule base is again a relation, and Figure 2.7 is a picture of the relation with its associated membership values. The rule base links the output y to the input x, much like a function $y = f(x)$. Formally, the rule base is a fuzzy relation on the (classical) Cartesian product of all input–output combinations.

The example rule base associates negative values of x to negative values of y, and positive values of x to positive values of y. It is thus a rough approximation of the function $y = x$. Each rule constitutes a *granule* of the function (Zadeh 1994), and the whole rule base is the union of the granules. If, in general, a rule i is represented by a relational matrix $\boldsymbol{R}_i$, then the whole rule base of m rules is the element-by-element union

$$\boldsymbol{R} = \bigcup_{i=1}^{m} \boldsymbol{R}_i$$

A granulated function can be seen as a fuzzy discretization of a continuous function. The discretization is coarse, but short; it may nevertheless be sufficient. In daily conversation, for instance, the sentence 'John is tall' rather than 'John is 196 cm (6 ft 5 in.)' is often sufficient.

Both the left-hand side and the right-hand side of a rule can accommodate several variables. It is thus straightforward to generalize to multi-variable relations.

Once the rule base exists, a fuzzy controller *infers* a control action from a given measured input x_0. The verb 'to infer' means to decide that something is true based on information that is already available. The inference mechanism is similar to finding a specific instance $y_0 = f(x_0)$ given the function $f(x)$ and a specific instance $x = x_0$.

2.4 Fuzzy Logic

The study of logic began as a study of language in arguments and persuasion, and it can be used to judge the correctness of a chain of reasoning – in a mathematical proof for instance. The goal of formal logic is to reduce principles of reasoning to a code. Similar to a program, a logical system builds on truth-values (constants), connectives (operators), and rules of inference (programming constructs).

2.4.1 *Truth-Values*

The 'truth' or 'falsity' assigned to a proposition is its *truth-value*. In two-valued logic, a proposition can be either true (1) or false (0), that is, the truth-values belong to the set {0, 1}. If we try to count the number of possible ways to combine two logical variables, the result is $(2^2)^2 = 16$ tables of truth-values, the *truth-tables*.

A possible extension to classical logic is to include an intermediate truth-value 'undecided' (0.5). The result is a three-valued logic. If one tries to identify all possible binary combinations, it would result in $(3^2)^3 = 729$ truth-tables, which is impractical. There are, nevertheless, a multitude of three-valued logics, differing in the specification of truth-tables. Extensions to more truth-values, finite in number, lead to multi-valued logics (Rescher in Nguyen and Walker 2000).

Fuzzy logic extends the range of truth-values to the continuous interval [0, 1] of real numbers. In *fuzzy logic* a proposition may be true or false, or any intermediate truth-value. For instance, the sentence 'John is tall' may assume an intermediate truth-value of, let us say, 0.8.

Originally, Zadeh interpreted a truth-value in fuzzy logic as a fuzzy set, for instance, *Very true* (Zadeh 1988). Thus, Zadeh based fuzzy logic on the treatment of *Truth* as a linguistic variable that takes words or sentences as values (Zadeh 1975). Our approach differs, as it is built on scalar truth-values rather than membership functions.

2.4.2 *Classical Connectives*

In daily conversation and mathematics, sentences are connected with the words *and*, *or*, *if–then* (or *implies*), and *if and only if*. These are called *connectives*. A sentence that is modified by the word 'not' is called the *negation* of the original sentence. The word 'and' is used to join two sentences to form the *conjunction* of the two sentences. The word 'or' is used to join two sentences to form the *disjunction* of the two sentences. From two sentences we may construct one, of the form 'If ... then ...'; this is called an *implication*. The sentence following 'If' is the *antecedent*, and the sentence following 'then' is the *consequent*. Other phrases which we shall regard as having the same meaning are 'p implies q', 'p only if q', 'q if p', etc.

Letters and special symbols make the connective structure stand out. Our choice of symbols, or *language*, is

$\neg$	for ‘not’
$\wedge$	for ‘and’
$\vee$	for ‘or’
$\Rightarrow$	for ‘if-then’ (implication)
$\Leftrightarrow$	for if and only if’ (equivalence)

A *propositional variable*, denoted by a letter, takes truth-values as values. An assertion that contains at least one propositional variable is called a *propositional formula*. The main difference between *proposition* and propositional formula is that every proposition has a truth-value, whereas a propositional formula is an assertion whose truth-value cannot be determined until propositions are substituted for its propositional variables. But when no confusion results, we will refer to propositional formulae as propositions.

The next example illustrates how the symbolic forms expose the underlying logical structure.

Example 2.14 *Baseball betting (after Stoll 1979)*

Consider the assertion about four baseball teams: If either the Pirates or the Cubs lose and the Giants win, then the Dodgers will be out of first place, and I will lose a bet. Transform the assertion into a formula.

▶ *Solution*

Since the assertion is an if–then statement, it is an implication, symbolized by $\Rightarrow$. *The antecedent is composed from the three sentences* $\neg p$ *(The Pirates lose),* $\neg c$ *(The Cubs lose), and* g *(The Giants win). The consequent is the conjunction of* $\neg d$ *(The Dodgers will be out of first place) and* $\neg b$ *(I will lose a bet). The original sentence may thus be represented by* $[(\neg p \vee \neg c) \wedge g] \Rightarrow (\neg d \wedge \neg b)$.

A *truth-table* summarizes the possible truth-values of an assertion. Take for example the truth-table for the formula $p \vee q$ below (left). The table lists all possible combinations of (classical) truth-values – the Cartesian product – of the variables p and q in the two leftmost columns. The rightmost column holds the truth-values of the formula. Alternatively, the truth-table can be rearranged into a two-dimensional array, a so-called *Cayley table* (below, right).

p	q	$p \vee q$
0	0	0
0	1	1
1	0	1
1	1	1

is equivalent to

Or

$p \vee q$	0	1	$\rightarrow q$
0	0	1	
1	1	1	
$\downarrow$			
p			

Along the vertical axis in the Cayley table, symbolized by arrow $\downarrow$, are the possible values 0 and 1 of the first argument p. Along the horizontal axis, symbolized by arrow $\rightarrow$, are the possible values 0 and 1 of the second argument q. Above the table is the symbolic form $p \vee q$ for disjunction. At the intersection of row i and column j (only counting the inside of the box) is the truth-value of the expression $p_i \vee q_j$.

By inspection, one entry renders $p \vee q$ false, while three entries render $p \vee q$ true. Truth-tables for binary connectives are thus given by two-by-two matrices. A total of 16 such tables can be constructed, and each is associated with a connective.

In order to proceed to other truth-tables, it is necessary to define *negation*. Classical negation is a function that maps the truth-value 0 to 1 and vice versa. Thus, if the variable p is 1, then $\neg p$ is 0. Double negation, that is, the negation of a negation, leaves the truth-value unchanged. Negation therefore is its own inverse according to the *law of involution*,

$$\neg(\neg p) = p \tag{2.5}$$

Having defined 'or' and 'negation' we can derive the truth-table for $(\neg p) \vee (\neg q)$, because we just have to negate the values of p and q first, and then combine those using the truth-table for 'or'. Before doing so, we introduce *De Morgan's laws* as follows,

$$\neg(p \wedge q) \equiv (\neg p) \vee (\neg q) \tag{2.6}$$

$$\neg(p \vee q) \equiv (\neg p) \wedge (\neg q) \tag{2.7}$$

The two laws define how to distribute negation over a parenthesis, much like multiplying a sum by −1. They furthermore provide a link between 'and' and 'or'. We observe that the previously mentioned expression $(\neg p) \vee (\neg q)$ appears on the right-hand side of Equation (2.6). Its left-hand side is clearly the negation of 'and', which we usually designate 'nand' (for 'not and'). We thus derive the truth-table for 'nand' as

Nand
$(\neg p) \vee (\neg q)$

$p \downarrow$ \ $q \rightarrow$	0	1
0	1	1
1	1	0

From 'nand' we derive 'and' by negating the entries inside the table (not nand = and),

And
$\neg((\neg p) \vee (\neg q))$

$p \downarrow$ \ $q \rightarrow$	0	1
0	0	0
1	0	1

Notice that it was possible to define 'and' by means of 'or' and 'not', rather than assume it given as an *axiom*, that is, a rule or principle most people believe to be true. It is straightforward to derive 'nor' by means of 'or' and negation.

2.4.3 *Fuzzy Connectives*

It is possible to define truth-tables for fuzzy connectives by analogy. Assume for the sake of clarity that the universe of truth-values is restricted to three truth-values, that is $\langle 0, 0.5, 1 \rangle$. The middle truth-value 0.5 can be interpreted as 'undecided' if the usual truth-values are interpreted as 'true' (1) and 'false' (0). If we start again by defining negation and disjunction, we can derive the truth-tables of other connectives from that point of departure.

Let us define *negation* as set complement, that is,

$$\neg p \equiv 1 - p$$

By insertion, we can immediately ascertain the law of involution: $\neg(\neg p) = p$. If we define *disjunction* as set union, that is,

$$p \vee q \equiv \max(p, q) \tag{2.8}$$

then the truth-table for the fuzzy connective 'or' is the following,

Or

$p \vee q$

$p \downarrow$ / $\rightarrow q$	0	0.5	1
0	0	0.5	1
0.5	0.5	0.5	1
1	1	1	1

As before, the p-axis is vertical and the q-axis horizontal. At the intersection of row i and column j (only considering the inside of the box) we have the value of the expression $\max(p_i, q_j)$ in accordance with Equation (2.8).

It is reasonable to demand that fuzzy connectives agree with their classical counterparts defined in the truth-domain $\langle 0, 1 \rangle$. That is, in terms of truth-tables, the values in the four corners of the fuzzy Cayley table should agree with the Cayley table for the classical connective. Indeed, the four corners of 'fuzzy or' check with the table for two-valued 'or'.

Now derive the truth-table for 'nand' from 'or' by the definition $(\neg p) \vee (\neg q)$ by negating the variables. Moving further, 'and' is the negation of the entries in the truth-table for 'nand'. We have thus acquired two more truth-tables,

Nand

$(\neg p) \vee (\neg q)$

$p \downarrow$ / $\rightarrow q$	0	0.5	1
0	1	1	1
0.5	1	0.5	0.5
1	1	0.5	0

And

$\neg((\neg p) \vee (\neg q))$

$p \downarrow$ / $\rightarrow q$	0	0.5	1
0	0	0	0
0.5	0	0.5	0.5
1	0	0.5	1

Observe that even though the truth-table for 'and' is derived from the truth-table for 'or', the truth-table for 'and' is identical to one generated using the min operation: set intersection.

The derivations above are certainly valid for the three-valued domain of truth-values, as shown, but they are also valid for *all* truth-values in [0, 1]. Assuming a truth-domain $\langle 0, u, 1\rangle$, where u varies between 0 and 1, all truth-values are accounted for. With our definitions of fuzzy 'or' and 'not', any expression built from the operations $\vee, \wedge, \neg$ and valid for $\langle 0, 0.5, 1\rangle$, is also valid for $\langle 0, u, 1\rangle$ (Nguyen and Walker 2000). That is, we only have to check the validity regarding the three truth-values rather than the whole continuous range.

Example 2.15 *Proof by truth-table*

Use truth-tables to prove that $\neg p \vee \neg q$ is the same as $\neg(p \wedge q)$.

► *Solution*

The propositions contain two variables p and q. Restricting each variable to the three-valued truth-domain, the number of possible combinations is $3^2 = 9$. By an exhaustive investigation of all these possibilities, we can check whether the two propositions result in the same truth-table. Below are two tables, one for each proposition:

p	q	$\neg p$	$\neg q$	$\neg p \vee \neg q$
0	0	1	1	1
0	0.5	1	0.5	1
0	1	1	0	1
0.5	0	0.5	1	1
0.5	0.5	0.5	0.5	0.5
0.5	1	0.5	0	0.5
1	0	0	1	1
1	0.5	0	0.5	0.5
1	1	0	0	0

p	q	$p \wedge q$	$\neg(p \wedge q)$
0	0	0	1
0	0.5	0	1
0	1	0	1
0.5	0	0	1
0.5	0.5	0.5	0.5
0.5	1	0.5	0.5
1	0	0	1
1	0.5	0.5	0.5
1	1	1	0

In each table the two leftmost columns hold all possible combinations of truth-values for p and q. The following columns stepwise develop the truth-values for the entire proposition.

Clearly, the rightmost columns of the two tables are identical. Moreover, since the propositions are valid for the three truth-values, they are also valid for all truth-values of fuzzy logic (at least as long as we use min *for $\wedge$ and* max *for $\vee$). Therefore $\neg p \vee \neg q$ is the same as $\neg(p \wedge q)$.*

The example shows that it is possible to use truth-tables to prove a logical proposition. It is even possible to do it automatically by computer, which is much easier and faster than traditional proof techniques.

To summarize, it is possible to derive truth-tables for the connectives 'nand', 'and' and 'nor' (not shown). In order to do so, it is necessary to assume that 'or' and 'negation' are defined beforehand. Furthermore, it is necessary to assume that De Morgan's two laws are valid. In other words, using the defined language of symbols, we started from four axioms and derived the other connectives.

Such an axiomatic approach is useful in practice, because it ensures a logically consistent system of operations. For instance, we are certain that we can apply De Morgan's laws when the connectives are defined as above. It is possible to prove that there are other laws, which are valid as well.

2.4.4 *Triangular Norms*

Above, we assumed that 'or' is the max operator, and consequently 'and' is min, but there are other alternatives. The operations 'and' and 'or' are always defined in pairs, and another such pair is,

$$\text{Product:} \quad x \wedge y \equiv xy \tag{2.9}$$

$$\text{Probabilistic sum:} \quad x \vee y \equiv x + y - xy \tag{2.10}$$

Example 2.16 *Product and probabilistic sum*

If we use product for 'and' and probabilistic sum for 'or', prove that De Morgan's laws are valid.

► *Solution*

Take the first De Morgan law $\neg(p \wedge q) \equiv (\neg p) \vee (\neg q)$. *With our usual definition of negation, the right-hand side is*

$$\begin{aligned}(\neg p) \vee (\neg q) &= (1 - p) + (1 - q) - (1 - p)(1 - q) \\ &= 1 - p + 1 - q - (1 - q - p + pq) \\ &= 1 - pq\end{aligned}$$

The left-hand side is,

$$\neg(p \wedge q) = 1 - pq$$

Thus the two sides are equivalent, and the law holds. The second De Morgan law can be proved in a similar way.

All so-called *t-norms* and *t-conorms* are valid candidates for the fuzzy connectives 'and' and 'or'. For conjunction, a *triangular norm* (*t-norm*) is an operation Δ satisfying (Nguyen and Walker 2000)

$$\begin{aligned}1 \,\Delta\, x &= x \\ x \,\Delta\, y &= y \,\Delta\, x \\ x \,\Delta\, (y \,\Delta\, z) &= (x \,\Delta\, y) \,\Delta\, z \\ \text{If } w \leq x \text{ and } y \leq z \text{ then } w \,\Delta\, y &\leq x \,\Delta\, z\end{aligned}$$

It is called triangular, because it satisfies the triangle inequality when viewed as a distance norm. Some examples of t-norms and their names are

$$\text{Nilpotent:} \quad \begin{cases} \min(x, y) & \text{if } x + y > 1 \\ 0 & \text{otherwise} \end{cases}$$

$$\text{Lukasiewicz:} \quad \max(0, x + y - 1)$$

$$\text{Drastic:} \quad \begin{cases} y & \text{if } x = 1 \\ x & \text{if } y = 1 \\ 0 & \text{otherwise} \end{cases}$$

$$\text{Product:} \quad xy$$

$$\text{Hamacher:} \quad \frac{xy}{x + y - xy}$$

$$\text{Minimum (Gödel):} \quad \min(x, y)$$

For disjunction, a *t-conorm* is a dual operation $\triangledown$ satisfying

$$\begin{aligned} 0 \triangledown x &= x \\ x \triangledown y &= y \triangledown x \\ x \triangledown (y \triangledown z) &= (x \triangledown y) \triangledown z \end{aligned}$$

$$\text{If } w \leq x \text{ and } y \leq z \text{ then } w \triangledown y \leq x \triangledown z$$

We can use De Morgan's second law to generate a t-conorm $\triangledown$ from a t-norm $\triangle$, because

$$\neg(x \triangledown y) = (\neg x) \triangle (\neg y) \Leftrightarrow$$

$$x \triangledown y = \neg[(\neg x) \triangle (\neg y)]$$

With our definition of negation the formula is

$$x \triangledown y = 1 - [(1 - x) \triangle (1 - y)] \tag{2.11}$$

Example 2.17 *Product conjunction*

Given that logical 'and' is defined as a product in the truth-domain $\langle 0, 0.5, 1 \rangle$, *use Equation (2.11) to define 'or'.*

► *Solution*

Insertion into Equation (2.11) yields

$$\begin{aligned} x \triangledown y &= 1 - [(1 - x)(1 - y)] \\ &= 1 - (1 - y - x + xy) \\ &= x + y - xy \end{aligned}$$

The result checks with the definition stated above in Equation (2.10).

Example 2.18 *Min conjunction (Gödel).*
Same as before, but now logical 'and' is defined as min. *How is then 'or' defined?*
► *Solution*
Insertion into Equation (2.11) yields

$$\begin{aligned} x \bigtriangledown y &= 1 - \min\left[(1-x), (1-y)\right] \\ &= 1 - \left[\min(1,1) - \min(1,y) - \min(x,1) + \min(x,y)\right] \\ &= x + y - \min(x,y) \\ &= \max(x,y) \end{aligned}$$

The result checks with our previous derivations by truth-tables.

In fuzzy control systems the definition of the fuzzy connectives 'and' and 'or' may vary from application to application depending on the problem area.

2.5 Summary

Fuzzy reasoning is based on fuzzy logic, which is in turn based on fuzzy set theory. The idea of a fuzzy set is basic and simple: an object is allowed to have a gradual membership of a set. The idea pervades all derived mathematical aspects of set theory. In fuzzy logic an assertion is allowed to be more or less true. A truth-value in fuzzy logic is a real number in the interval [0, 1], rather than the set of two truth-values {0, 1} of classic logic.

Fuzzy logic has seen many applications, one among them being fuzzy control. There is more to be said about fuzzy logic, and fuzzy control applies just a subset of the arsenal of operations and definitions.

2.6 Theoretical Fuzzy Logic*

This section can be skipped on a first reading, since it is an excursion into the theory of logic without immediate application in fuzzy control systems.

In a broad sense, fuzzy logic has several facets, including mathematical logic. There are difficult, unresolved issues around the fuzzy implication connective, which is why this section points to some opportunities for research.

The *implication* connective has always troubled fuzzy logicians. If we define it in the same way as *material implication* above, then it causes several useful logical laws to break down. We must make a design choice at this point, in order to proceed with the definition of implication and equivalence.

2.6.1 Tautologies

We have to select a set of logical laws, which we would like to use in a given logical system. We cannot select any set of laws, however; some laws known from two-valued logic do not hold, whatsoever, in fuzzy logic.

Example 2.19 *Excluded middle*

Take the formula

$$p \vee \neg p \Leftrightarrow 1 \tag{2.12}$$

which is equivalent to the law of the excluded middle. *Show that it does not hold in fuzzy logic.*

► *Solution*

In order to show that the law does not hold, it is sufficient to find a single instance where it breaks down. Testing with the truth-value $p = 0.5$ *(fuzzy logic) the left-hand side of the equivalence symbol* $\Leftrightarrow$ *yields*

$$p \vee \neg p = 0.5 \vee \neg 0.5 = \max(0.5, 1 - 0.5) = 0.5.$$

The right-hand side is 1, *which is different, and thus the law of the excluded middle does not hold in fuzzy logic.*

If a proposition is true with a truth-value of 1, for *any* combination of truth-values assigned to the variables, we shall say it is *valid*. Such a proposition is a *tautology*. If the proposition is true for some, but not all, combinations, we shall say it is *satisfiable*. Thus Equation (2.12) in the previous example is only satisfiable; it is valid in two-valued logic, but not in three-valued logic.

Many tautologies exist in classic logic, and Table 2.1 lists seven that are very useful to have. Whether these propositions are valid in fuzzy logic depends on how we define the connectives, or rather, we must define the connectives – implication in particular – such that those propositions become valid. The table thus constitutes a test battery against which we can assess the quality of our definitions of the connectives.

We have to be careful, though, because testing with just three truth values $\langle 0, 0.5, 1 \rangle$ may be insufficient to establish validity everywhere. On the other hand, if we just find a single instance where a law breaks down, then the law is invalid.

Table 2.1 Useful tautologies.

No.	Definition	Name
1	$[p \wedge (p \Rightarrow q)] \Rightarrow q$	modus ponens
2	$[(p \Rightarrow q) \wedge (q \Rightarrow r)] \Rightarrow (p \Rightarrow r)$	hypothetical syllogism
3	$[p \wedge (p \Rightarrow q)] \Leftrightarrow (p \wedge q)$	weak conjunction
4	$\neg(p \wedge q) \Leftrightarrow (\neg p) \vee (\neg q)$	De Morgan 1
5	$\neg(p \vee q) \Leftrightarrow (\neg p) \wedge (\neg q)$	De Morgan 2
6	$[\neg q \wedge (p \Rightarrow q)] \Rightarrow p$	modus tollens
7	$[\neg p \wedge (p \vee q)] \Rightarrow q$	disjunctive syllogism

2.6.2 *Fuzzy Implication*

We would like to proceed to define the connective 'implication' by means of 'or' and 'not'. Classical logic defines implication as *material implication*, $\neg p \vee q$. We negate the p-axis of the 'or' table, which is equivalent to permuting the rows of the 'or' table,

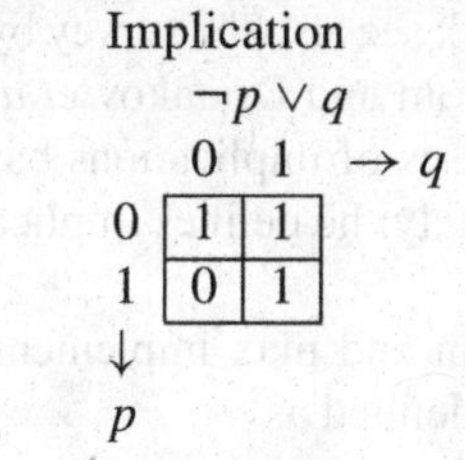

Implication

$\neg p \vee q$

$p \downarrow$ \ $q \rightarrow$	0	1
0	1	1
1	0	1

Equivalence ($\Leftrightarrow$) is taken to mean $(p \Rightarrow q) \wedge (q \Rightarrow p)$. We find its truth-table from the truth-table of implication by taking the conjunction of each entry with the elements of the transposed table, element by element:

Equivalence

$(p \Rightarrow q) \wedge (q \Rightarrow p)$

$p \downarrow$ \ $q \rightarrow$	0	1
0	1	0
1	0	1

It is possible to evaluate, in principle at least, a formula by an exhaustive test of all combinations of truth-values of the variables, as the next example illustrates.

Example 2.20 *Array based logic*

In the baseball example, we derived the relation $((\neg p \vee \neg c) \wedge g) \Rightarrow (\neg d \wedge \neg b)$. Find the winning combinations.

► *Solution*

The proposition contains five variables, and each variable can take two truth-values. The total number of possible combinations is therefore $2^5 = 32$. Only 23 are legal in the sense that the implication ($\Rightarrow$) is true for these combinations. The remaining $32 - 23 = 9$ cases are illegal, in the sense that the implication is false for those combinations. We are interested only in those legal combinations for which 'I win the bet' ($b = 1$), and the following table is obtained:

p	c	g	d	b
0	0	0	0	1
0	0	0	1	1
0	1	0	0	1
0	1	0	1	1
1	0	0	0	1
1	0	0	1	1
1	1	0	0	1
1	1	0	1	1
1	1	1	0	1
1	1	1	1	1

There are thus 10 *winning outcomes out of* 32 *possible.*

Such an exhaustive test of all possible combinations is the idea behind array based logic *(Franksen 1979).*

Many researchers have proposed implication connectives (e.g. Zadeh 1973, Mizumoto *et al.* 1979, Fukami *et al.* 1980, Wenstøp 1980; see also the survey by Lee 1990). In fact, Kiszka *et al.* (1985) list 72 alternatives to choose from and Driankov *et al.* (1996) test nine implications. Nguyen and Walker (2000) define classes of implications based on three basic forms. Hájek (1998) approaches the problem oppositely: he defines implication and negation first, then the others follow.

One candidate, that matches the min and max implementations of 'and' and 'or', is the so-called *Gödel implication*. It can be defined as

$$p \Rightarrow q \equiv (p \leq q) \vee q \tag{2.13}$$

The truth-tables for implication and equivalence are

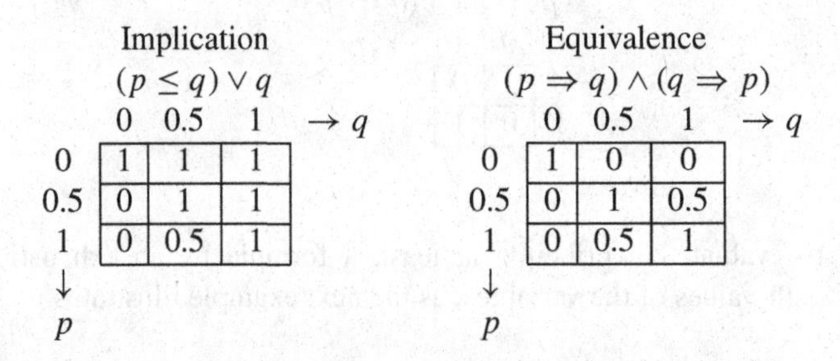

Implication $(p \leq q) \vee q$

$p \downarrow$ / $\rightarrow q$	0	0.5	1
0	1	1	1
0.5	0	1	1
1	0	0.5	1

Equivalence $(p \Rightarrow q) \wedge (q \Rightarrow p)$

$p \downarrow$ / $\rightarrow q$	0	0.5	1
0	1	0	0
0.5	0	1	0.5
1	0	0.5	1

The truth-table for equivalence ($\Leftrightarrow$) is derived from implication and conjunction, once it is agreed that $p \Leftrightarrow q$ is the same as $(p \Rightarrow q) \wedge (q \Rightarrow p)$. Further truth-tables can be built on the previously defined operations, for example 'exclusive or', $\neg(p \Rightarrow q)$, and 'nor' = 'not or', $\neg(p \vee q)$.

Example 2.21 *Fuzzy baseball*

The baseball example illustrates what difference three-valued logic makes. Try the baseball example with three fuzzy truth-values.

► *Solution*

The proposition $((\neg p \vee \neg c) \wedge g) \Rightarrow (\neg d \wedge \neg b)$ *contains five variables. Now that each can take three truth-values, there are* $3^5 = 243$ *possible combinations;* 148 *of these are legal in the sense that the proposition is true (truth-value* 1*). If we are interested again in the legal combinations for which 'I win the bet'* ($b = 1$)*, then there are* 33 *winning outcomes out of* 148*. Instead of listing them all, we show one for illustration,*

$$(p, c, g, d, b) = (0.5, 0.5, 0, 1, 1)$$

With two-valued logic, we found 10 *winning outcomes out of* 32 *possible.*

Table 2.2 Proof of tautology 1 in three-valued logic. The tautology is valid since the rightmost column contains purely 1s.

p	q	$p \Rightarrow q$	$[p \wedge (p \Rightarrow q)]$	$[p \wedge (p \Rightarrow q)] \Rightarrow q$
0	0	1	0	1
0	0.5	1	0	1
0	1	1	0	1
0.5	0	0	0	1
0.5	0.5	1	0.5	1
0.5	1	1	0.5	1
1	0	0	0	1
1	0.5	0.5	0.5	1
1	1	1	1	1

The example indicates that fuzzy logic provides more solutions, compared to two-valued logic, and requires more computational effort.

It is straightforward to test whether tautologies are valid in three-valued logic, because it is possible to perform an exhaustive test of all combinations of truth-values of the variables by computer.

Example 2.22 *Proof of tautology 1*

Is tautology 1 valid in three-valued logic?

► *Solution*

Tautology 1 is

$$[p \wedge (p \Rightarrow q)] \Rightarrow q$$

Since the proposition contains two variables p and q, and each variable can take three truth-values, there will be $3^2 = 9$ possible combinations to test. Table 2.2 has nine rows accordingly. Columns 1 and 2 are the input combinations of p and q. Column 3 is the result of Gödel implication, and column 4 is the left-hand side of tautology 1. Since the rightmost column is all ones, the proposition is a tautology.

In fact, the example suggests a new tautology. Compare the truth-values for $[p \wedge (p \Rightarrow q)]$ (column 4 in Table 2.2) with the truth-table for conjunction – they are identical. We thus have

$$[p \wedge (p \Rightarrow q)] \Leftrightarrow p \wedge q \tag{2.14}$$

This is in fact the so-called *weak conjunction* (tautology 3).

Example 2.23 *Mamdani 'implication'*

Let $\mathcal{A}$ and $\mathcal{B}$ be fuzzy sets defined on $\mathcal{X}$ and $\mathcal{Y}$ respectively, then the so-called Mamdani 'implication' *(Mamdani 1977) is a fuzzy set in $\mathcal{X} \times \mathcal{Y}$ with the membership function*

$$\mu_{\mathcal{A}' \Rightarrow' \mathcal{B}}(x, y) = \min(\mu_{\mathcal{A}}(x), \mu_{\mathcal{B}}(y))$$

The Mamdani 'implication' is often used in fuzzy control. Is it a valid implication?

► *Solution*

Notice that the definition is similar to the definition of the fuzzy Cartesian product. Its Cayley table, in two-valued logic, is

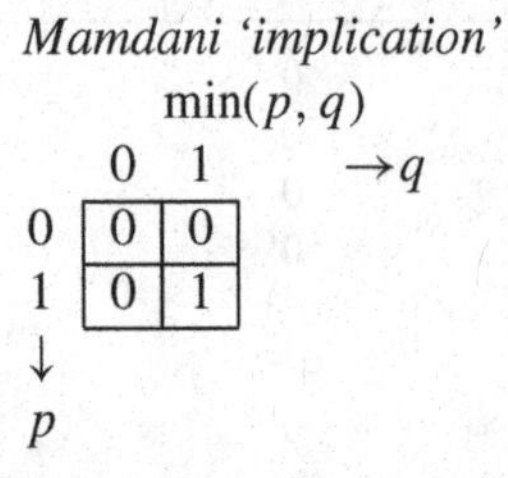

Mamdani 'implication'
min(p, q)
0 1 →q
0 0 0
1 0 1
↓
p

Only one out of the four truth-values matches the truth-table for two-valued implication.

The example shows, that, despite its name, the Mamdani 'implication' is not an implication in the sense of material implication in two-valued logic.

2.6.3 Rules of Inference

Logic provides means of reasoning by *inference,* which is the drawing of conclusions from assertions. The verb 'to infer' means to decide that something is true based on information that is available. *Rules of inference* specify conclusions drawn from assertions known or assumed to be true.

One such rule of inference is *modus ponens*. It is often presented in the form of an *argument*:

$$\frac{\begin{array}{c} P \\ P \Rightarrow Q \end{array}}{Q}$$

In words, if (1) P is known to be true, and (2) we assume that $P \Rightarrow Q$ is true, then (3) Q must be true. Restricting for a moment to two-valued logic, we see from the truth-table for implication,

Implication

$p \Rightarrow q$

$p \downarrow$ \ $\rightarrow q$	0	1
0	1	1
1	0	1

that whenever both $P \Rightarrow Q$ and P are true, then so is Q; in P true we consider only the second row, which contains only a single 1 leaving Q true as the only solution.

In such an argument, the assertion P is the *premise*, the assertion $P \Rightarrow Q$ is the *implication*, and the assertion below the line is the *conclusion*. Notice that the premise and the implication are considered as true and only true.

On the other hand, underlying the modus ponens is tautology 1, which expresses the same, but for *all* truth-values. Only if tautology 1 is valid in fuzzy logic, can we rely on modus ponens inference in fuzzy logic.

Example 2.24 *Four useful rules of inference*
There are several useful rules of inference, which can be represented in tautological form. Give examples from daily life.
► *Solution*
(a) Modus ponens. *Its tautological form is (tautology 1)*

$$[p \wedge (p \Rightarrow q)] \Rightarrow q$$

Let p stand for 'altitude sickness', and let $p \Rightarrow q$ *stand for 'altitude sickness causes a headache'. If it is known that John suffers from altitude sickness, p is true. Furthermore, if* $p \Rightarrow q$ *is assumed to be true for the case of illustration, then the conclusion q is true, that is, John has a headache.*
(b) Modus tollens. *Its tautological form is (tautology 6)*

$$[\neg q \wedge (p \Rightarrow q)] \Rightarrow \neg p$$

Let p and q be as in (a). Thus, if John does not have a headache, then we may infer that John is not suffering from altitude sickness.
(c) Disjunctive syllogism. *Its tautological form is (tautology 7)*

$$[(p \vee q) \wedge \neg p] \Rightarrow q$$

Let p stand for 'altitude sickness' as previously, but let q stand for 'dehydration'. Thus, if it is known for a fact that John's headache is due to either altitude sickness or dehydration, and it is not altitude sickness, then we may infer that John is suffering from dehydration.
(d) Hypothetical syllogism. *Its tautological form is (tautology 2):*

$$[(p \Rightarrow q) \wedge (q \Rightarrow r)] \Rightarrow (p \Rightarrow r).$$

Let p stand for 'high altitude and fast ascent', let q stand for 'altitude sickness', and let r stand for 'a headache'. Further, assume that high altitude and fast ascent together cause altitude sickness, and in turn that altitude sickness causes a headache. Thus, we may infer that John will get a headache at high altitude if John ascends fast.

Provided the tautological forms are valid in fuzzy logic, the inference rules may be applied in fuzzy logic as well. Testing with our previous definitions of $\neg$, $\wedge$, $\vee$, and $\Rightarrow$ shows that tautologies 1, 2, 3, 4, 5, and 7 are valid, while tautology 6 (modus tollens) is only satisfiable. Further logical relationships (31 in total) have been tested earlier (Jantzen 1995).

In the case of discrete membership functions, the modus ponens inference mechanism is recast into arrays. The next example shows the mechanism.

Example 2.25 *Array based modus ponens*

Switching to vector-matrix representation, how is modus ponens performed?

► *Solution*

With **p** *a (column) vector and* ***R*** *a two-dimensional truth-table, with the p-axis vertical, the inference is defined as*

$$\mathbf{q}^t = \mathbf{p}^t \circ \boldsymbol{R}$$

The operation $\circ$ *is an inner* $\vee - \wedge$ *product (read 'or-and product'). The* $\wedge$ *operation is the same as in* $p \wedge (p \Rightarrow q)$ *and the* $\vee$ *operation along the columns yields what can possibly be implied about q; compare the rightmost implication in* $[p \wedge (p \Rightarrow q)] \Rightarrow p$*. Assuming that p is true corresponds to setting*

$$\mathbf{p} = \begin{bmatrix} 0 \\ 1 \end{bmatrix}$$

But the scheme is more general, because we could also assume **p** *is false, compose with* **R** *and study what can be inferred about* **q**. *In the case of modus ponens,*

$$\boldsymbol{R} = \begin{bmatrix} 1 & 1 \\ 0 & 1 \end{bmatrix}$$

which is the truth-table for $p \Rightarrow q$*. Assigning* **p** *as above,*

$$\mathbf{q}^t = \mathbf{p}^t \circ \boldsymbol{R} = \begin{bmatrix} 0 & 1 \end{bmatrix} \circ \begin{bmatrix} 1 & 1 \\ 0 & 1 \end{bmatrix} = \begin{bmatrix} 0 & 1 \end{bmatrix}$$

The outcome $\mathbf{q}^t$ *is a truth-vector pointing at q true as the only possible conclusion, as expected.*

Trying $\mathbf{p} = (1\ 0)^t$ *yields*

$$\mathbf{q}^t = \mathbf{p}^t \circ \boldsymbol{R} = \begin{bmatrix} 1 & 0 \end{bmatrix} \circ \begin{bmatrix} 1 & 1 \\ 0 & 1 \end{bmatrix} = \begin{bmatrix} 1 & 1 \end{bmatrix}$$

Thus q could be anything, true or false, as expected.

The inference could even proceed in the reverse direction, from q to p, but then we must compose from the right side of ***R*** *to match the axes. Assume for instance that q is true, or* $\mathbf{q} = (10)^t$*, then*

$$\mathbf{p} = \boldsymbol{R} \circ \mathbf{q} = \begin{bmatrix} 1 & 1 \\ 0 & 1 \end{bmatrix} \circ \begin{bmatrix} 1 \\ 0 \end{bmatrix} = \begin{bmatrix} 1 \\ 0 \end{bmatrix}$$

To interpret: if q is false and $p \Rightarrow q$*, then p is false (modus tollens).*

The array based inference mechanism is even more general, because **R** *can be any dimension n* ($n > 0$ *and integer). Given values of* $n - 1$ *variables, the possible outcomes of the remaining variable can be inferred by an n-dimensional inner product. Furthermore, given the values of* $n - d$ *variables (d integer and* $0 < d < n$*), then the truth-array connecting the remaining* *d variables can be inferred. The mechanism is the basis of array based technology (Møller 1998).*

2.6.4 Generalized Modus Ponens

Consider now the argument

$$\frac{\begin{array}{c}A' \\ A \Rightarrow B\end{array}}{B'} \tag{2.15}$$

It is similar to modus ponens, but the premise A' is slightly different from A and thus the conclusion B' is slightly different from B. Mizumoto and Zimmermann give an example (in Zimmermann 1993):

This tomato is very red
If a tomato is red, then the tomato is ripe
This tomato is very ripe

To make the inference rule operational, we use the following definition.

Definition 2.7 *Generalized modus ponens. Let $\mathcal{A}$ and $\mathcal{A}'$ be fuzzy sets defined on $\mathcal{X}$, and $\mathcal{B}$ a fuzzy set defined on $\mathcal{Y}$. Then the fuzzy set $\mathcal{B}'$, induced by 'x is $\mathcal{A}'$' from the fuzzy rule*

$$\text{if } x \text{ is } \mathcal{A} \text{ then } y \text{ is } \mathcal{B},$$

represented by the relation $\mathcal{R}$, is given by

$$\mu_{\mathcal{B}'}(y) = \mu_{\mathcal{A}'}(x) \circ \mathcal{R}$$

The operation $\circ$ is $\vee - \wedge$ composition.

The generalized modus ponens is thus closely tied to relational composition. Notice also that x and y are scalars, but the definition applies vectorially as well, if taken element by element. The next example illustrates the calculations numerically.

Example 2.26 *Generalized modus ponens*
Given the rule 'if altitude is High, then oxygen is Low'. Let the fuzzy set $High$ be defined on a range of altitudes from 0 to 4000 metres (about 12 000 feet),

$$High = \{\langle 0, 0\rangle, \langle 1000, 0.25\rangle, \langle 2000, 0.5\rangle, \langle 3000, 0.75\rangle, \langle 4000, 1\rangle\}$$

and Low be defined on a set of percentages of normal oxygen content,

$$Low = \{\langle 0, 1\rangle, \langle 25, 0.75\rangle, \langle 50, 0.5\rangle, \langle 75, 0.25\rangle, \langle 100, 0\rangle\}$$

Perform a generalized modus ponens using vector-matrix notation.

► *Solution*

As a shorthand notation we write the rule as a logical proposition $High \Rightarrow Low$, where it is understood that the proposition concerns altitude on the left side and oxygen on the right side. We construct the relation $\boldsymbol{R}$, connecting $High$ and Low, using the Gödel implication $(\mu_{High}(x) \leq \mu_{Low}(y)) \vee \mu_{Low}(y)$:

$\boldsymbol{R} =$

	1	0.75	0.5	0.25	0
0	1	1	1	1	1
0.25	1	1	1	1	0
0.5	1	1	1	0.25	0
0.75	1	1	0.5	0.25	0
1	1	0.75	0.5	0.25	0

The boxes and axis annotations make the construction of the table clearer: each element r_{xy} is the evaluation of $\mu_{High}(x) \Rightarrow \mu_{Low}(y)$. The numbers on the vertical axis correspond to μ_{High} and the numbers on the horizontal axis correspond to μ_{Low}. Assuming altitude is $High$, we find by modus ponens

$$\begin{aligned}
\mu^t &= \mu^t_{High} \circ \boldsymbol{R} \\
&= \begin{bmatrix} 0 & 0.25 & 0.5 & 0.75 & 1 \end{bmatrix} \circ \begin{bmatrix} 1 & 1 & 1 & 1 & 1 \\ 1 & 1 & 1 & 1 & 0 \\ 1 & 1 & 1 & 0.25 & 0 \\ 1 & 1 & 0.5 & 0.25 & 0 \\ 1 & 0.75 & 0.5 & 0.25 & 0 \end{bmatrix} \\
&= \begin{bmatrix} 1 & 0.75 & 0.5 & 0.25 & 0 \end{bmatrix}
\end{aligned}$$

The result is identical to Low. Thus modus ponens returns a result as expected.

Assume instead that altitude is Very High,

$$\mu^t_{VeryHigh} = \begin{bmatrix} 0 & 0.06 & 0.25 & 0.56 & 1 \end{bmatrix},$$

That is, the square of μ^t_{High}. Modus ponens yields

$$\begin{aligned}
\mu^t &= \mu^t_{VeryHigh} \circ \boldsymbol{R} \\
&= \begin{bmatrix} 0 & 0.06 & 0.25 & 0.56 & 1 \end{bmatrix} \circ \begin{bmatrix} 1 & 1 & 1 & 1 & 1 \\ 1 & 1 & 1 & 1 & 0 \\ 1 & 1 & 1 & 0.25 & 0 \\ 1 & 1 & 0.5 & 0.25 & 0 \\ 1 & 0.75 & 0.5 & 0.25 & 0 \end{bmatrix} \\
&= \begin{bmatrix} 1 & 0.75 & 0.5 & 0.25 & 0 \end{bmatrix}
\end{aligned}$$

The result is not *identical to the square of* μ_{Low}. *Written as an argument, we have in fact*

$$\frac{\begin{array}{l}\textit{Very High}\\ \textit{High} \Rightarrow \textit{Low}\end{array}}{\textit{Low}}$$

This is not as desired, but in fact all implication operators have some inconvenient feature, and we rest the case.

2.7 Notes and References*

The notes below contain recommendations for beginners as well as for more advanced reading.

Fuzzy logic For an introduction to fuzzy logic and its applications, see the book by Ross (2010). It gives an overview of the mathematics, without being overly mathematical. For an advanced study of pure fuzzy logic, the book by Nguyen and Walker (2000) includes algebras, relations, possibility theory, and fuzzy integrals. The original papers by Zadeh are still relevant and accessible. An article in IEEE Computer is a good starting point (Zadeh 1988) before reading the two central papers (Zadeh 1973, 1975). The first original paper is historically interesting (Zadeh 1965). The array based approach to logic is founded on the work by Franksen (1979) and developed into commercial use by Møller (1986, 1998). The book by Hájek develops a consistent version of fuzzy logic starting from implication and negation (Hájek 1998).

Applications There are now many collections describing applications of fuzzy logic. Two remarkable books by Constantin von Altrock describe case studies related to control and business (von Altrock 1995, 1996). A collection of contributions, edited by Zimmermann, gives an overview of the applications to engineering, medicine, management, and psychology, among others (Zimmermann 1999). An important collection of articles regarding image processing and pattern recognition is Bezdek and Pal, published by IEEE (1992). Regarding automatic control, see the overview article by Lee (1990).

Tools The Fuzzy Logic Toolbox is a Matlab toolbox for membership functions, connectives, inference systems, adaptation of the membership functions, Simulink support, and C code generation. The toolbox includes a tutorial, which is another excellent starting point for learners. Together with the book by Jang *et al.* (1997) students of fuzzy logic and fuzzy control are well equipped. There are many software tools; since software products develop quickly, the World Wide Web is the best reference for these.

3

Fuzzy Control

Fuzzy controllers possess a unique rule based interface to the end-user and the designer. The control strategy is in terms of rules, which makes it readable for process operators. The rules are executed by a computer, by means of an inference engine. Modern fuzzy controllers use the Sugeno type of rule base, in which the right-hand side of a rule is a function of the controller inputs. The rules interpolate between the right-hand sides. In practice, it may be convenient to discretize the rule base and inference, such that the controller is reduced to a table-lookup. With a certain choice of definitions, the rule base is a linear combination of its inputs, and thus the fuzzy controller contains a conventional PID controller as a special case. There are, in general, many design choices to consider, but if the goal is to design a linear controller, most of the choices are made beforehand. In the simplest of cases, the rule base is equivalent to a linear combination of membership values.

If fuzzy logic is 'computing with words', then fuzzy control can be described as 'automatic control with rules'. A fuzzy controller includes not only words, but also sentences that form empirical rules. Fuzzy controllers are especially useful in connection with operator-controlled plants. Consider, for instance, a typical fuzzy controller:

1. If *error* is Neg and *change in error* is Neg then *control* is NB
2. If *error* is Neg and *change in error* is Zero then *control* is NM

⋮

The rules are in the familiar if–then format, with the *premise* on the *if*-side and the *conclusion* on the *then*-side. The term 'Neg' is a *linguistic term* short for the word 'negative', the term 'NB' stands for 'negative big' and 'NM' for 'negative medium'. The collection of rules is a *rule base*. A computer can execute the rules and compute a control action depending on the measured values of *error* and *change in error*.

Foundations of Fuzzy Control: A Practical Approach, Second Edition. Jan Jantzen.

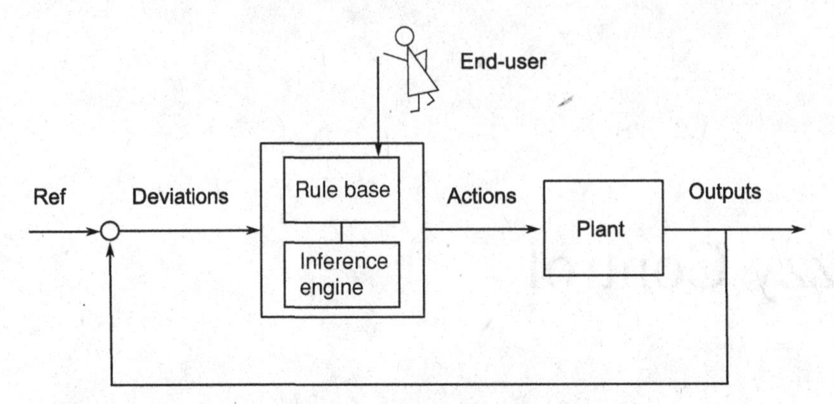

Figure 3.1 Direct control.

A designer is faced with more design questions than usual when designing a fuzzy controller. This chapter will identify and explain those design choices. The last part of the chapter, which consists of *-marked sections that can be skipped on a first reading, contains further details that help to understand the components in commercial software products, and it also contains material suitable for research and development.

3.1 The Rule Based Controller

In a rule based controller the control strategy is in a more or less natural language. A rule based controller is intelligible and maintainable for a non-specialist. An equivalent controller could be implemented using conventional techniques – it is just more convenient to isolate the control strategy in a rule base when operators control the plant.

In the *direct control* scheme in Figure 3.1, the fuzzy controller is in the forward path of a feedback control system. The plant output y is compared with a reference *Ref*, and if the deviation $e = Ref - y$ is non-zero, the controller takes action according to the control strategy embedded in the rule base. In the figure, the arrows can be understood as hyper-arrows containing several signals at a time for multi-loop control.

3.1.1 Rule Base Block

A rule allows for several variables both in the premise and the conclusion. A controller can therefore be multi-input–multi-output (MIMO) or single-input–single-output (SISO). The typical SISO controller regulates a control signal according to an error signal. A controller may actually apply the *error*, the *change in error*, and the *integral error*, but we will still call it SISO control, because the change in error and the integral error are derived from the error, and the loop is a single feedback loop. This section assumes that the control objective is to regulate a plant output around a prescribed *setpoint* (*reference*) using a SISO controller.

A linguistic controller contains rules in the *if–then* format (although they can appear in other formats), for example:

$$\begin{aligned}
&1.\ \text{If } \textit{error} \text{ is Neg and } \textit{change in error} \text{ is Neg then } \textit{control} \text{ is NB}\\
&2.\ \text{If } \textit{error} \text{ is Neg and } \textit{change in error} \text{ is Zero then } \textit{control} \text{ is NM}\\
&3.\ \text{If } \textit{error} \text{ is Neg and } \textit{change in error} \text{ is Pos then } \textit{control} \text{ is Zero}\\
&4.\ \text{If } \textit{error} \text{ is Zero and } \textit{change in error} \text{ is Neg then } \textit{control} \text{ is NM}\\
&5.\ \text{If } \textit{error} \text{ is Zero and } \textit{change in error} \text{ is Zero then } \textit{control} \text{ is Zero}\\
&6.\ \text{If } \textit{error} \text{ is Zero and } \textit{change in error} \text{ is Pos then } \textit{control} \text{ is PM}\\
&7.\ \text{If } \textit{error} \text{ is Pos and } \textit{change in error} \text{ is Neg then } \textit{control} \text{ is Zero}\\
&8.\ \text{If } \textit{error} \text{ is Pos and } \textit{change in error} \text{ is Zero then } \textit{control} \text{ is PM}\\
&9.\ \text{If } \textit{error} \text{ is Pos and } \textit{change in error} \text{ is Pos then } \textit{control} \text{ is PB}
\end{aligned} \tag{3.1}$$

The rule base is an example for the sake of illustration, but we shall use it throughout the book. The names Zero, Pos, Neg are labels of fuzzy sets as well as NB, NM, PB, and PM (negative big, negative medium, positive big, and positive medium respectively). The designer assigns the names. The rule base has two inputs, *error* and *change in error*. The latter is based on the time derivative of the former. These are the inputs to the controller. The conclusion side of each rule prescribes a value for the variable *control*, which is the output of the controller.

The two inputs are combined by means of the logical connective 'and', but this could be any valid combination of fuzzy logic connectives and modifiers – a rule such as 'If *error* is very Neg and not Zero or *change in error* is Zero then ...' is also possible. The most prominent connective, however, is the 'and' connective, often implemented as multiplication instead of minimum.

The connectives 'and' and 'or' are always defined in pairs, as the earlier chapter on fuzzy logic demonstrated. For example, loosely written,

$$\mathcal{A} \wedge \mathcal{B} \equiv \min(\mu_{\mathcal{A}}(x), \mu_{\mathcal{B}}(x)) \text{ (minimum)}$$

$$\mathcal{A} \vee \mathcal{B} \equiv \max(\mu_{\mathcal{A}}(x), \mu_{\mathcal{B}}(x)) \text{ (maximum)}$$

or

$$\mathcal{A} \wedge \mathcal{B} \equiv \mu_{\mathcal{A}}(x) * \mu_{\mathcal{B}}(x) \text{ (product)}$$

$$\mathcal{A} \vee \mathcal{B} \equiv \mu_{\mathcal{A}}(x) + \mu_{\mathcal{B}}(x) - \mu_{\mathcal{A}}(x) * \mu_{\mathcal{B}}(x) \text{ (probabilistic sum)}$$

There are other, more complex, definitions to choose from (Section 2.4.4).

Membership functions can be flat on the top, piece-wise linear and triangular, trapezoidal, or ramps with horizontal shoulders. Figure 3.2 shows some typical shapes of membership functions.

A constant in the conclusion is theoretically a *singleton conclusion*. For example, let $NB = -200$, $NM = -100$, $Zero = 0$, $PM = 100$, $PB = 200$ in the rule base (3.1). There are at least

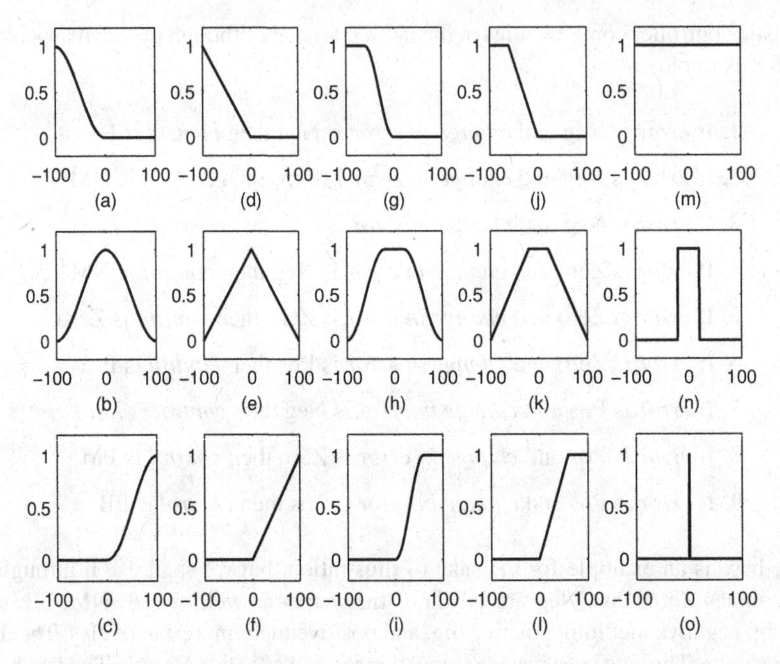

Figure 3.2 Examples of primary sets. Read column-wise: families of smooth triangular functions (a–c), triangular functions (d–f), smooth trapezoidal functions (g–i), and trapezoidal functions (j–l). The last column is the set *anything* (m), a crisp set (n), and a singleton set (o). (figsets.m)

three advantages to using singletons rather than full fuzzy sets on the conclusion side: (1) the computations are simpler; (2) it is possible to drive the control signal to the extremes of the universe; and (3) it is natural for an operator to think in terms of signal magnitude. The constant can be viewed as a fuzzy singleton $\langle x, \mu_A(x)\rangle$ placed in position x. For instance, 200 would be equivalent to the fuzzy membership function $\langle 0, 0, 0, 0, 1\rangle$ defined on the universe $\langle -200, -100, 0, 100, 200\rangle$.

3.1.2 Inference Engine Block

Figure 3.3 is a graphical construction of the inference, where each of the four top rows represents one rule. Consider for instance the first row: if the *error* is negative (row 1, column 1) and the *change in error* is negative (row 1, column 2) then the control action is negative big, NB (row 1, column 3). The chart corresponds to rules 1, 3, 7, and 9 of the rule base (3.1). Since the controller combines the *error* and the *change in error*, the controller is a fuzzy version of a proportional-derivative (PD) controller.

The inference mechanism follows a stepwise calculation procedure, and there is a certain terminology associated with each step (IEC 2000).

1. *Fuzzification*. The measured instances of the *error* and the *change in error* are indicated by the vertical dashed lines through the first and second columns of the chart starting at the points (1) and (2) in the figure. For each rule, the inference engine looks up the fuzzy membership value, which is where the vertical dashed line intersects a membership

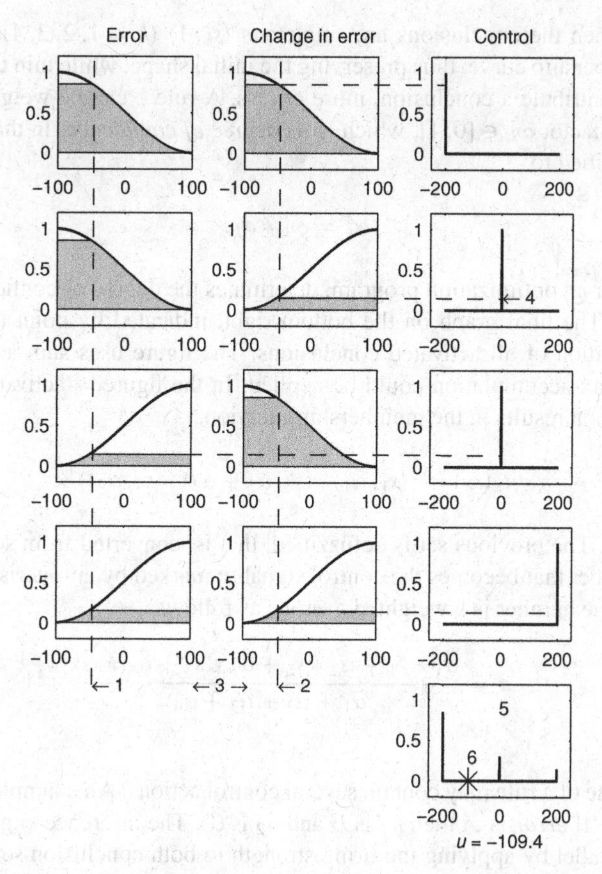

Figure 3.3 Graphical construction of the control signal in a fuzzy PD controller. (figinfer2in)

function. This is called fuzzification, since it associates a fuzzy membership value to a crisp measurement.

2. *Aggregation*. The *firing strength* α_k of a rule k is the degree of fulfilment of the rule premise. Rule k causes a fuzzy membership value $\mu_{A,k}(error)$ corresponding to the *error* measurement, and a membership value $\mu_{B,k}(change\ in\ error)$ corresponding to the *change in error* measurement. Their aggregation, indicated by point (3) in the figure, is the combination of the two,

$$\alpha_k = \mu_{A,k}(error) \wedge \mu_{B,k}(change\ in\ error) \tag{3.2}$$

The $\wedge$-operation is the 'and' connective combining the two propositions in rule base (3.1); in general, it could be a combination of many propositions connected by $\wedge$ or $\vee$. In the figure, $\wedge$ is implemented as the min operation.

3. *Activation*. The activation of a rule is the derivation of a local conclusion depending on the firing strength as indicated by point (4) in the figure. Only a portion of each singleton is activated, and min or $*$ (multiplication) is applied as the *activation operator*. The result

is the same, when the conclusions are singletons $\langle s_k, 1\rangle$ $(k = 1, 2, 3, 4)$, but in general, $*$ scales a membership curve, thus preserving the initial shape, while min clips it. In Figure 3.3, all rules contribute a conclusion, more or less. A rule k can be weighted beforehand by a weighting factor $\omega_k \in [0, 1]$, which is its *degree of confidence.* In that case the firing strength is modified to

$$\alpha_k^* = \omega_k * \alpha_k. \tag{3.3}$$

The designer, or an optimization program, determines the degree of confidence ω_k.

4. *Accumulation.* The final graph on the bottom right, indicated by point (5) in the figure, is the accumulation of all activated conclusions. The figure uses sum-accumulation, but alternatively max-accumulation could be applied. In the figure, $*$-activation followed by sum-accumulation results in the membership function,

$$\mu_c(s_k) = (\langle s_1, \alpha_1\rangle, \langle s_2, \alpha_2 + \alpha_3\rangle, \langle s_4, \alpha_4\rangle) \tag{3.4}$$

5. *Defuzzification.* The previous set is defuzzified, that is, converted from several singletons to a single number that becomes the control signal u, marked by an asterisk (*) at point (6) in the figure. The number is a weighted average as follows,

$$u = \frac{\alpha_1 * s_1 + \alpha_2 * s_2 + \alpha_3 * s_3 + \alpha_4 * s_4}{\alpha_1 + \alpha_2 + \alpha_3 + \alpha_4} \tag{3.5}$$

The conclusion side of a rule may contain several control actions. An example of a one-input–two-output rule is 'If *error* is $\mathcal{A}$ then u_1 is $\mathcal{B}$ and u_2 is $\mathcal{C}$'. The inference engine executes two conclusions in parallel by applying the firing strength to both conclusion sets $\mu_{\mathcal{B}}$ and $\mu_{\mathcal{C}}$. In practice, one would implement this situation as two rules rather than one: 'If *error* is $\mathcal{A}$ then u_1 is $\mathcal{B}$' and 'If *error* is $\mathcal{A}$ then u_2 is $\mathcal{C}$'.

Example 3.1 *Two inputs and one output inference*

Take the rule base in Figure 3.3. The first input is error $= -50$*, and the second input is* change in error$= -50$. *Find the final control signal.*

►*Solution*

1. *Fuzzification. The membership functions yield the following memberships regarding* error*:* $\mu_{Neg,k}(-50) = 0.85$, $\mu_{Pos,k}(-50) = 0.15$. *Regarding* change in error *we get:* $\mu_{Neg,k}(-50) = 0.85$, $\mu_{Pos,k}(-50) = 0.15$.
2. *Aggregation, Equation (3.2). For the four rules* $(k = 1, 2, 3, 4)$ *we get:*

$$\alpha_1 = \mu_{Neg,1}(-50) \wedge \mu_{Neg,1}(-50) = \min(0.85, 0.85) = 0.85$$
$$\alpha_2 = \mu_{Neg,2}(-50) \wedge \mu_{Pos,2}(-50) = \min(0.85, 0.15) = 0.15$$
$$\alpha_3 = \mu_{Pos,3}(-50) \wedge \mu_{Neg,3}(-50) = \min(0.15, 0.85) = 0.15$$
$$\alpha_4 = \mu_{Pos,4}(-50) \wedge \mu_{Pos,4}(-50) = \min(0.15, 0.15) = 0.15$$

3. *Activation. From the four rules we get the weighted singletons*

$$k = 1: \ \langle 0.85, -200 \rangle$$
$$k = 2: \ \langle 0.15, 0 \rangle$$
$$k = 3: \ \langle 0.15, 0 \rangle$$
$$k = 4: \ \langle 0.15, 200 \rangle$$

4. *Accumulation, Equation (3.4). The accumulated fuzzy set consists of three weighted singletons*

$$\mu_c(s_k) = \langle \langle -200, 0.85 \rangle, \langle 0, 0.30 \rangle, \langle 200, 0.15 \rangle \rangle$$

5. *Defuzzification, Equation (3.5). The final control signal is*

$$u = \frac{0.85 * (-200) + 0.15 * 0 + 0.15 * 0 + 0.15 * 200}{0.85 + 0.15 + 0.15 + 0.15} = \frac{-140}{1.30} = -108$$

The result is slightly off the number given in the figure, but that is just because of rounding. Once we find the firing strength of each rule, then the final result is the weighted average of the local conclusions.

3.2 The Sugeno Controller

We saw that conclusions can be singletons, but they can also be linear combinations of the inputs, or even a complex function of the inputs (Takagi and Sugeno 1985). The general *Sugeno* rule structure is

$$\text{If } f(e_1 \text{ is } \mathcal{A}_1, e_2 \text{ is } \mathcal{A}_2, \ldots, e_k \text{ is } \mathcal{A}_k) \text{ then } y = g(e_1, e_2, \ldots, e_k)$$

Here f is a logical function that connects the sentences forming the premise, y is the conclusion, and g is a function of the inputs. A simple example is

$$\text{If E is Pos and CE is Pos then control } y = c,$$

where E and CE are abbreviations for *error* and *changein error* respectively, and c is a constant, or in other words, a singleton $\langle x_i, \mu_y(x_i) \rangle = \langle c, 1 \rangle$. This is a *zero*-order conclusion. An example of a *first*-order conclusion is

$$\text{If } E \text{ is } \mathcal{A} \text{ and } CE \text{ is } \mathcal{B} \text{ then } u = a * E + b * CE + c$$

where a, b, and c are all constants. Inference with several rules proceeds as before, but each control action is linearly dependent on the inputs. The control action from each rule

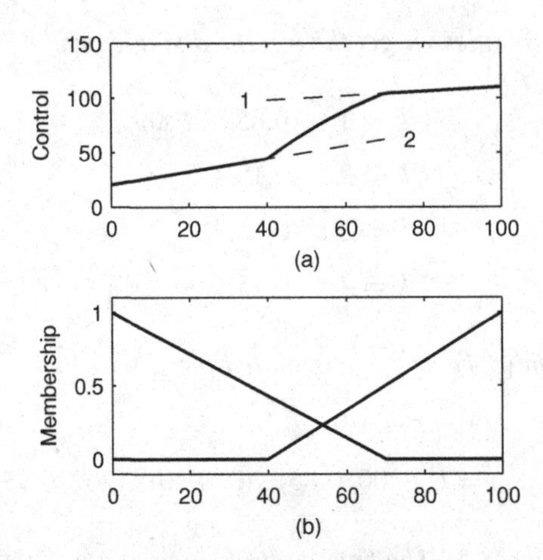

Figure 3.4 Interpolation between two lines (a), in the interval of overlap of two membership functions (b). (figsug2.m)

is a singleton, and the final control signal is the weighted average of the contributions from each rule.

Such a controller can interpolate between linear PD controllers – where each controller is dominated by one rule – with a weighting depending on the overlap of the premise membership functions. It can be useful, in a nonlinear control system, to have each controller operate in its own subspace of the operating space.

Example 3.2 *Interpolation between lines (after Takagi and Sugeno 1985)*
Suppose we have two rules

1. *If error is Large then output is Line1*
2. *If error is Small then output is Line2*

Line 1 is defined as $0.2 * error + 90$ *and line 2 is defined as* $0.6 * error + 20$*. Draw the input–output mapping.*
►*Solution*
The rules interpolate between the two lines in the interval where the membership functions overlap; see Figure 3.4. Outside of that interval the conclusion is a linear function of error.

Example 3.3 *Interpolation between points*
Suppose we have two Sugeno rules of the order zero,

1. *If error is Pos then Control is PB*
2. *If error is Neg then Control is NB*

Let $PB = 100$ *and* $NB = -100$*, draw the input–output mapping.*

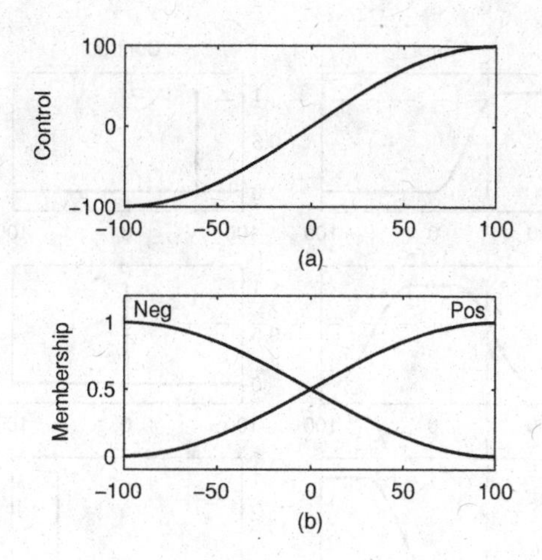

Figure 3.5 Interpolation between two singletons 100 and −100 (a), in the interval of overlap of the two membership functions Pos and Neg (b). (figsug3.m)

►*Solution*

The rules interpolate between the two singletons where the membership functions overlap – see Figure 3.5. The interpolated function inherits a shape more or less like the input membership function Pos.

If the membership functions had been linear (triangular membership functions), the interpolated function would be a straight line. This can be seen as follows. The output signal control is the firing strength weighted average of PB and NB, or, in symbols,

$$Control = \frac{\mu_{Pos}(error) * PB + \mu_{Neg}(error) * NB}{\mu_{Pos}(error) + \mu_{Neg}(error)}$$

However, we have defined the membership functions such that the denominator is always 1. We can therefore rewrite the expression as follows,

$$\begin{aligned} Control &= \mu_{Pos}(error) * PB + \mu_{Neg}(error) * NB \\ &= \mu_{Pos}(error) * PB + [1 - \mu_{Pos}(error)] * NB \\ &= (PB - NB) * \mu_{Pos}(error) + NB \end{aligned}$$

It is now clear that Control will be a linear function of error if – and only if – $\mu_{Pos}(error)$ is a linear function of error, that is, a triangular membership function.

The previous little example may seem insignificant, but it is the foundation of linear fuzzy control and the following chapters of this book. It provides insight into the interpolative nature of fuzzy rules as well.

Figure 3.6 illustrates a single input, zero-order Sugeno controller. There are three premise membership functions in the figure, one for each rule, and the conclusions are singletons. The

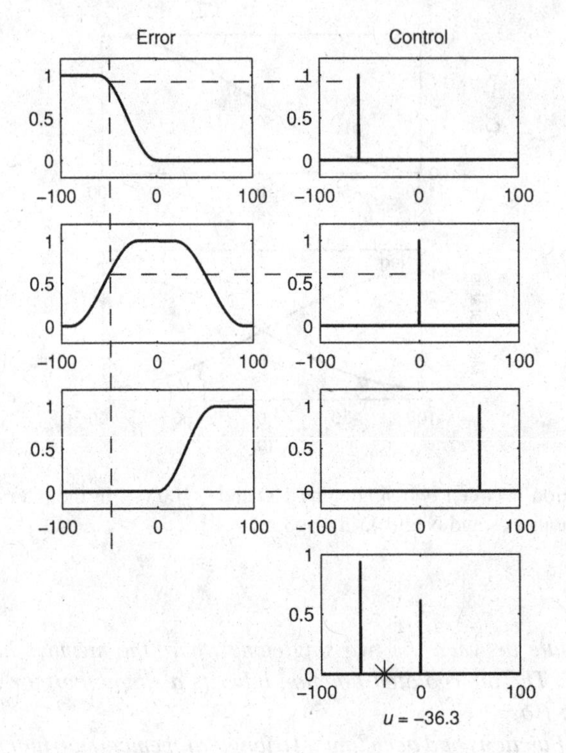

Figure 3.6 Sugeno inference with singleton output. (figrbase.m)

control signal, indicated by an asterisk (*) in the figure, is a firing strength weighted average of the singleton conclusions, $u = -36.3$.

3.3 Autopilot Example: Four Rules

Let us use the Autopilot simulator, mentioned in the first chapter, to illustrate a rule based Sugeno type of controller.

A train car drives on a track shaped like a parabola, and the initial position of the car is -15 m from the stop line; see Figure 3.7 for definitions. The track curves in the vertical plane according to the equation $(x, y) = \left(x, 0.02x^2\right)$. The car is initially standing still (initial

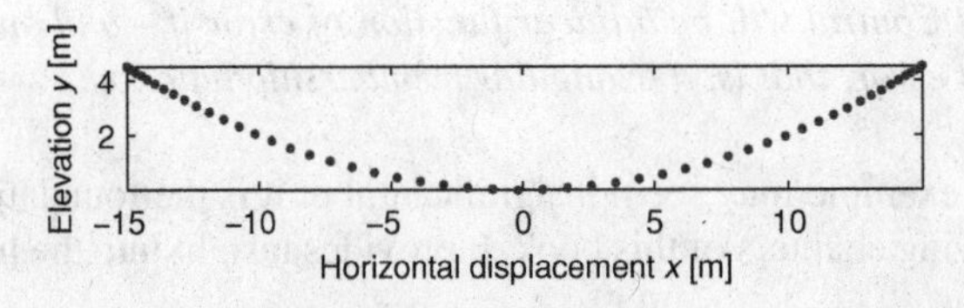

Figure 3.7 A plot of the car positions at evenly spaced time-instants along the trajectory. (figrunautopilot2.m)

velocity is 0 m/s), and it is only the gravitational acceleration that makes it roll. How would a fuzzy controller look, if it was supposed to stop the train car in the bottom at $x = 0$?

Given the initial conditions (and disregarding friction), we first note that the uncontrolled car will roll downwards, pass the stop line at $x = 0$, and roll up again to the same elevation as the starting level. Without a controller, the car will oscillate between the two end points, because the total energy is conserved. Incidentally, the car will always rise to the same elevation in the right end point regardless of the shape of the track.

Now assume that a controller drives an electric motor in the car, thus adding a tangential force either forward or in the reverse driving direction. A first guess on a control strategy could be to let the motor run forward (towards right) when the car is on the left side of the stop line, and backward (towards left) when it is on the right side. Such a position controller will speed up the response. But position alone is not enough to stabilize the motion, that is, the controller must also use a velocity measurement in order to brake the car. An intuitive control strategy could be the following:

1. If car is left and moves right then set motor signal to zero
2. If car is right and moves left then set motor signal to zero
3. If car is left and moves left then set motor signal to forward
4. If car is right and moves right then set motor signal to reverse

This set of rules covers all possible combinations of states, as long as we only consider the fuzzy terms 'left' and 'right'. The next step would be to define universes and membership functions for the position (left, right), the velocity (left, right), and the control signal (zero, forward, reverse). With a proper choice of membership functions, the rules will interpolate between the control signals zero, forward, and reverse. Even if 'forward' refers to full speed ahead and 'reverse' to full speed reverse, the controller will apply moderate control actions, when it is between the extreme states.

Alternatively, we can apply a more control theoretic approach. Given that the stop line is at $x = 0$ we can define this position as the reference point, that is, $Ref = 0$. Furthermore, for a given position x we can define the deviation from the reference as the error, that is, $error = Ref - x$. When the car is on the left side of the reference point, the error will thus be positive, and it will be negative on the right side. If we then define the change in error as the time derivative of error, we can use rules 1, 3, 7, and 9 from the previously defined rule base (3.1). It is now clear that the controller is a proportional–derivative (PD) type of controller, although it is possibly nonlinear.

3.4 Table Based Controller

With discrete premise universes, it is possible to infer all possible control actions offline, *before* putting the controller into operation. In a *table based controller* the relation between all combinations of the premise universe points (Cartesian product) and their corresponding control actions are arranged in a table. With two controller inputs and one control action, the table is a two-dimensional *lookup table*. With three inputs, the table becomes a three-dimensional array. Pre-calculating the possible outcomes improves execution speed at runtime,

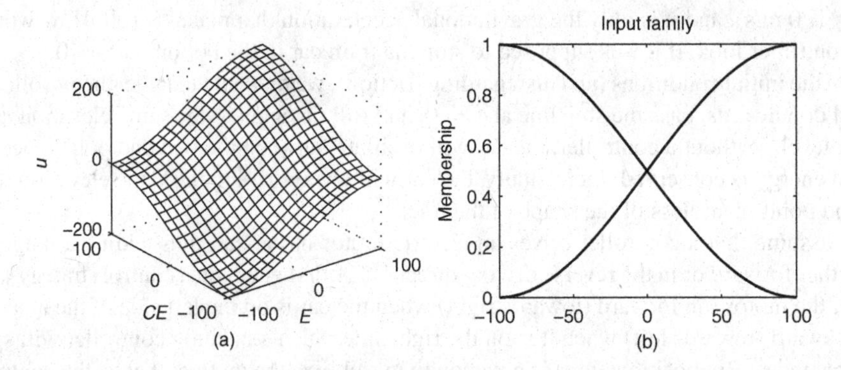

Figure 3.8 The control surface (a) corresponds to the rule base in Fig. 3.3. The two membership functions in the right pane (b) correspond to Neg and Pos. (figsurfs.m)

as the inference is reduced to a table lookup, which is normally faster than a rule based inference. Below is a small example of a lookup table derived from the rule base in Figure 3.3,

		Change in error				
		−100	−50	0	50	100
	100	0	29	100	171	200
	50	−29	0	71	141	171
Error	0	−100	−71	0	71	100
	−50	−171	-141	−71	0	29
	−100	−200	−171	−100	−29	0

A typical application for the table based controller is to embed it in a processing unit – in a car, for instance – where the table is downloaded to the processing unit that performs the table look-up.

With two inputs and one output, the input–output mapping is a surface, which we call the *control surface*. Figure 3.8 is a mesh plot of the relationship between *error* and *change in error* on the premise side, and *control* action u on the conclusion side, resulting from the four rules in Figure 3.3. The horizontal tangents of the premise sets cause horizontal plateaus in the corners of the surface. When the plant is on the reference and in a steady state, the controller takes the control signal from the centre of the surface. Small deviations result in relatively large changes in the control signal, because the slope is relatively steep here. The steep slope helps to keep an otherwise unstable plant steady on the reference. In case noise sensitivity must be kept low when the plant is near the reference, we would prefer another kind of control surface with a flat plateau around the centre. The designer must therefore consider carefully the choice of control surface.

When the plant output y is above the reference *Ref* the value of *error* becomes negative, and when the plant output y is below the reference *Ref* the value of *error* becomes positive. A negative value of *change in error* is the result of an increasing plant output y (for constant *Ref*), while a positive value is the result of a decreasing plant output y.

The table in the preceding text contains several regions of interest. The cell in the centre of the table has *error* equal to zero, that is, the plant is on the reference. Furthermore, the *change in error* is zero here, that is, the plant is in a steady state. Thus the centre cell is the stable point where the plant has settled on the reference. The diagonal is zero in the example; all these are favourable states, where the plant is either stable on the reference or approaching the reference. Should the plant move away a little from the zero diagonal – because of noise or a disturbance – the table values there have a small magnitude, and the controller will make small corrections to get it back in place. Should the plant be far off the reference and, furthermore, heading away from it, we are in the upper left and lower right corners; here the controller calls for drastic changes. Generally, the numerical values on the two sides of the zero diagonal may be any values, as a result of the rule based control strategy whether it be symmetric or asymmetric. The table governs the behaviour of the controller, and the surface is a convenient picture of the table.

For a stable system, and after a positive step in the reference, the point (*error*, *change in error*) will follow a trajectory in the table which spirals clockwise from the lower left corner of the table towards the centre. It is similar to a *phase plane* trajectory, which is the plot of a variable against its own derivative. A skilled designer, or an optimization algorithm, may adjust the numbers during a tuning session to obtain a favourable response.

If the resolution in the table is too coarse *limit cycles* will appear, that is, oscillations about the reference. This is because the table allows the *error* to drift away from the centre cell until it jumps into a neighbouring cell with a non-zero control action. This effect can be removed with *bilinear interpolation* between the cells instead of rounding to the nearest cell.

Example 3.4 *Bilinear interpolation*

Considering a two-dimensional table of discrete values, how does the controller interpolate between the values?

►*Solution*

An observed error E may fall between two neighbouring discrete values in the universe, E_i and E_{i+1}, such that $E_i < E < E_{i+1}$. Likewise, the observed change in error CE may fall between two neighbouring discrete values in its universe, CE_j and CE_{j+1}, such that $CE_j < CE < CE_{j+1}$. The resulting control signal is found by interpolating linearly in the E-axis direction between the first pair,

$$u_1 = g(E; F(i, j), F(i+1, j))$$

and the second pair,

$$u_2 = g(E; F(i, j+1), F(i+1, j+1))$$

and then in the CE-axis direction,

$$u = g(CE; u_1, u_2)$$

The function g is linear interpolation, and F_{ij} is the fuzzy lookup table ($i, j = 1, 2, \ldots$).

Table 3.1 Sketch of a 3D look-up table in relational representation.

x	y	z	u
−100	−100	−100	−100
−100	−100	−67	−89
−100	−100	0	−67
−100	−100	67	−44
−100	−100	100	−33
−100	−67	−100	−89
−100	−67	−67	−78
−100	−67	0	−56
⋮	⋮	⋮	⋮
100	100	100	100

A three-input controller implies a three-dimensional lookup table. Assuming a resolution of, say, 21 points in each universe, the table holds $21^3 = 9261$ elements. The number 21 comes from choosing a standard universe $[-100, 100]$ with discrete steps of 10, but the choice of resolution is arbitrary. It would be a tremendous task to fill these in manually, but it is manageable with rules.

Example 3.5 *Two-dimensional implementation*

If only two-dimensional tables are available in the programming language at hand, how can we then implement a three-dimensional table?

►*Solution*

Reshape the table into four columns: one for each of the three inputs (x, y, z), *and one for the control action u; see Table 3.1 for an example. The table lookup is now a question of finding the correct row, and picking the corresponding u value.*

3.5 Linear Fuzzy Controller

There are several sources of nonlinearity in a fuzzy controller. The position, shape and number of membership functions on the premise side, as well as nonlinear input scaling, cause a nonlinear control surface. Even the rules themselves most often express a nonlinear control strategy. Furthermore, if the connectives $\wedge$ and $\vee$ are implemented as min and max respectively, they are nonlinear functions themselves.

It is possible to construct a rule base with a linear control surface (Siler and Ying 1989; Mizumoto 1992, 1995; Qiao and Mizumoto 1996):

- The premise universes must be large enough for the inputs to stay within the limits, in other words, to avoid saturation. Each premise family should contain a number of terms, with an overlap such that the sum of membership values for any particular input instance is 1. This is achieved with duplicates of symmetric, triangular sets that cross their neighbour sets at the membership value $\mu = 0.5$. Their peaks will thus be equidistant. Any input instance can thus be a member of at most two sets simultaneously, and the membership of each is a piece-wise linear function of the input.

- The number of terms in each family determines the number of rules, since the rule base must consist of the $\wedge$-combination of all terms to ensure completeness.
- We must choose multiplication for the connective $\wedge$.
- Using a weighted average of the control signals the denominator vanishes, because all firing strengths add up to 1.

We can force the rule base to be equivalent to a plain *summation* of the inputs. With singleton conclusion sets $\langle s_i, 1\rangle$, the s_i must be the sum of the peak positions of the premise sets. Take for example the first of the nine rules in rule base (3.1),

if *error* is Neg and *change in error* is Neg, then *control* is NB.

Assuming the peak of Neg is in -100, then the constant NB must equal $-100 + (-100) = -200$, if the rule base is to act like a summation. The following list summarizes five design choices for achieving a fuzzy rule base equivalent to a summation.

Theorem 3.1 *Summing point fuzzy controller*
In order to design a fuzzy controller $u = F(E, CE)$, which is equivalent to the summation $u = E + CE$, the following conditions are necessary and sufficient:

1. *Use triangular premise sets that cross at $\mu = 0.5$.*
2. *Build a rule base containing all possible $\wedge-$ combinations of the premise terms.*
3. *Use product conjunction $(*)$ for the $\wedge-$ connective.*
4. *Use conclusion singletons, positioned at the sum of the peak positions of the premise sets.*
5. *Use the firing strength weighted average to find the resulting control signal.*

With these design choices the control surface is a diagonal plane, and the output universe is the sum of the input universes (Figure 3.9).

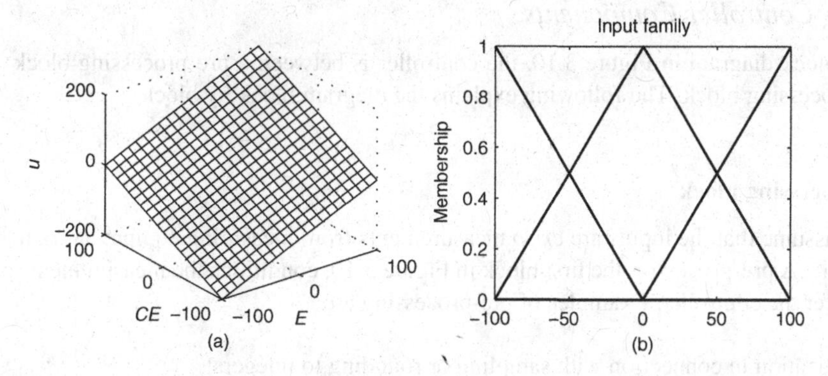

Figure 3.9 Linear control surface (a) and the membership functions that generated it (b). Inside the universes, the controller acts like a summation: $u = E + CE$. (figsurfs.m)

The theorem follows from the analytical simplification in Section 3.9. Such a fuzzy controller is two controllers in one: (1) it has the design of a fuzzy controller, and (2) it is equivalent to a summation. Therefore it has a transfer function, and the usual methods for tuning and calculating stability of the closed loop system apply.

3.6 Summary

In a fuzzy controller the measurement data pass through fuzzy membership functions, and the firing strength of each rule determines the local contribution to the final control signal. With all other choices being equal, it is recommended to apply continuous premise membership functions, but discrete conclusion membership functions, preferably singletons in a Sugeno controller.

When designing the rule base, the designer must consider the number of term sets, their shape, and their overlap. The rules themselves must be established, unless more advanced means such as adaptation are available. There are many design choices to make, some of which must always be considered, while others may not play a role in a particular design.

The control surface provides insight, because it is a picture of the input–output characteristics of the controller; it is the central object that governs the behaviour of the controller. Changing the membership functions causes the control surface to change shape. A linear control surface settles several design choices and opens the way for a fuzzy proportional–integral–derivative (PID) type of control, as the next chapter shows.

3.7 Other Controller Components*

This section contains descriptions and explanations that will help a designer to understand various options in a commercial software package such as the Fuzzy Logic Toolbox for MATLAB® (MathWorks 2012). The section is a slight detour, as the material is not strictly necessary for the rest of the book, therefore it can be skipped on a first reading (hence the *-mark in the heading).

3.7.1 Controller Components

In the block diagram in Figure 3.10, the controller is between a pre-processing block and a post-processing block. The following explains the diagram block by block.

Pre-processing block

Let us assume that the inputs are *crisp* measurements from measuring equipment, rather than linguistic. A pre-processor, the first block in Figure 3.10, conditions the measurements before they enter the controller. Examples of pre-processing are

- quantization in connection with sampling or rounding to integers;
- normalization or scaling onto a particular, standard range;
- filtering in order to remove noise;

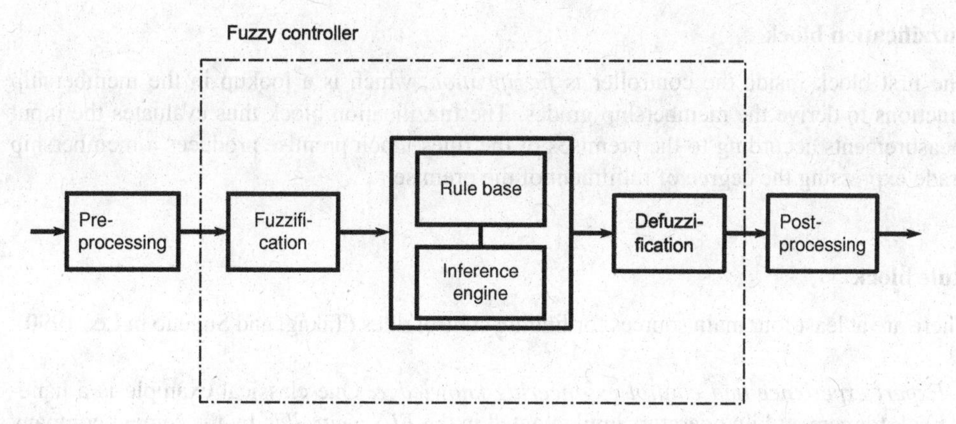

Figure 3.10 Fuzzy controller components. The rule base with its inference engine performs most of the work.

- averaging to obtain long-term or short-term tendencies;
- a combination of several measurements to obtain key indicators;
- differentiation and integration, or their approximations in discrete time.

A *quantizer* converts an inbound measurement in order to fit it to a discrete universe. Assume, for instance, that the variable *error* is 4.6, but the universe is $\mathcal{U} = \langle -5, -4, \ldots, 0, \ldots, 4, 5 \rangle$. The quantizer thus rounds it to 5, the nearest level. Quantization is a means to reduce data, but if the quantization is too coarse the controller may oscillate around the reference or even become unstable.

The FLSmidth (FLS) controller applies nonlinear *scaling* (Figure 3.11). The operator supplies a value for a typical small measurement, a typical normal measurement, and a typical large measurement according to experience (Holmblad and Østergaard 1982). The combined effect of the scaling and the membership functions, is a distortion of the membership functions. In fact, over the years, only the break points have been adjusted, while the primary sets have remained unchanged.

A dynamic controller has additional time-related inputs: derivatives, integrals, or previous values of measurements backwards in time. The pre-processor forms these.

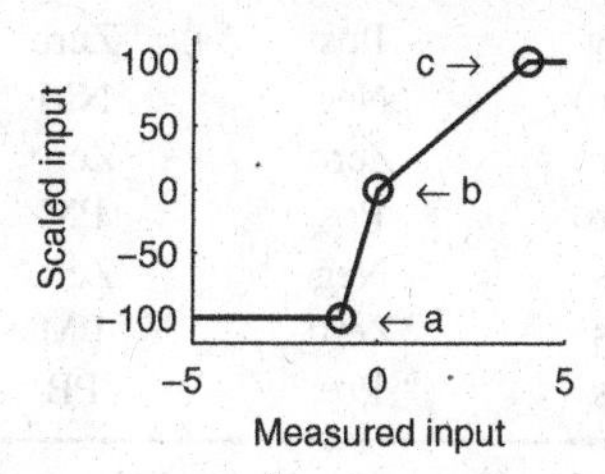

Figure 3.11 Example of nonlinear scaling of an input measurement to a standard universe [−100, 100]. The x-coordinates of the circled points indicate a typical small measurement (a), a typical normal measurement (b), and a typical large measurement (c), as decided by a skilled operator.

Fuzzification block

The first block inside the controller is *fuzzification*, which is a lookup in the membership functions to derive the membership grades. The fuzzification block thus evaluates the input measurements according to the premises of the rules. Each premise produces a membership grade expressing the degree of fulfilment of the premise.

Rule block

There are at least four main sources for finding control rules (Takagi and Sugeno in Lee 1990):

- *Expert experience and control engineering knowledge.* One classical example is a handbook for cement kiln operators implemented in the *FLS controller,* by the cement company FLSmidth (Holmblad and Østergaard 1982, 1995). The most common approach is to question experts or operators with a carefully organized questionnaire.
- *Based on the operator's control actions.* Observing an operator's control actions or examining a log book may reveal fuzzy *if–then* rules of input–output relationships.
- *Based on a fuzzy model of the plant.* A linguistic rule base may be viewed as an inverse model of the controlled plant. Thus the fuzzy control rules can perhaps be obtained by inverting a fuzzy model of the plant, if fuzzy models of the open and closed loop systems are available (Braae and Rutherford in Lee 1990). This method is restricted to low-order systems. Another approach is *fuzzy identification* (Tong; Takagi and Sugeno; Sugeno – all in Lee 1990; Pedrycz 1993; Babuška 1998, 1999).
- *Based on learning.* The self-organizing controller is an example of a controller that finds the rules itself. It is an example of the more general control scheme: model reference adaptive control.

The set of rules in the rule base (3.1) can be represented in other formats, such as a *relational* format.

Error	Change in error	Control
Neg	Neg	NB
Neg	Zero	NM
Neg	Pos	Zero
Zero	Neg	NM
Zero	Zero	Zero
Zero	Pos	PM
Pos	Neg	Zero
Pos	Zero	PM
Pos	Pos	PB

The top row is a heading. It is understood that the two leftmost columns constitute the premises, the rightmost the conclusions, and each row represents one rule. This format is more compact, and it provides an overview of the rule base quickly. The relational format is certainly suitable for storing in a relational database. The relational format implicitly assumes that the premise variables are connected by a connective – logical 'and' or logical 'or'. The same connective applies to all rules, not a mixture of connectives. Incidentally, a fuzzy rule with an 'or' combination of terms can be converted into an equivalent 'and' combination of terms using the laws of logic (foremost the two De Morgan laws).

Another, even more compact format is the *tabular* format,

		Change in error		
		Neg	Zero	Pos
	Pos	Zero	PM	PB
Error	Zero	NM	Zero	PM
	Neg	NB	NM	Zero

The premise variables *error* and *change in error* are laid out along the axes, and the conclusions are inside the table. Symmetries can be discovered readily, and an empty cell is an indication of a missing rule; thus the format is useful for checking completeness. When the premise variables are *error* and *change in error*, that format is also called a *linguistic phase plane*. If the number of premise variables is $n > 2$, the table grows to an n-dimensional array.

A nested arrangement can accommodate several conclusions. A rule with several conclusions can alternatively be broken down into several rules, each having one conclusion.

Example 3.6 *Membership functions*

Fuzzy rule bases use a variety of membership functions. Are there any functional definitions?

► *Solution*

A common example is the Gaussian *curve based on the exponential function,*

$$\mu_{Gauss}(x) = \exp\left[\frac{-(x - x_0)^2}{2\sigma^2}\right]$$

This is a standard Gaussian curve with a maximum value of 1; *x is the independent variable on the universe, x_0 is the position of the peak relative to the universe, and σ is the standard deviation.*

The bell *membership function does not use the exponential,*

$$\mu_{Bell}(x) = \left[1 + \left(\frac{x - x_0}{\sigma}\right)^{2a}\right]^{-1}$$

The extra parameter a, usually positive, affects the width of the membership function and the slope of the sides.

The FLS *controller uses the equation*

$$\mu_{FLS}(x) = 1 - \exp\left[-\left(\frac{\sigma}{x_0 - x}\right)^a\right]$$

It is also possible to use other functions, for example the sigmoid *known from neural networks, or the cosine based smooth trapezoids from the previous chapter.*

Universes

Before designing the membership functions it is necessary to consider the universes for the premises and conclusions. Take, for example, the rule

If *error* is Neg and *change in error* is Pos then *control* is Zero

The membership functions for Neg and Pos must be defined for all acceptable measurements of *error* and *change in error.* Nevertheless, a *standard universe* may be convenient.

Premise membership functions can be continuous or discrete. A continuous membership function is a function defined on a continuous universe. A discrete membership function is a vector with a finite number of elements. This case requires specification of the range of the universe and the value at each discrete sampling point. The choice between fine or coarse resolution is a compromise between accuracy, computational speed, and memory space. The quantizer takes time to execute, and if this time is too precious, continuous membership functions will make the quantizer obsolete.

Example 3.7 *Standard universes*

Does the universe change during runtime or is it fixed once and for all?

▶ *Solution*

In most practical cases, the universes are fixed. Authors and commercial products use a variety of standard universes.

- *The FLS controller uses the real number interval* $[-1, 1]$.
- *Authors of the early papers on fuzzy control used the short integer range* $[-6, 6]$, *because computer memory was scarce at that time.*
- *Another possibility is the interval* $[-100, 100]$ *corresponding to the percentage of full range of a measurement.*
- *Yet another possibility is the integer range* $[0, 4095]$ *arising from a 12-bit conversion of an analogue signal to digital representation.*
- *A variant is the integer range* $[-2047, 2048]$, *where the interval is shifted in order to accommodate negative numbers.*

The choice of data type may govern the choice of universe. For example, the voltage range $[-5, 5]$ *volts could be represented as an integer range* $[-50, 50]$, *or as a floating point range* $[-5.0, 5.0]$; *a signed byte data type has an allowable integer range of* $[-128, 127]$.

Scaling is a means of expanding the range of operation of a variable. If a controller input mostly operates within a small interval, increasing the scaling factor increases the range of operation within the universe.

The designer is faced with the problem of how to design the term sets, for example the *family* of terms: Neg, Zero, and Pos. There are two specific questions to consider:

(1) How does one determine the shape of the sets? (2) How many sets are necessary and sufficient?

According to fuzzy set theory the choice of shape and width is subjective, thus a solution is to ask the plant operators to draw their personal preferences for the membership curves; but operators will probably find it difficult to settle on particular curves. A few rules of thumb apply, however.

- A term set should be sufficiently wide to allow for noise in the measurement.
- A certain amount of overlap is desirable otherwise the controller may run into poorly defined states, where it does not return a well-defined output.
- If there is a gap between two neighbouring sets, no rules fire for values in the gap, so the controller is undefined in that gap.

Thus the necessary and sufficient number of sets in a family depends on the width of the sets, and vice versa.

Defuzzification block

The fuzzy set μ_c resulting from the inference (Figure 3.3, bottom right) must be converted to a single number in order to form a control signal to the plant. This is *defuzzification*. In the figure the defuzzified control signal is the x-coordinate marked by an asterisk (*). Several defuzzification methods exist.

The crisp control value u_{COG} is the abscissa of the *centre of gravity* of the fuzzy set. For discrete sets, its name is *centre of gravity for singletons*, COGS,

$$u_{COGS} = \frac{\sum_i \mu_c(x_i)\, x_i}{\sum_i \mu_c(x_i)}, \tag{3.6}$$

where x_i is a point in the universe $\mathcal{U}$ of the conclusion ($i = 1, 2, \ldots$), and $\mu_c(x_i)$ its membership of the resulting conclusion set. The expression is the membership weighted average of the elements of the set. For continuous sets, replace summations by integrals and call it COG. The method is much used, although its computational complexity is relatively high.

With singleton conclusions (Figure 3.3) and sum-accumulation, the resulting defuzzified value is

$$u = \frac{\sum_k \alpha_k^* s_k}{\sum_k \alpha_k^*} \tag{3.7}$$

Here s_k is the position of the singleton in rule k in $\mathcal{U}$, and α_k^* is the firing strength of rule k. It has the advantage that u is differentiable with respect to the singletons s_k, which is a useful property for optimization algorithms.

The *bisector of area* method, BOA, finds the abscissa x of the vertical line that partitions the area under the membership function into two areas of equal size. For discrete sets, u_{BOA} is the abscissa x_j that minimizes

$$\left|\sum_{i=1}^{j} \mu_c(x_i) - \sum_{i=j+1}^{i_{\max}} \mu_c(x_i)\right|, \qquad 1 < j < i_{\max} \tag{3.8}$$

Here $i_{\max}$ is the index of the largest abscissa $x_{i_{\max}} \in \mathcal{U}$. Its computational complexity is relatively high. There may be several solutions x_j.

An intuitive approach is to choose the point of the universe with the highest membership. Several such points may exist, and it is common practice to take the *mean of maxima* (MOM),

$$u_{MOM} = \frac{\sum_{i \in \mathcal{I}} x_i}{|\mathcal{I}|}, \qquad \mathcal{I} = \{i \mid \mu_c(x_i) = \mu_{\max}\}$$

where $\mathcal{I}$ is the (crisp) set of indices i where $\mu_c(x_i)$ reaches its maximum $\mu_{\max}$, and $|\mathcal{I}|$ is its cardinality (the number of members). This method disregards the shape of the fuzzy set, but the computational complexity is relatively good.

Another possibility is to choose the position in the universe of the *leftmost maximum* (LM),

$$u_{LM} = x_{\min(\mathcal{I})}$$

or the position in the universe of the *rightmost maximum* (RM)

$$u_{RM} = x_{\max(\mathcal{I})}$$

A robot, for example, must choose between left or right to avoid an obstacle in front of it; thus the defuzzifier must choose one (LM) or the other (RM), not something in between. These defuzzification methods are indifferent to the shape of the fuzzy set, but the computational complexity is relatively small.

Figure 3.12 shows the result of the five methods on a particular fuzzy set. Each method gives a different result.

Post-processing block

If the inferred control value is defined on a standard universe, it must be scaled to *engineering units,* for instance: volts, metres, or tons per hour. An example is the scaling from the standard universe $[-1, 1]$ to the physical units $[-10, 10]$ volts. The post-processing block contains an output gain that can be tuned.

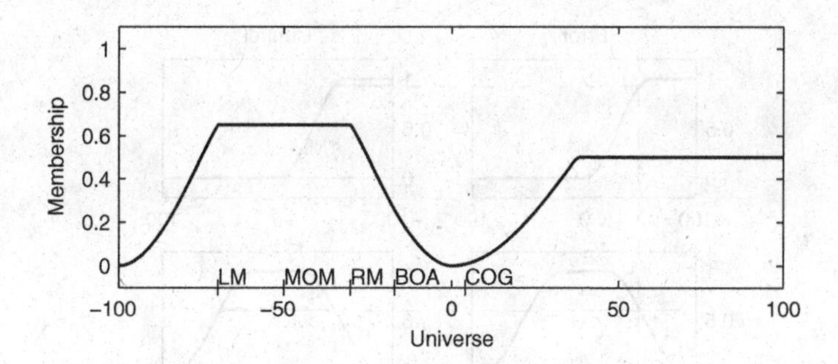

Figure 3.12 Different defuzzification methods give different results. From left to right: leftmost maximum (LM), mean of maxima (MOM), rightmost maximum (RM), bisector of area (BOA), and centre of gravity (COG). (figdefuz.m)

3.8 Other Rule Based Controllers*

Three distinct variants of controller have evolved historically: the *Mamdani*, *FLS*, and *Sugeno* controllers. They use the same general inference scheme, but they differ with respect to activation method and conclusion membership functions.

The following SISO rule base pinpoints the essential differences,

If *error* is Neg then *control* is Neg

If *error* is Zero then *control* is Zero

If *error* is Pos then *control* is Pos

The input is *error*, proportional to the deviation from the setpoint. The *control* action has the same sign as *error*, linguistically. Thus the rule base expresses a fuzzy proportionality between *error* and *control*, and the controller is a fuzzy version of a proportional (P) controller. The following paragraphs explain the characteristics of the controller variants by means of graphical construction.

3.8.1 The Mamdani Controller

Figure 3.13 illustrates the Mamdani controller. Each of the three rows refers to one rule. A particular instance of an *error* measurement ($error = -50$) is indicated by a vertical dashed line intersecting all three rules. Firing strengths are indicated by horizontal dashed lines.

The Mamdani controller applies activation function min, resulting in a clipping of the conclusion sets. The accumulated conclusion on the bottom right of the figure thus contains sharp breakpoints. Defuzzification method COG results in the control signal $u = -25.7$ in the figure.

The inference mechanism is motivated by intuitive clarity: it is evident how gradual fulfilment of a rule contributes to the accumulated conclusion.

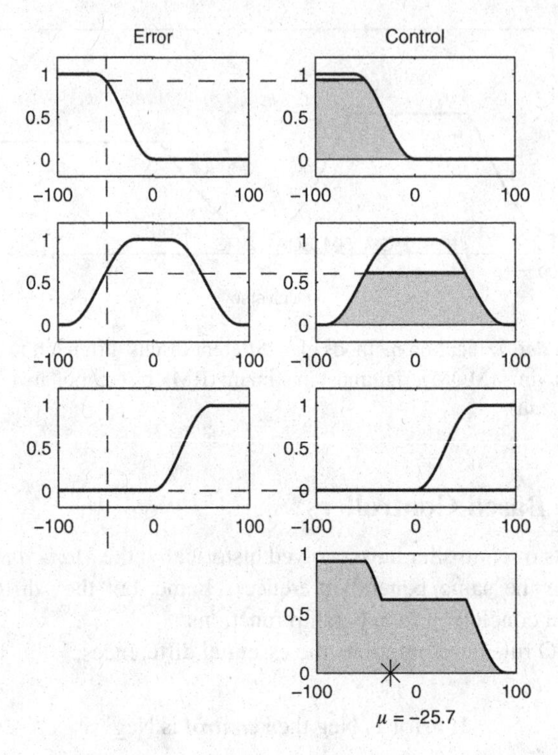

Figure 3.13 Mamdani inference. (figrbase.m)

Example 3.8 *Mamdani inference*

Take the inference in Figure 3.13; how is it implemented?

▶ *Solution*

Behind the scenes the premise membership functions are continuous, while the conclusion functions are discrete, divided into 201 *integer points in* $[-100, 100]$. *But for brevity, let us just use five points here. Assume the conclusion universe is defined by the vector*

$$\mathbf{u} = \begin{bmatrix} -100 & -50 & 0 & 50 & 100 \end{bmatrix}$$

A smooth trapezoid function $\mu_{STrapezoid}(x; a, b, c, d)$ *defines the membership functions with*

$$\mu_{Neg} = \mu_{STrapezoid}(x; -100, -100, -60, 0)$$
$$\mu_{Zero} = \mu_{STrapezoid}(x; -90, -20, 20, 90)$$
$$\mu_{Pos} = \mu_{STrapezoid}(x; 0, 60, 100, 100),$$

where a *is the left foot-point,* b *is the left shoulder-point,* c *is the right shoulder-point, and* d *is the right foot-point. Inserting* $\mathbf{u}$ *for* x *the conclusion term set is represented by three vectors*

$$\mathbf{Neg} = \begin{bmatrix} 1 & 0.93 & 0.05 & 0 & 0 \end{bmatrix}$$
$$\mathbf{Zero} = \begin{bmatrix} 0 & 0.61 & 1 & 0.61 & 0 \end{bmatrix}$$
$$\mathbf{Pos} = \begin{bmatrix} 0 & 0 & 0.05 & 0.93 & 1 \end{bmatrix}$$

Here we inserted the whole vector $\mathbf{u}$ *in place of the running point* x*; the result is thus a vector for each set. In the figure* $error = -50$*, and the unit is a percentage of full range. Thus the firing strength of the first rule is* $\alpha_1 = \mu_{Neg}(-50) = 0.93$*. Using* min *as the activation function, the conclusion is*

$$\min(\alpha_1, \mathbf{Neg}) = \begin{bmatrix} 0.93 & 0.93 & 0.05 & 0 & 0 \end{bmatrix}$$

The firing strength of the second rule is $\alpha_2 = \mu_{Zero}(-50) = 0.61$*, and the firing strength of the third rule is* $\alpha_3 = \mu_{Pos}(-50) = 0$*. Now stack all three contributions on top of each other,*

$$\begin{bmatrix} \min(\alpha_1, \mathbf{Neg}) \\ \min(\alpha_2, \mathbf{Zero}) \\ \min(\alpha_3, \mathbf{Pos}) \end{bmatrix} = \begin{bmatrix} 0.93 & 0.93 & 0.05 & 0 & 0 \\ 0 & 0.61 & 0.61 & 0.61 & 0 \\ 0 & 0 & 0 & 0 & 0 \end{bmatrix}$$

Accumulation using max *down each column yields the vector*

$$\mu_\mathbf{c} = \begin{bmatrix} 0.93 & 0.93 & 0.61 & 0.61 & 0 \end{bmatrix}$$

COGS defuzzification yields

$$\begin{aligned} u_{COGS} &= \frac{\sum_i \mu_c(x_i)\, x_i}{\sum_i \mu_c(x_i)} \\ &= \frac{0.93 * (-100) + 0.93 * (-50) + 0.61 * 0 + 0.61 * 50 + 0 * 100}{0.93 + 0.93 + 0.61 + 0.61 + 0} \\ &= -35.4 \end{aligned}$$

This is the defuzzified control signal. The number differs from the previously quoted result ($u = -25.7$)*, because we chose a lower resolution of five element vectors rather than* 201 *element vectors.*

3.8.2 The FLS Controller

Figure 3.14 illustrates the FLS controller. The membership functions are the same as earlier, and the error measurement is the same.

But the FLS controller applies the activation function 'product', causing a scaling of the conclusion sets. The defuzzification method is again COGS for the sake of comparison, although the real FLS controller applies the BOA method. The defuzzified control signal is $u = -29.7$, which is less negative than the Mamdani control signal.

The $*$-activation is motivated by a wish to preserve the shape of the conclusion sets.

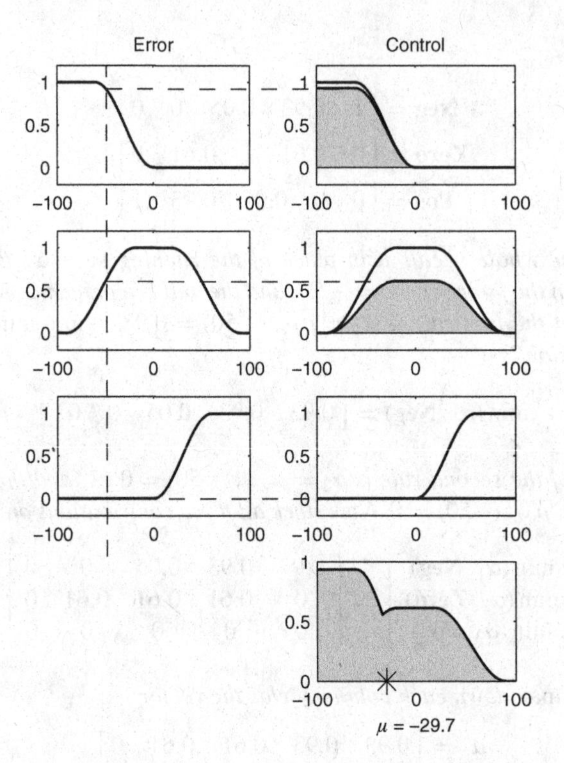

Figure 3.14 FLS inference. (figrbase.m)

3.9 Analytical Simplification of the Inference*

This section can be skipped at first reading (*-marked), because it is heavily symbolic. It contains, however, a simplification of the Sugeno controller which may be of practical use since it speeds up the computations. It is also useful for a theoretical analysis of stability, because under certain conditions the controller is reduced to a simple linear expression.

Assuming the conditions for linearity are fulfilled (Theorem 3.1), the inferred output has a simple analytical expression, even for the nonlinear case. We shall use the shorthand notation

N_E for $\mu_{Neg}(E)$ corresponding to 'error is Neg',

Z_E for $\mu_{Zero}(E)$ corresponding to 'error is Zero',

P_E for $\mu_{Pos}(E)$ corresponding to 'error is Pos',

N_{CE} for $\mu_{Neg}(CE)$ corresponding to 'change in error is Neg',

Z_{CE} for $\mu_{Zero}(CE)$ corresponding to 'change in error is Zero'

P_{CE} for $\mu_{Pos}(CE)$ corresponding to 'change in error is Pos', and

S_i for the singleton of rule i.

3.9.1 Four Rules

Take the rule base with four rules,

1. If *error* is Neg and *change in error* is Neg then *control* is NB
3. If *error* is Neg and *change in error* is Pos then *control* is Zero
7. If *error* is Pos and *change in error* is Neg then *control* is Zero
9. If *error* is Pos and *change in error* is Pos then *control* is PB

It corresponds to rules 1, 3, 7 and 9 in (3.1). The rule base is constructed in accordance with design choice 2. The controller output is by design choice 5 and Equation (3.7),

$$u = \frac{\sum_k \alpha_k^* S_k}{\sum_k \alpha_k^*}$$

where S_k is the position of the conclusion singleton belonging to rule k, and α_k^* is the firing strength of rule k. The expression is the activation weighted average of the conclusion singletons. For example, the first rule ($k = 1$) is activated to the degree

$$\alpha_1^* = N_E * N_{CE}$$

by design choice 3. Singleton $S_1 = s_{NB}$, and the contribution from the first rule to the numerator is

$$\alpha_1^* * s_{NB}$$

by design choices 4 and 5. Similar results can be derived for the remaining rules. We can in fact write the inferred controller output directly. The numerator is,

$$\begin{aligned}\sum_{k=1}^{4} \alpha_k^* S_k = N_E * N_{CE} * s_{NB} + N_E * P_{CE} * s_{Zero} \\ + P_E * N_{CE} * s_{Zero} + P_E * P_{CE} * s_{PB}\end{aligned}$$

Singleton s_{Zero} is 0 by design choice 4, therefore two terms vanish. Furthermore, $s_{NB} = -s_{PB}$, and by design choice 1, $N(x) = 1 - P(x)$ where x is either E or CE. The numerator can therefore be reduced,

$$\begin{aligned}\sum_{k=1}^{4} \alpha_k^* S_k &= N_E * N_{CE} * s_{NB} + P_E * P_{CE} * s_{PB} \\ &= (1 - P_E)(1 - P_{CE}) * (-s_{PB}) + P_E * P_{CE} * s_{PB} \\ &= (P_E + P_{CE} - 1) * s_{PB}\end{aligned}$$

The denominator is

$$\sum_{k=1}^{4} \alpha_k^* = N_E * N_{CE} + N_E * P_{CE} + P_E * N_{CE} + P_E * P_{CE}$$

Or,

$$\begin{aligned}\sum_{k=1}^{4} \alpha_k^* &= (1 - P_E) * (1 - P_{CE}) + (1 - P_E) * P_{CE} \\ &\quad + P_E * (1 - P_{CE}) + P_E * P_{CE} \\ &= 1\end{aligned}$$

In summary, the controller output can be written

$$u = (P_E + P_{CE} - 1) * s_{PB}$$

Singleton $s_{PB} = 200$ when using standard universes, and the final expression is clearly linear when P_E and P_{CE} are linear. The expression is valid even for nonlinear P_{CE} and P_E, that is, nonlinear fuzzy membership functions μ_{Pos} and μ_{Neg}, as long as $\mu_{Pos}(x) + \mu_{Neg}(x) = 1$. But, in order to depend linearly on E and CE, the function P must be triangular.

3.9.2 Nine Rules

For the rule base with nine rules (3.1) we can derive a similar result, only slightly more complex.

Again we note that the rule base is constructed in accordance with design choice 2. The numerator is, in this case,

$$\begin{aligned}\sum_{k=1}^{9} \alpha_k^* S_k &= N_E * N_{CE} * s_{NB} + N_E * Z_{CE} * s_{NM} + N_E * P_{CE} * s_{Zero} \\ &\quad + Z_E * N_{CE} * s_{NM} + Z_E * Z_{CE} * s_{Zero} + Z_E * P_{CE} * s_{PM} \\ &\quad + P_E * N_{CE} * s_{Zero} + P_E * Z_{CE} * s_{PM} + P_E * P_{CE} * s_{PB}\end{aligned}$$

Singleton s_{Zero} is 0 by design choice 4, therefore three terms vanish. Furthermore, $s_{NB} = -s_{PB}$, $s_{NM} = -0.5 * s_{PB}$, and $s_{PM} = 0.5 * s_{PB}$. By design choice 1, we have $Z(x) = 1 - (P(x) + N(x))$, where x is either E or CE. The numerator reduces to

$$\begin{aligned}\sum_{k=1}^{9} \alpha_k^* S_k &= N_E * N_{CE} * s_{NB} + N_E * Z_{CE} * s_{NM} \\ &\quad + Z_E * N_{CE} * s_{NM} + Z_E * P_{CE} * s_{PM} \\ &\quad + P_E * Z_{CE} * s_{PM} + P_E * P_{CE} * s_{PB}\end{aligned}$$

$$
\begin{aligned}
&= N_E * N_{CE} * (-s_{PB}) \\
&\quad + N_E * (1 - (P_{CE} + N_{CE})) * \left(-\frac{1}{2} * s_{PB} \right) \\
&\quad + (1 - (P_E + N_E)) * N_{CE} * \left(-\frac{1}{2} * s_{PB} \right) \\
&\quad + (1 - (P_E + N_E)) * P_{CE} * \frac{1}{2} * s_{PB} \\
&\quad + P_E * (1 - (P_{CE} + N_{CE})) * \frac{1}{2} * s_{PB} \\
&\quad + P_E * P_{CE} * s_{PB} \\
&= \frac{1}{2} (P_E - N_E + P_{CE} - N_{CE}) s_{BP}
\end{aligned}
$$

The denominator is

$$
\begin{aligned}
\sum_{k=1}^{9} \alpha_k^* &= N_E * N_{CE} + N_E * Z_{CE} + N_E * P_{CE} \\
&\quad + Z_E * N_{CE} + Z_E * Z_{CE} + Z_E * P_{CE} \\
&\quad + P_E * N_{CE} + P_E * Z_{CE} + P_E * P_{CE} \\
&= N_E * N_{CE} + N_E * (1 - (P_{CE} + N_{CE})) + N_E * P_{CE} \\
&\quad + (1 - (P_E + N_E)) * N_{CE} \\
&\quad + (1 - (P_E + N_E)) * (1 - (P_{CE} + N_{CE})) \\
&\quad + (1 - (P_E + N_E)) * P_{CE} \\
&\quad + P_E * N_{CE} + P_E * (1 - (P_{CE} + N_{CE})) + P_E * P_{CE} \\
&= 1
\end{aligned}
$$

In summary the controller output can be written

$$
u = \frac{1}{2} (P_E - N_E + P_{CE} - N_{CE}) s_{PB}
$$

Singleton $s_{PB} = 200$ when using standard universes, and the final expression is clearly linear when when P_{CE}, P_E, N_{CE}, and N_E are linear. The expression is valid even for nonlinear P_{CE}, P_E, N_{CE}, and N_E, that is, nonlinear fuzzy membership functions μ_{Pos}, μ_{Zero} and μ_{Neg}, as long as $\mu_{Pos}(x) + \mu_{Zero}(x) + \mu_{Neg}(x) = 1$. But, in order to depend linearly on E and CE, the functions P and N must be triangular.

3.10 Notes and References*

Theory The overview by Lee (1990) is still a good starting point to gain insight into the inner workings of fuzzy controllers. The most efficient way to learn is to study the Fuzzy Logic Toolbox for MATLAB®, especially the Fuzzy Inference System (MathWorks 2012). The book by Driankov *et al.* (1996) is explicitly aimed at the control engineering community, in particular, engineers in industry and university students, with the intention of covering just the relevant parts of the theory and focusing on principles rather than particular applications or tools. Another central reference is the book by Passino and Yurkovich (1998), which provides case studies and a wide coverage of the control area, including identification, adaptive control, and supervisory control. Another wide coverage of the control area is the book by Wang (1997). The more recent book by Michels *et al.* (2006) is a comprehensive reference book that includes the current state of the art at the time of writing. Fuzzy model identification is treated thoroughly by Babuška (1998), who also offers a related MATLAB® toolbox for download.[1] Another central reference for modelling is Jang *et al.* (1997), which combines neural networks and machine learning in the so-called soft computing approach. Model-based fuzzy control with gain scheduling and sliding mode control is treated by Palm *et al.* (1997). Fuzzy control has been merged with the learning capabilities of artificial neural networks into neuro–fuzzy control (e.g. Nauck *et al.* 1997, Lin and Lee 1996) or intelligent control systems (Gupta and Sinha 1996), and with further techniques such as genetic algorithms into soft computing (Jang *et al.* 1997). For an overview of theoretically oriented work, see the collection edited by Farinwata *et al.* (2000). For future perspectives see the article by Sala *et al.* (2005).

Applications The book by Constantin von Altrock (1995) describes case studies related to appliances (air conditioning, heating, washing machine, clothes dryer), the automotive industry (brakes, engine, transmission, skidding, air conditioning), process control (decanter, incineration, ethylene production, cooling, wastewater, food processing), and other applications (battery charger, optical disk drive, camcorder, climate control, elevator, camera, anaesthesia, aircraft landing). Many related reports are easily accessible from the Fuzzy Application Library on the web site[2] of the software product fuzzyTECH. The collection of intelligent control systems (Gupta and Sinha 1996) mentioned here presents applications within robot control, adaptive control, knowledge based systems, robust control, expert systems, and discrete event systems. There is an international standard for programmable controllers defining programming methodology, environment and functional characteristics (IEC 2000). A survey paper presents industrial applications of fuzzy control reported after the year 2000 (Precup and Hellendoorn 2011).

[1] www.dcsc.tudelft.nl/~babuska

[2] www.fuzzytech.com

4

Linear Fuzzy PID Control

The linear fuzzy controller opens for a range of PID type fuzzy controllers: a fuzzy P controller, a fuzzy PD controller, a fuzzy PID controller, and a fuzzy incremental controller. The fuzzy versions can be made to perform exactly the same as the crisp, linear versions. The usual tuning methods apply, including hand-tuning, the Ziegler–Nichols rules, and other tuning methods known from linear control theory. An example with a linear third-order process demonstrates the design, and a nonlinear example with a stable equilibrium demonstrates the design in the nonlinear domain. As with classical PID control, there are practical obstacles, such as derivative spikes and integrator windup, that have to be overcome. Since the linear fuzzy controller is equivalent to a crisp controller, its stability can be analysed. The Nyquist plot conveniently shows the stability margin of the closed-loop system in a single picture.

A fuzzy PID controller is a fuzzified proportional-integral-derivative (PID) controller. It acts on the same input signals, but the control strategy is formulated as fuzzy rules.

If a control engineer changes the rules, or the tuning gains to be discussed later, it is difficult to predict the effect on rise time, overshoot, and settling time of a closed-loop step response, because the controller is generally nonlinear and its structure is complex.

In contrast, a PID controller is a simple linear combination of three signals: the P-action proportional to the error e, the I-action proportional to the integral of the error $\int e dt$, and the D-action proportional to the time derivative of the error de/dt, or $\dot{e}$ for short. This chapter introduces a systematic tuning procedure for fuzzy PID type controllers.

Assume that the control objective (Figure 4.1) is to control the controlled output $c = x + n$, which includes measurement noise n, around a reference input r, after a change in the reference, a change in the load l, or in the presence of noise. Assume for simplicity that the process has a monotonic transfer function such that the process output x increases when the process input $u + l$ increases. The controller must increase the control signal u when the controlled output c is below the reference, and decrease u when the controlled output c is above the reference. This is negative feedback, reflected by the minus sign under the leftmost summing junction, that is, the control signal moves in the opposite direction to the process output.

Foundations of Fuzzy Control: A Practical Approach, Second Edition. Jan Jantzen.

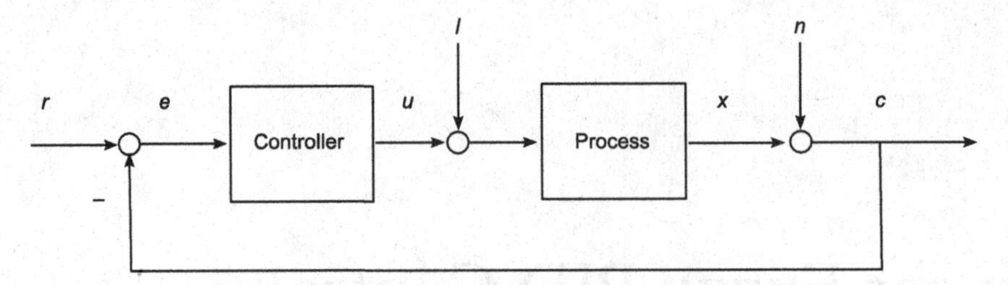

Figure 4.1 Feedback loop with load l and noise n.

It is natural to choose the error $e = r - c$ as an input to a fuzzy controller, as in PID control, and, furthermore, let the controller act on the magnitude and the sign of e. It follows that the integral of the error and the derivative of the error are useful signals to act on as well. A simple fuzzy control strategy with only four rules, based on error and its derivative, is

1. If *error* is Neg and *change in error* is Neg then *control* is NB
2. If *error* is Neg and *change in error* is Pos then *control* is Zero
3. If *error* is Pos and *change in error* is Neg then *control* is Zero
4. If *error* is Pos and *change in error* is Pos then *control* is PB (4.1)

The premise variable *error* is proportional to e, the premise variable *change in error* is proportional to $\mathrm{d}e/\mathrm{d}t$, and the conclusion variable *control* is proportional to the control signal. We shall distinguish between *error* and e, and *change in error* and $\mathrm{d}e/\mathrm{d}t$; when possible we reserve the terms *error* and *change in error* for rule bases, and e and $\mathrm{d}e/\mathrm{d}t$ for the signals derived from $e = r - c$. The difference in each case is a gain factor. The names Pos and Neg are labels of fuzzy sets, and likewise NB (negative big), Zero (control signal is zero), and PB (positive big).

There are methods for tuning PID controllers, for example: hand-tuning, Ziegler–Nichols tuning, optimal design, pole placement design, and auto-tuning (Åström and Hägglund 2006). There is much to gain if these methods are carried forward to fuzzy controllers.

Fuzzy PID controllers are similar to PID controllers under certain assumptions about the shape of the membership functions and the inference method (Siler and Ying 1989, Mizumoto 1992, Qiao and Mizumoto 1996, Tso and Fung 1997). A design procedure for fuzzy controllers of the PID type, based on PID tuning, is the procedure in Algorithm 4.1.

Algorithm 4.1 Design fuzzy PID.

1. Build and tune a conventional PID controller first.
2. Replace it with an equivalent linear fuzzy controller.
3. Make the fuzzy controller nonlinear.
4. Fine-tune it.

The idea is to start the controller design with a crisp PID controller, stabilize the closed-loop system, and tune it to a satisfactory performance. With a linear controller, and given a linear model of the process, it is even possible at this stage to carry out stability calculations, for instance: gain margins, eigenvalues, and Nyquist plots. From the solid foundation of linear control theory, it is safer to move to fuzzy control, rather than starting from scratch. The scope of such a procedure is limited by the scope of PID control, therefore the procedure is relevant whenever PID control is possible, or already implemented.

Our starting point is the ideal continuous PID controller,

$$u = K_p \left(e + \frac{1}{T_i} \int e(t)\, dt + T_d \frac{\mathrm{de}}{\mathrm{dt}} \right) \tag{4.2}$$

The control signal u is a linear combination of the error e, its integral, and its derivative. The parameter K_p is the *proportional gain*, T_i is the *integral time*, and T_d the *derivative time*.

In digital controllers, the equation must be approximated. Replacing the derivative term by a backward difference and the integral by a sum using rectangular integration, and given a constant – preferably small – sampling time T_s, the simplest approximation is

$$u(n) = K_p \left(e(n) + \frac{1}{T_i} \sum_{j=1}^{n} e(j) T_s + T_d \frac{e(n) - e(n-1)}{T_s} \right) \tag{4.3}$$

Index n refers to the time instant. By *tuning* we shall mean the activity of adjusting the parameters K_p, T_i, and T_d in order to achieve a good closed-loop performance.

4.1 Fuzzy P Controller

In discrete time, a proportional controller is defined by

$$u(n) = K_p e(n) \tag{4.4}$$

It is derived from the PID controller in Equation (4.3) with the I-action set to zero ($1/T_i = 0$) and the D-action set to zero ($T_d = 0$). The *fuzzy proportional* (FP) controller in the block diagram in Figure 4.2 accordingly acts on the error e, and its control signal is U.

Signals are represented by lower case symbols before gains and upper case symbols after gains. Thus the notation E represents the term *error,* and $E = GE * e$ (the symbol $*$ is multiplication); the symbol u represents *control,* where $GU * u = U$.

The FP controller has two tuning gains GE and GU, where the crisp proportional controller has just one, K_p. The control signal $U(n)$, at the time instant n is generally a nonlinear function of the input $e(n)$,

$$U(n) = f(GE * e(n)) * GU \tag{4.5}$$

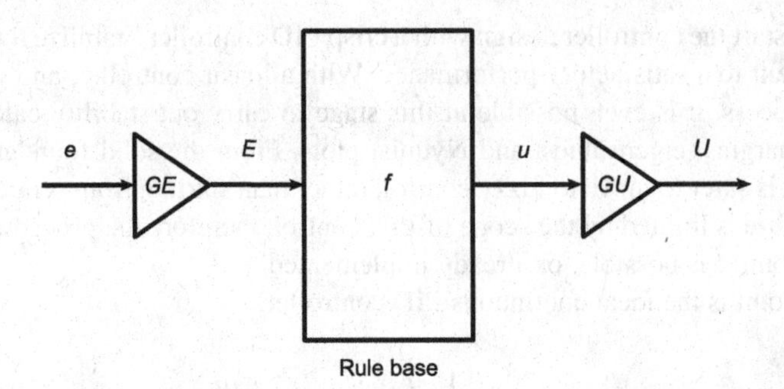

Figure 4.2 Fuzzy proportional controller, FP.

The function f denotes the rule base mapping. It is generally nonlinear, as mentioned; however, with a favourable choice of design, a linear approximation is

$$f(GE * e(n)) \approx GE * e(n) \tag{4.6}$$

Insertion into Equation (4.5) yields the control signal

$$U(n) = GE * e(n) * GU = GE * GU * e(n) \tag{4.7}$$

Comparing with Equation (4.4), the product of the gain factors for the linear controller corresponds to the proportional gain,

$$GE * GU = K_p. \tag{4.8}$$

The linear approximation is exact if, firstly, we choose the same universe for premise sets and conclusion sets, for example percentages of full scale [−100, 100]. Secondly, the rule base

1. If $E(n)$ is Pos then $u(n)$ is 100 (4.9)

2. If $E(n)$ is Zero then $u(n)$ is 0 (4.10)

3. If $E(n)$ is Neg then $u(n)$ is −100 (4.11)

with Pos, Zero, and Neg implemented appropriately – that is, according to the conditions in the previous chapter – provides a linear input–output mapping. The controller is thus equivalent to a crisp P controller.

Given a target proportional gain K_p – from a tuned, crisp P controller – Equation (4.8) determines one fuzzy gain factor when the other is chosen. The equation has one degree of freedom, since the fuzzy P controller has one more gain factor to adjust than the crisp P controller. This can be used to exploit the full range of the premise universe. For example, assume the maximal reference step is 1, whereby the maximal $e(n)$ is 1, and assume the universe for $E(n)$ is [−100, 100], then GE should be close to 100. When GE is chosen, Equation (4.8) determines GU.

4.2 Fuzzy PD Controller

Because of the process dynamics, it will take some time before a change in the control signal is noticeable in the process output, and the proportional controller will be equally late in correcting for an error. Derivative action helps to predict the future error, and the proportional-derivative controller uses the derivative action to attenuate oscillations. The discrete time PD controller is,

$$u(n) = K_p \left(e(n) + T_d \frac{e(n) - e(n-1)}{T_s} \right) \tag{4.12}$$

by Equation (4.3) with the I-action set to zero ($1/T_i = 0$).

The term in the parenthesis is proportional to an estimate of the error, T_d seconds ahead of the time instant n, where the estimate is obtained by linear extrapolation of the straight line connecting $e(n-1)$ and $e(n)$.

With $T_d = 0$ the controller is purely proportional, but when T_d is gradually increased, it will dampen possible oscillations. If T_d is increased too much the step response of the closed-loop system becomes *overdamped,* and it will start to oscillate again.

Input to the *fuzzy proportional-derivative* (FPD) controller in Figure 4.3 is $e(n)$ and $\dot{e}(n)$, where

$$\dot{e}(n) \approx \frac{e(n) - e(n-1)}{T_s} \tag{4.13}$$

The backward difference is a simple discrete approximation to the differential quotient, and more accurate computer implementations are available (e.g. Åström and Hägglund 2006 p. 412). The notation CE represents the term *change in error,* and $CE = GCE * \dot{e}$. Notice that Equation (4.13) deviates from the straight difference $e(n) - e(n-1)$ used in the early days of fuzzy control.

The control signal $U(n)$, at the time instant n, is a nonlinear function of *error* and *change in error*,

$$U(n) = f(GE * e(n), GCE * \dot{e}(n)) * GU \tag{4.14}$$

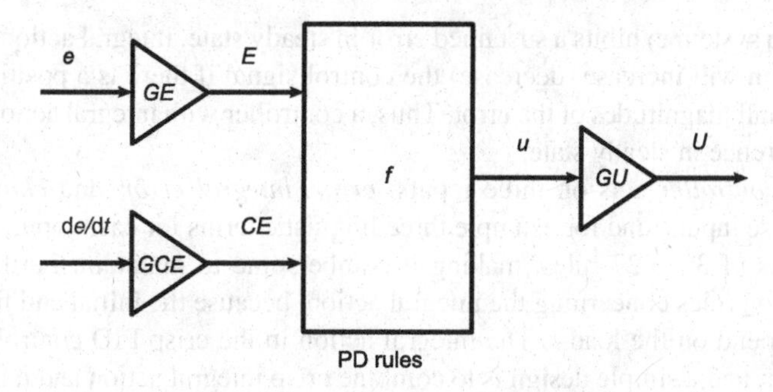

Figure 4.3 Fuzzy PD controller, FPD.

Again the function f is the rule base mapping, only this time it is a surface depending on two variables. It is usually nonlinear, but with a favourable choice of design, a linear approximation is

$$f(GE * e(n), GCE * \dot{e}(n)) \approx GE * e(n) + GCE * \dot{e}(n) \tag{4.15}$$

Insertion into Equation (4.14) yields the control action for the linear controller,

$$U(n) = (GE * e(n) + GCE * \dot{e}(n)) * GU \tag{4.16}$$

$$= GE * GU * \left(e(n) + \frac{GCE}{GE} \dot{e}(n) \right) \tag{4.17}$$

Comparing Equations (4.12) and (4.17) term by term, the gains are related as follows,

$$GE * GU = K_p \tag{4.18}$$

$$\frac{GCE}{GE} = T_d \tag{4.19}$$

The linear approximation is exact when the fuzzy control surface is a plane acting like a summation; compare the conditions in the previous chapter (Theorem 3.1). Thus the rule base (4.1), with Pos and Neg implemented appropriately, provides a linear input–output mapping. The conclusion universe must be defined as the sum of the premise universes. Assume, for instance, that the premise universes are both $[-100, 100]$, and we choose singleton conclusions $NB = -200$ and $PB = 200$, then the control surface will be the plane $u(n) = E(n) + CE(n)$. By that choice, the controller is equivalent to a crisp PD controller, and we can exploit Equations (4.18) and (4.19).

The fuzzy PD controller may be applied when proportional control is inadequate. The derivative term reduces overshoot, but it may be sensitive to noise as well as abrupt changes of the reference causing *derivative kick* in Equation (4.13).

4.3 Fuzzy PD+I Controller

If the closed-loop system exhibits a sustained error in steady state, integral action is necessary. The integral action will increase (decrease) the control signal if there is a positive (negative) error, even for small magnitudes of the error. Thus, a controller with integral action will always return to the reference in steady state.

A *fuzzy PID controller* acts on three inputs: *error*, *integral error*, and *change in error*. With three premise inputs, and for example three linguistic terms for each input, the complete rule base consists of $3^3 = 27$ rules, making it cumbersome to maintain. Furthermore, it is difficult to settle on rules concerning the integral action, because the initial and final values of the integrator depend on the load l. The integral action in the crisp PID controller serves its purpose, however, and a simple design is to combine crisp integral action and a fuzzy PD rule base in the *fuzzy PD+I* (FPD+I) controller; see Figure 4.4.

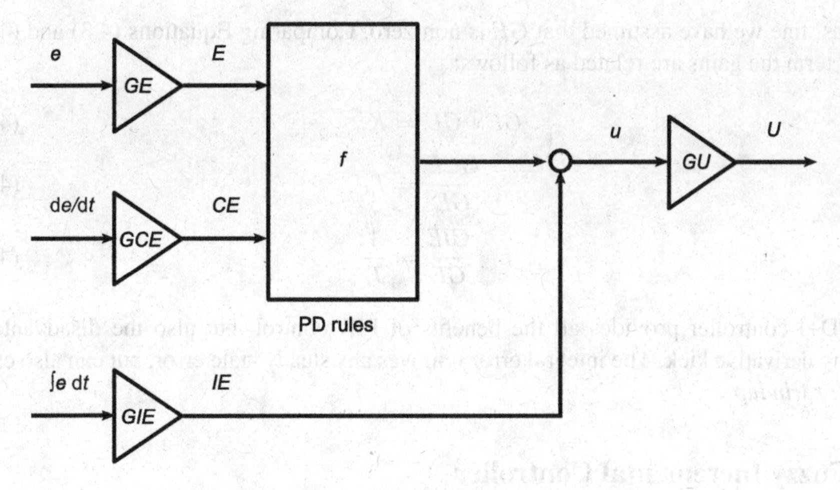

Figure 4.4 Fuzzy PID controller, FPD+I.

The integral error $IE = GIE * \int e\mathrm{d}t$ is proportional to the accumulation of all previous error measurements in discrete time, with

$$\int edt \approx \sum_{j=1}^{n} e(j)T_s$$

Rectangular integration is a simple approximation to the integral, and more accurate approximations exist (e.g. Åström and Hägglund 2006 p. 412). The control signal $U(n)$ after the gain GU, at the time instant n, is a nonlinear function of *error*, *change in error*, and *integral error*,

$$U(n) = \left[f(GE * e(n), GCE * \dot{e}(n)) + GIE \sum_{j=1}^{n} e(j)T_s \right] * GU \tag{4.20}$$

The function f is again the control surface of a PD rule base. The mapping is usually nonlinear, but with a favourable choice of design, Equation (4.15) is a linear approximation. Insertion into Equation (4.20) yields the control action,

$$U(n) \approx \left[GE * e(n) + GCE * \dot{e}(n) + GIE \sum_{j=1}^{n} e(j)T_s \right] * GU \tag{4.21}$$

$$= GE * GU * \left[e(n) + \frac{GCE}{GE} * \dot{e}(n) + \frac{GIE}{GE} \sum_{j=1}^{n} e(j)T_s \right] \tag{4.22}$$

In the last line we have assumed that GE is non-zero. Comparing Equations (4.3) and (4.22) term by term the gains are related as follows:

$$GE * GU = K_p \tag{4.23}$$

$$\frac{GCE}{GE} = T_d \tag{4.24}$$

$$\frac{GIE}{GE} = \frac{1}{T_i} \tag{4.25}$$

The FPD+I controller provides all the benefits of PID control, but also the disadvantages regarding derivative kick. The integral error removes any steady state error, but can also cause *integrator windup*.

4.4 Fuzzy Incremental Controller

An incremental controller adds a *change* in control signal Δu to the current control signal,

$$u(n) = u(n-1) + \Delta u(n)T_s \Rightarrow$$
$$\Delta u(n) = K_p \left(\frac{e(n) - e(n-1)}{T_s} + \frac{1}{T_i} e(n) \right)$$

using Equation (4.3) with $T_d = 0$. The controller output is an increment to the current control signal.

The *fuzzy incremental* (FInc) controller in Figure 4.5 is of almost the same configuration as the FPD controller, except for the added integrator. The conclusion in the rule base is now called *change in output* (cu), and the gain on the output is, accordingly, GCU. The control signal $U(n)$ at time instant n is the sum of all previous increments,

$$U(n) = \sum_{j=1}^{n} (cu(j) * GCU * T_s) \tag{4.26}$$

$$= \sum_{j=1}^{n} (f(GE * e(j), GCE * \dot{e}(j)) * GCU * T_s) \tag{4.27}$$

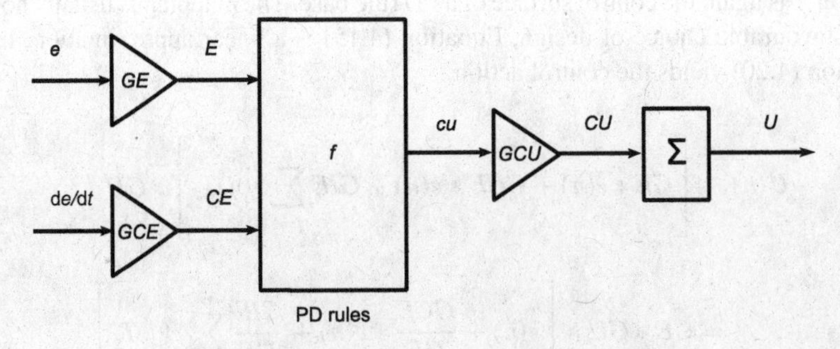

Figure 4.5 Incremental fuzzy controller, FInc.

Notice again that this definition deviates from the historical fuzzy controllers, where the sampling period T_s was left out. The function f is again the control surface of a PD rule base. The mapping is usually nonlinear, but with the usual favourable choice of design, Equation (4.15) is a linear approximation. Insertion into Equation (4.27) yields the control action,

$$U(n) \approx \sum_{j=1}^{n}(GE * e(j) + GCE * \dot{e}(j)) * GCU * T_s \tag{4.28}$$

$$= GCU * \sum_{j=1}^{n}\left[GE * e(j) + GCE * \frac{e(j) - e(j-1)}{T_s}\right] * T_s \tag{4.29}$$

$$= GCU * \left[GE * \sum_{j=1}^{n} e(j) * T_s + GCE * \sum_{j=1}^{n}(e(j) - e(j-1))\right] \tag{4.30}$$

$$= GCE * GCU * \left[\frac{GE}{GCE}\sum_{j=1}^{n}(e(j) * T_s) + e(n)\right] \tag{4.31}$$

By comparing Equations (4.3) and (4.31) it is clear that the linear controller is equivalent to a crisp PI controller ($T_d = 0$), and the gains are related as follows,

$$GCE * GCU = K_p$$

$$\frac{GE}{GCE} = \frac{1}{T_i}$$

Notice that the proportional gain K_p now depends on *GCE*. The gain $1/T_i$ is determined by the ratio between the two fuzzy input gains, and is the inverse of the derivative gain T_d in FPD control; thus, the gains *GE* and *GCE* change roles in FPD and FInc controllers.

It is an advantage that the controller output is driven directly from an integrator, because: (1) simply limiting the integrator prevents integrator windup, and (2) the integrator cancels noise to an extent which smooths the control signal.

To summarize, Table 4.1 shows for each of the four controller types the relationships between the PID tuning parameters and fuzzy gain factors valid for fuzzy linear controllers acting like a summation.

Table 4.1 Relationships between linear fuzzy and PID gains.

Controller	K_p	$1/T_i$	T_d
FP	$GE * GU$	—	—
FInc	$GCE * GCU$	GE/GCE	—
FPD	$GE * GU$	—	GCE/GE
FPD+I	$GE * GU$	GIE/GE	GCE/GE

4.5 Tuning

Several tuning aspects may be illustrated by static considerations (Åström and Hägglund 2006). For purely proportional control, consider the feedback loop in Figure 4.1, where the controller has the proportional gain K_p and the process has the gain K in steady state. The process output x is related to the reference r, the load l, and the measurement noise n by the equation

$$x = \frac{K_p K}{1 + K_p K}(r - n) + \frac{K}{1 + K_p K} l$$

If n and l are zero, then K_p should be high in order to ensure that the process output x is close to the reference r, such that the controller follows the reference well, as in a *servomechanism*. Furthermore, if the load l is non-zero, a high value will make the system less sensitive to changes in the load, such that the controller can maintain a constant controlled output as in a *regulator*. But if the noise n is non-zero, K_p should be moderate – otherwise the system will be too sensitive to noise. If the process dynamics are considered, the closed-loop system will normally be unstable if K_p is high. Obviously the tuning of K_p is a balance between the control objectives: stability, noise sensitivity, reference following, and load regulation.

4.5.1 Ziegler–Nichols Tuning

A PID controller may be tuned using the *Ziegler–Nichols frequency response method* (Ziegler and Nichols in Åström and Hägglund 2006). The procedure is quoted in Algorithm 4.2.

Algorithm 4.2 Ziegler–Nichols frequency response method.

1. Increase the proportional gain until the system oscillates; the resulting gain is the ultimate gain K_u.
2. Read the time between peaks T_u at this setting.
3. Use Table 4.2 to acquire approximate values for the controller gains.

The sample period may be related to the derivative gain T_d. Åström and Wittenmark (1984) suggest that the sample period should be between $1/10$ and $1/2$ of T_d. Combining with the Ziegler–Nichols rules, T_s should be approximately equal to 1–5% of the ultimate period T_u.

Table 4.2 The Ziegler–Nichols rules (frequency response method).

Controller	K_p	T_i	T_d
P	$0.5K_u$		
PI	$0.45K_u$	$T_u/1.2$	
PID	$0.6K_u$	$T_u/2$	$T_u/8$

Another rule of thumb is that T_s should be chosen to be slightly smaller than the dominating time constant in the process, for instance, between $1/10$ and $1/5$ of that time constant.

Ziegler and Nichols also give another method called the *reaction curve* or *step response* method (see for example Åström and Hägglund 2006). That method uses the open-loop step response to find the gains, and this is an advantage if oscillations in the closed-loop system must be avoided for safety reasons.

Example 4.1 *Ziegler–Nichols frequency response method*
Assume the process in Figure 4.1 has the transfer function

$$G(s) = \frac{1}{(s+1)^3}$$

Use the Ziegler–Nichols recommendations (Table 4.2) to tune a controller for it.
► *Solution*

Insert a PID controller with differential and integral action removed by setting $T_d = 0$ and $1/T_i = 0$. Gradually increase the proportional gain until the closed-loop system reaches a stable oscillation (Figure 4.6). This gain is $K_u = 8$ and the ultimate period is $T_u = 15/4$ (the plot shows four peaks in about 15 seconds).

There is a unit load on the system, so, the controller must employ integral action. The third row in Table 4.2 implies $K_p = 0.6 * K_u = 4.8$, $T_i = T_u/2 = 15/8$, and $T_d = T_u/8 = 15/32$.

Figure 4.7 shows the closed-loop response after a step in the reference at time equal to zero, and a step in the load at time equal to 20 seconds. The initial overshoot is fairly large.

Ziegler and Nichols aimed at a response to a load change with a *decay ratio* of one quarter. Decay ratio is the ratio between two consecutive peaks of the error after a step change in reference or load; thus in a quarter-decay response the second overshoot is 25% of the first – a compromise between a fast response and a small overshoot.

The relationships of Table 4.2 therefore do not fit all situations. The results are poor for systems with a time lag much greater than the dominating time constant. In general, the rules often result in rather poor damping. The table generally works better for PID control than for PI control, and it does not give guidance for PD control. A related, more accurate, method is called AMIGO (Åström and Hägglund 2006 p. 233).

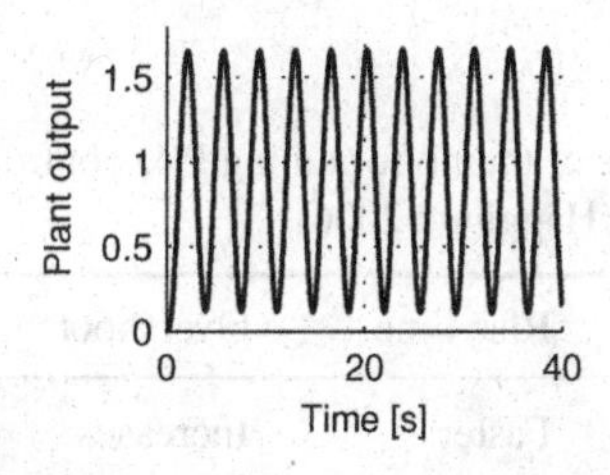

Figure 4.6 Ziegler–Nichols oscillation of process $1/(1+s)^3$. (figzn.m)

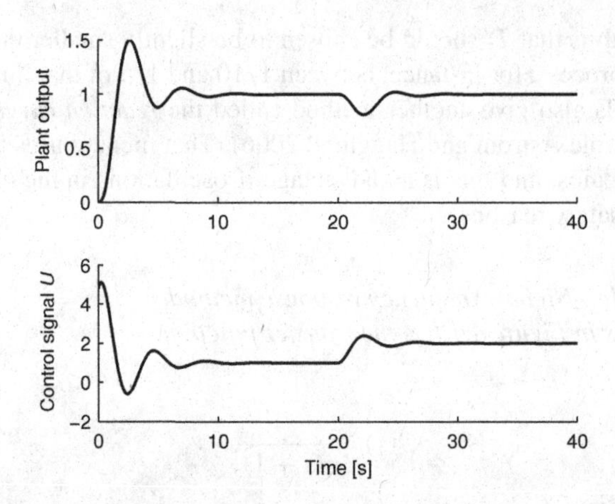

Figure 4.7 PID control of the process $1/(1+s)^3$. A reference step at 0 seconds is followed by a load step at 20 seconds. Ziegler–Nichols settings: $K_p = 4.8$, $T_i = 15/8$, and $T_d = 15/32$. (figzn.m)

4.5.2 Hand-Tuning

Hand-tuning is based on the rules of thumb in Table 4.3. There are exceptions to the rules in the table; if the process contains an integrator, an increase in K_p often results in more stable control.

Algorithm 4.3 is a more rigorous procedure for hand-tuning. The tuning result is a compromise between fast reaction and stability. The procedure adjusts the derivative gain before the integral gain, but in practice the sequence may be reversed. A process engineer can use the procedure right away, online, and develop a feel for how the closed-loop system behaves. A disadvantage is that it may take a long time to develop this feel, and it is difficult to sense whether the final settings are optimal.

Algorithm 4.3 Hand-tuning procedure (adapted from Smith 1979).

1. Remove all integral and derivative action by setting $T_d = 0$ and $1/T_i = 0$.
2. Tune the proportional gain K_p to give the desired response, ignoring any final value offset from the setpoint.
3. Increase the proportional gain further and adjust the derivative gain T_d to dampen the overshoot.
4. Adjust the integral gain $1/T_i$ to remove any final value offset.
5. Repeat from step 3 until the proportional gain K_p is as large as possible.

Table 4.3 Rules of thumb for tuning PID controllers (adapted from Åström and Hägglund 2006).

Action	Rise time	Overshoot	Stability
Increase K_p	Faster	Increases	Decreases
Increase T_d	Slower	Decreases	Increases
Increase $1/T_i$	Faster	Increases	Decreases

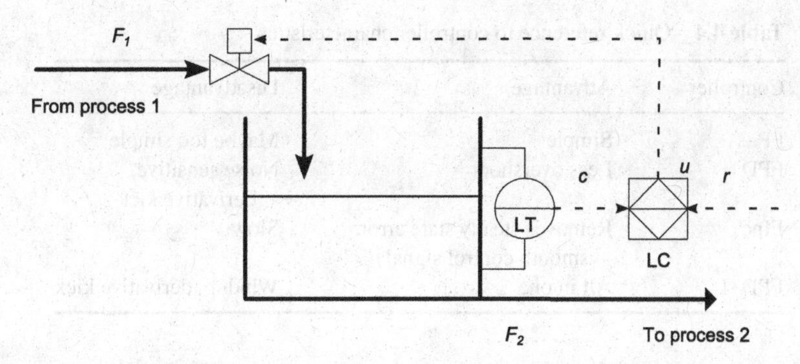

Figure 4.8 Piping and instrumentation diagram of a surge tank system. The level transmitter (LT) sends the measured level c of the liquid to the level controller (LC). The LC compares the measured level to the reference r and sends a control signal u to the valve. The valve adjusts the input flow rate F_1.

Example 4.2 *Choice of controller*

Surge tanks provide intermediate storage for liquid flows between process units. Consider the diagram in Figure 4.8 where a liquid stream from process 1 flows into a surge tank. What kind of controller do we need to control the level?

► *Solution*

(a) Assume first that the outlet is closed, that is, $F_2 = 0$. *For this problem, a P controller is sufficient. As soon as the liquid reaches the setpoint level, the error becomes zero, and the controller will close the valve.*

(b) If there is overshoot – maybe the valve reacts sluggishly – the controller should slow the feed stream down well before the liquid reaches the setpoint. A PD controller is then appropriate.

(c) Assume instead that the outlet is open, that is, $F_2 \neq 0$. *The controller must try and reach the setpoint and therefore keep* F_1 *flowing to compensate for the liquid leaving the tank. A sustained control signal in steady state is necessary to keep the valve open, enough to balance the outflow. Thus, integral action is necessary, and a PI or PID controller will be appropriate.*

In our design procedure we replace the initial PID controller with an equivalent linear fuzzy controller – in accordance with the requirements for linearity from the previous chapter, and gain factors in accordance with Table 4.1. The closed-loop system should show exactly the same step response. As a proof of correct implementation, replace the fuzzy rule base by a pure summation, and check that the performance is exactly the same.

Table 4.4 summarizes advantages and disadvantages of the four fuzzy controllers. The fuzzy P controller may be used as a starting point. To improve the settling time and reduce overshoot, fuzzy PD is the choice. If there is a steady state error, an FInc controller or an FPD+I is the choice.

To emphasize, the controllers with f replaced by a summation are linear approximations to the corresponding fuzzy configurations; the relations hold for the *approximations only*. Also, for fixed universe controllers, the conclusion universe must be the sum of the premise universes. With premise universes $[-100, 100]$, for instance, the conclusion universe of an FPD controller should be $[-200, 200]$.

Table 4.4 Quick reference to controller characteristics.

Controller	Advantage	Disadvantage
FP	Simple	Maybe too simple
FPD	Less overshoot	Noise sensitive, derivative kick
FInc	Removes steady state error, smooth control signal	Slow
FPD+I	All in one	Windup, derivative kick

Example 4.3 *Gain transfer in other implementations*

What if we come across controller implementations other than the above? How are the gains related then?

► *Solution*

(a) In a particular fuzzy PD controller $\dot{e}$ is implemented as a straight difference $\Delta e = e(n) - e(n-1)$. Comparing Equations (4.3) and (4.16) implies $(GCE/GE) * \Delta e = T_d * \Delta e/T_s$, and this implies in turn $T_d = (GCE/GE) * T_s$. Similarly with the FPD+I controller. Then the last column in Table 4.1 should be multiplied by T_s. As a consequence, an increase in the sampling period will increase the differential time.

(b) In a particular FInc controller $\dot{e} = \Delta e/T_s$, but $U(n) = \sum u_i * GCU$ (without the multiplication by T_s). Then (4.31) must be modified to

$$U(n) = \frac{GCE}{T_s} * GCU * \left[\frac{GE}{GCE}\sum_{i=1}^{n} e(i) * T_s + e(n)\right]$$

Comparison with (4.3) yields $K_p = GCE * GCU/T_s$, and the integral time is unchanged. As a consequence, an increasing sampling period implies a decreasing proportional gain.

(c) A particular fuzzy PD has the premise and conclusion universes [−100, 100]. The linear controller is equivalent to the usual linear approximation, but with half the output gain. Thus Table 4.1 must be used with $GU/2$ instead of GU. The general rule is to use the table with GU/r (or GCU/r) where r is the number of inputs when the conclusion universe equals the premise universes.

(d) The PID controller can be given on the so-called parallel form,

$$u = K_p e + K_i \int e(t) * \mathrm{d}t + K_d \frac{\mathrm{d}e}{\mathrm{d}t}$$

where the control signal is a linear combination of three terms. The proportional gain K_p has been multiplied through in the parenthesis, such that $K_i = K_p * 1/T_i$ and $K_d = K_p * T_d$. By inspection, the gains of the linear FPD+I controller are related in the following manner: $K_p = GE * GU$, $K_i = GIE * GU$, and $K_d = GE * GU$.

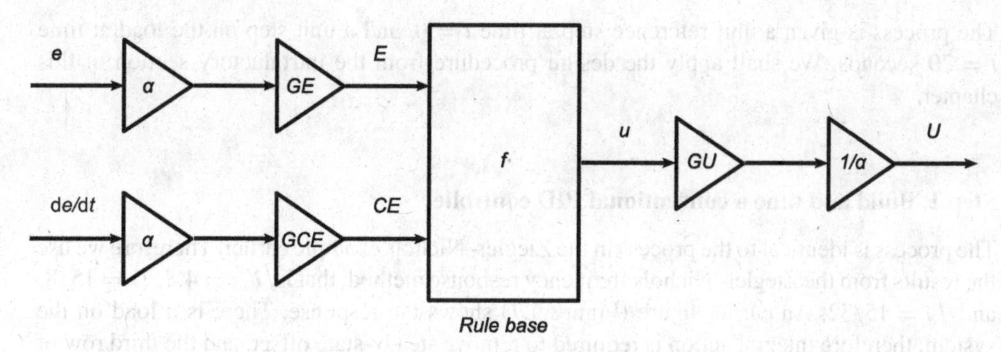

Figure 4.9 Scaling of gains in an FPD controller by means of a scaling factor α.

4.5.3 Scaling

Saturation of the input signals in the premise universes upsets the linearity of the fuzzy controller. Take the third-order process $1/(s+1)^3$ from Figure 4.7. A suitable value of the gain on *error* is $GE = 100$. But if the gain is increased to $GE = 400$, and all other gains adjusted according to Table 4.1 – whereby the proportional gain, differential time, and integral time remain invariant – the controller saturates in the premise universes.

Scaling is a means of avoiding saturation, but the relationships in Table 4.1 must still be observed in order to preserve the tuning. Consider, for example, the FPD controller in Equation (4.16). We may scale the input gains by a factor α without altering the tuning, because

$$(GE * e(n) + GCE * \dot{e}(n)) * GU \tag{4.32}$$

$$= (\alpha * GE * e(n) + \alpha * GCE * \dot{e}(n)) * GU * \frac{1}{\alpha} \tag{4.33}$$

That is, we multiply by α on the premise side, and cancel out by $1/\alpha$ on the conclusion side. This is *not* valid for nonlinear controllers, only their linear approximations. Figure 4.9 illustrates the scaling in a block diagram. Given a linear rule base, the values of K_p and T_d are independent of α. If saturation occurs in the premise universes, α must be decreased in magnitude until signals E and CE do not saturate during a step response. Scaling is introduced analogously in the other fuzzy PID controllers.

4.6 Simulation Example: Third-Order Process

When the process is of order higher than two, the PID controller starts to experience difficulties, and better responses can be achieved with more complex controllers. Consider therefore the third-order process

$$G(s) = \frac{1}{(s+1)^3}$$

The process is given a unit reference step at time $t = 0$, and a unit step on the load at time $t = 20$ seconds. We shall apply the design procedure from the introductory section of this chapter.

Step 1. Build and tune a conventional PID controller

The process is identical to the process in the Ziegler–Nichols example earlier. Therefore we use the results from the Ziegler–Nichols frequency response method, that is, $K_p = 4.8$, $T_i = 15/8$, and $T_d = 15/32$. An earlier figure (Figure 4.7) shows the response. There is a load on the system, therefore integral action is required to remove steady-state offset, and the third row of the Ziegler–Nichols table (Table 4.2) was used.

The response could possibly be improved by hand-tuning, but we shall settle on the Ziegler–Nichols settings for now.

Step 2. Replace it with an equivalent linear fuzzy controller

We will apply fuzzy PD+I control, since it is the only configuration of those previously mentioned which can be made equivalent to crisp PID control. The fuzzy PD sub-controller must be implemented as a linear Sugeno type controller, in accordance with the summing point conditions (Theorem 3.1).

The simplest rule base with two inputs consists of the four rules (4.1). The premise universes are arbitrarily chosen to be the generic percentages of full scale, $[-100, 100]$. They are continuous, since we will apply continuous membership functions on the premise side. The input families are already hinted at in the rule base: on the premise side we shall use membership functions Neg and Pos, and on the conclusion side we shall use NB, Zero, and PB. In order to achieve linearity and a rule base equivalent to a pure summation, Neg and Pos must be triangular sets that overlap by 50% stretching over the full width of the universe. Furthermore, the conclusion universe must be $[-200, 200]$, since there are two inputs, and the membership functions must be singletons at $NB = -200$, $PB = 200$.

The two premise variables *error* and *change in error* must be combined with the connective 'and'. Furthermore 'and' must be implemented as multiplication in order to achieve linearity. For activation we apply multiplication and for accumulation we apply sum. Since we are implementing a Sugeno type controller, the combined activation, accumulation, and defuzzification operation simplifies to weighted average, with the firing strengths weighting the singleton positions.

For keeping the design general, we shall implement the α-*scaling* from Equation (4.33), but keep $\alpha = 1$ as long as we are in the linear domain. Quantization is not relevant, because we are employing continuous premise membership functions, and it will not be necessary to implement a table based controller.

Regarding the choice of gain factors, we are guided by the PID settings. We choose $GE = 100$ since the error universe is $[-100, 100]$ and the maximal error is 1, according to the plot in Figure 4.7. By Equation (4.23) the gain factor GU is now fixed by the relation

$$GU = K_p/GE = 4.8/100$$

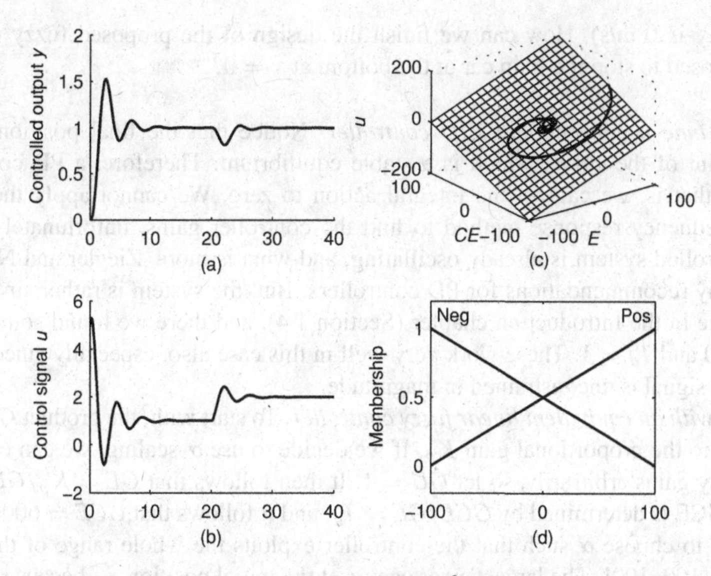

Figure 4.10 Fuzzy FPD+I control of the process $1/(s+1)^3$. The left column shows the response (a) and the control signal (b). The right column shows that the control surface is linear as well as the trajectory of the control signal (c). The bottom plot (d) shows the input family Neg and Pos. (figfpdi4.m)

The gain GCE is then determined by Equation (4.24),

$$GCE = GE * T_d = 100 * 15/32$$

The last gain is, by Equation (4.25),

$$GIE = GE * 1/T_i = 100 * 8/15$$

The sample time T_s is chosen at 0.05 seconds, since T_d is near 0.5 and one-tenth of that should be appropriate.

The step response with the linear FPD+I (Figure 4.10, left column) is exactly identical to that of the PID controller (Figure 4.7). The right-hand column of Figure 4.10 shows the trajectory of the control signal mapped onto the control surface. It shows that $|E_{\max}| \leq 100$ and $|CE_{\max}| \leq 55$ – thus there was no saturation in the universes.

The two final steps of the design procedure concern the nonlinear aspects of fuzzy control that are introduced in the next chapter.

4.7 Autopilot Example: Stable Equilibrium

We now have enough material to complete the fuzzy controller in the Autopilot example from the previous chapter (Section 3.3). Recall that the train car drives on a track shaped like a parabola. Its initial position is -15 m from the stop line. The car is initially standing still (initial velocity is 0 m/s). How can we finish the design of the proposed fuzzy controller, which is supposed to stop the train car at the bottom at $x = 0$?

1. *Build and tune a conventional PID controller*. Notice that the final position is at the bottom point of the curve, which is a stable equilibrium. Therefore, a PD controller is sufficient, that is, we can set the integral action to zero. We cannot apply the Ziegler–Nichols frequency response method to find the controller gains, unfortunately, because the uncontrolled system is already oscillating, and what is more, Ziegler and Nichols did not give any recommendations for PD controllers. But, the system is rather similar to the car example in the Introduction chapter (Section 1.4), and there we found some settings: $K_p = 6000$ and $T_d = 1$. These work very well in this case also, especially since this time the control signal is unconstrained in magnitude.
2. *Replace it with an equivalent linear fuzzy controller*. To start with, the product $GE * GU$ is equivalent to the proportional gain K_p. If we decide to use α-scaling, we can choose one of the fuzzy gains arbitrarily, so let $GU = 1$. It then follows that $GE = K_p/GU = 6000$. The gain GCE is determined by $GCE/GE = T_d$, and it follows that $GCE = 6000$ also. We would like to choose α such that the controller exploits the whole range of the premise universe $[-100, 100]$. The largest error occurs at the initial position x_0, because we expect to improve the uncontrolled oscillation. We choose, therefore,

$$\alpha = \frac{100}{K_p * \text{abs}(x_0)} = \frac{100}{6000 * 15} = 0.0011$$

The tuning is now completed. What remains is to build a linear FPD controller. We proposed earlier (Section 3.3) to take four rules out of the standard rule base with nine rules, that is,

1. If *error* is Neg and *change in error* is Neg then *control* is NB

3. If *error* is Neg and *change in error* is Pos then *control* is Zero

7. If *error* is Pos and *change in error* is Neg then *control* is Zero

9. If *error* is Pos and *change in error* is Pos then *control* is PB

With triangular membership functions, and other specific design choices (Theorem 3.1), the controller will act like a summation.

4.7.1 *Result*

Figure 4.11 shows the step response and the controller design. The response is fast, and there is no overshoot. The control signal is rather large at the beginning, which explains the fast response. The controller starts to brake before the train car reaches the stop line. The response is exactly the same as the one from the linear PD controller in the first step of the design procedure.

The figure also shows that the controller stays inside the control surface thanks to the α-scaling. The trajectory starts on the edge of the surface and turns clockwise towards the centre of the surface.

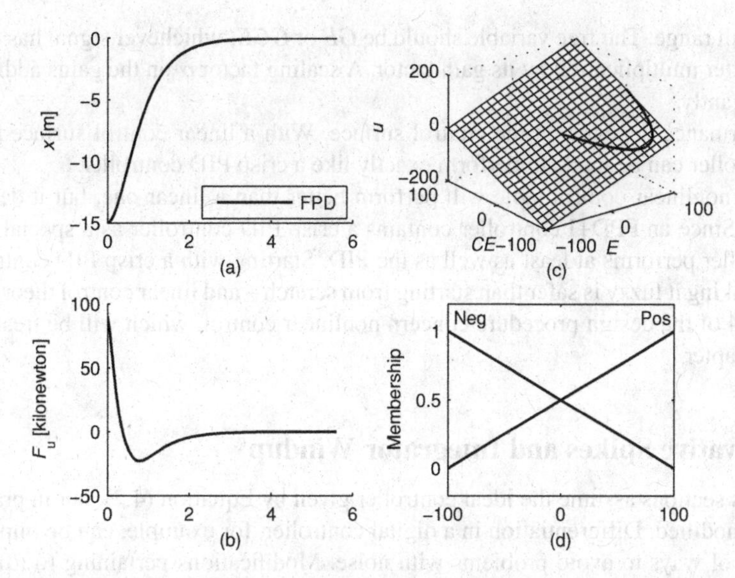

Figure 4.11 FPD control of Autopilot train car. The left column shows the response (a) and the control signal (b). The right column shows the control surface (c) and the membership functions that generated the surface (d). (figrunautopilot2.m)

The linear controller works well despite the nonlinear load forces caused by gravity and the geometry of the track. Admittedly, the reference is at a stable equilibrium point, which makes the control task relatively easy. The example demonstrates, nevertheless, that it is straightforward to build the fuzzy controller once the PD settings are known.

4.8 Summary

We have achieved a tuning procedure for fuzzy controllers. Referring to the FPD+I controller – because it covers both proportional, integral, and derivative action – the first steps of an algorithm are as follows:

1. *Design a crisp PID controller*. Use a PID tuning method to find K_p, $1/T_i$, and T_d.
2. *Replace it with a linear fuzzy*. Use the FPD+I configuration, transfer K_p, $1/T_i$, and T_d to *GE*, *GCE*, *GIE*, and *GU* (*GCU*) using Table 4.1. Run the controller, and check for saturation in the premise universes. When it is removed, by means of *α-scaling*, check that the closed-loop response is exactly the same with the fuzzy rule base replaced by a pure summation.
3. *Make it nonlinear* (see next chapter).
4. *Fine-tune it* (see next chapter).

The FPD+I controller has one degree of freedom, since it has one more gain factor than the crisp PID. This is used to exploit the full range of one premise universe. If, for example, the reference step is 1 and the universe for E is $[-100, 100]$, then fix *GE* at 100 in order to exploit the full range. The free variable should be GE or GCE, whichever signal has the larger magnitude after multiplication by its gain factor. A scaling factor α on the gains addresses the problem elegantly.

The performance depends on the control surface. With a linear control surface the fuzzy FPD+I controller can be made to perform exactly like a crisp PID controller.

Perhaps a nonlinear control table will perform better than a linear one, but it depends on the process. Since an FPD+I controller contains a crisp PID controller as a special case, the fuzzy controller performs at least as well as the PID. Starting with a crisp PID controller and gradually making it fuzzy is safer than starting from scratch – and linear control theory applies. Steps 3 and 4 of the design procedure concern nonlinear control, which will be treated in the following chapter.

4.9 Derivative Spikes and Integrator Windup*

The previous sections assume the ideal controller given by Equation (4.2), but in practice the equation is modified. Differentiation in a digital controller, for example, can be implemented in a number of ways to avoid problems with noise. Modifications pertaining to PID control can also be transferred to fuzzy controllers.

4.9.1 Setpoint Weighting

The ideal PID controller in Equation (4.3) is sensitive to abrupt changes in the reference. For example, a unit step in the reference causes the proportional action to jump by the amount K_p and the derivative action to jump to a large magnitude, when the sampling time is small. *Setpoint weighting* modifies the error signal used in the proportional action e_p and the derivative action e_d such that the effect of a sudden change in the reference signal will be attenuated. The modified controller based on e_p and e_d is (Åström and Hägglund 2006),

$$u(n) = K_p\left(e_p(n) + \frac{1}{T_i}\sum_{j=1}^{n} e(j)T_s + T_d\frac{e_d(n) - e_d(n-1)}{T_s}\right) \tag{4.34}$$

The error signal in the proportional action is

$$e_p(n) = b * r(n) - c(n) \tag{4.35}$$

and the error signal in the derivative action is

$$e_d(n) = c * r(n) - c(n) \tag{4.36}$$

The error signal in the integral action remains unmodified. For constant r, the closed-loop response to load changes will be independent of the values of the parameters b and c. The response to changes in the reference signal, however, will depend on b and c. For $b = 0$, the proportional action reacts to changes in the controller output only, which generally reduces the overshoot. For $c = 0$, the derivative action reacts to changes in the controlled output only, thus completely avoiding differentiation of a discontinuous jump in the reference signal. With $b = c = 1$, we achieve the original configuration. All simulations in this book use $b = 1, c = 0$.

The configuration of the equivalent linear fuzzy controller is unaffected, as long as it is understood that e is replaced by e_p from Equation (4.35) and $\dot{e}$ by e_d from Equation (4.36).

4.9.2 Filtered Derivative

In the presence of high frequency noise, the derivative action causes unwanted spikes in the control signal. A low-pass filter in combination with the pure derivative in the ideal controller in Equation (4.2) attenuates the spikes. The derivative action is modified to (Åström and Hägglund 2006),

$$u_d = -\frac{sK_pT_d}{1 + s\frac{T_d}{N}}c \tag{4.37}$$

where s is the Laplace operator for differentiation and $\left(-sK_pT_dc\right)$ is the ideal D-action. It is multiplied by a first-order transfer function $1/(1 + sT_d/N)$, which is a low-pass filter with the cut-off frequency N/T_d. High frequency noise ($s \to \infty$) is amplified at most by a factor K_pN, and in steady state ($s = 0$) the D-action vanishes. For $N \to \infty$ the expression in Equation (4.37) tends towards the ideal derivative action. Typical values in practice are $8 \le N \le 20$.

Replacing s with backward differences leads to a digital implementation. Note first that the factor K_pT_d in the numerator of Equation (4.37) is the gain on the derivative of the ideal differential, which is equivalent to $GCE * GU$ in the linear FPD controller. The remainder is

$$z = -\frac{s}{1 + s\frac{T_d}{N}}c$$

A discrete time approximation is

$$z(n) + \frac{T_d}{N}\frac{z(n) - z(n-1)}{T_s} = -\frac{c(n) - c(n-1)}{T_s} \Leftrightarrow$$

$$z(n)(T_s + \frac{T_d}{N}) = \frac{T_d}{N}z(n-1) - (c(n) - c(n-1)) \Leftrightarrow$$

$$z(n) = \frac{T_d}{NT_s + T_d}z(n-1) - \frac{c(n) - c(n-1)}{T_s + \frac{T_d}{N}}$$

Thus $z(n)$ is the signal to feed into the FPD controller in place of $\dot{e}(n)$, with T_d replaced by GCE/GE. It is a linear combination of the previous value $z(n-1)$ and the change in controlled output $c(n) - c(n-1)$. Again it is seen that for $N \to \infty, z(n)$ tends towards the unfiltered difference quotient $-(c(n) - c(n-1))/T_s$ arising from the combination of Equations (4.34) and (4.36).

4.9.3 Anti-Windup

The integrator in the I-action can sum up to a magnitude much larger than called for, that is, it *winds up*. Integrator windup occurs when the actuator after the controller along the signal path operates within limits, such as a maximum opening of a valve. The actuator remains at its limit corresponding to u_{lim}, while the integrator keeps integrating to a control signal $u(n) > u_{lim}$. When the error changes sign, the integrator starts to wind down, but the actuator stays at its limit until $u(n)$ passes u_{lim} as the control signal returns to the operating range of the actuator. Windup can cause a large overshoot, or an oscillation, where the actuator bounces from one extreme to the other, the so-called *chattering*.

One remedy (Rundqwist 1991) is to apply *conditional integration*, where the integrator is switched off at a prescribed condition, for example:

- to stop integrating when the control error is large, or $|e| > e_0$;
- to stop integrating when the controller saturates, or $u(n) > u_{lim}$;
- to stop integrating as before *and* the control error has the same sign as the control signal, $\text{sgn}(u(n)) = \text{sgn}(e(n))$;
- to limit the integrator I, such that $|I(n)| \leq I_0$; or
- to stop integrating and assign a prescribed value to the integrator when a condition is true.

Another approach is to use feedback from the control signal and the saturation value, such that their difference drives the controller. One final approach is to limit the controller input such that the control signal never saturates. This will often lead to a conservative, sluggish behaviour of the closed-loop system.

4.10 PID Loop Shaping*

Loop shaping is a design method in the linear domain. Given a pre-specified performance criterion, related to the robustness of the closed-loop system, it helps to choose the PID settings, and in some cases to calculate these automatically (Guzman *et al.* 2008 and the interactive learning modules by Calerga[1]). It would be very useful, if possible, to carry the ideas forward to the nonlinear fuzzy controllers.

Take the third-order process from the earlier simulation example (Section 4.6). Since the process and the controller are linear, we can plot the frequency response of the closed-loop system. The frequency response characterizes the dynamics by the way sine waves propagate through the loop. A Nyquist plot provides a complete description for the chosen frequencies.

In principle, it is determined by sending sinusoids of varying angular frequencies ω through the system, and then analysing the frequency phase shift $\phi(\omega)$ and amplitude gain $a(\omega)$ by comparing the input with the response. It is equivalent to Bode plots of $\phi(\omega)$ and $a(\omega)$, except that the information is merged into one plot, instead of two, with ω as an independent parameter. It is a polar plot of the complex transfer function $L(i\omega)$, with $a(\omega) = |L(i\omega)|$ the length of the locus vector and $\phi(\omega) = \angle L(i\omega)$ the angle.

[1] www.calerga.com/contrib/1/index.html

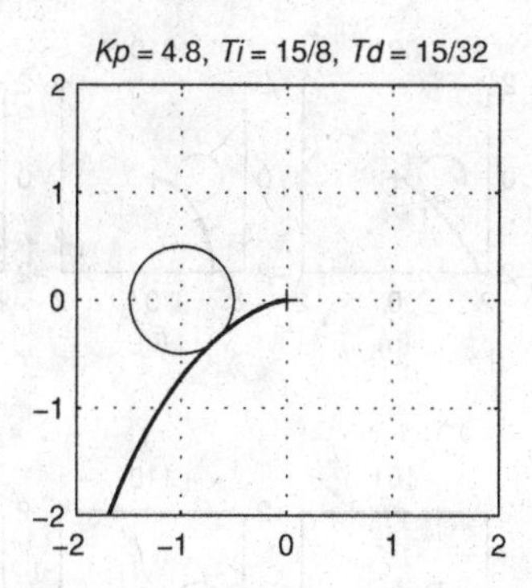

Figure 4.12 Partial Nyquist plot of the loop transfer function. The process $G(s) = 1/(s+1)^3$ is controlled by a PID controller with parameters $K_p = 4.8$, $T_i = 15/8$, and $T_d = 15/32$. The frequency ω increases from 0.1 to 20 towards the origin of the coordinate system. (fignyqs.m)

Figure 4.12 is a Nyquist plot of the *loop transfer function* $L = CG$, where G is the process and C the ideal controller transfer function (in the Laplace domain),

$$C(s) = K_p \left(1 + \frac{1}{T_i}\frac{1}{s} + T_d s\right)$$

$$= K_p + K_i \frac{1}{s} + K_d s \tag{4.38}$$

The last line used the change of variables $K_i = K_p/T_i$ and $K_d = K_p T_d$ for later convenience. The Nyquist plot is the map of $L(i\omega)$ – where $i\omega$ replaces the Laplace s – as $\omega \to \infty$. The controller parameters are the Ziegler–Nichols parameters found previously. The additional circle around the *critical point* $(-1 + i0)$ has diameter 0.5. In order to achieve a good attenuation of the closed-loop response, the Nyquist curve should pass by the critical point at a safe distance, which is equivalent to having closed-loop poles at a safe distance from the imaginary axis of a pole plot. A reasonable distance is in the range 0.5 to 0.8 (corresponding to the sensitivity range 2 to 1.2, Åström and Hägglund 2006 p. 127). Thus the circle marks the lower limit of the range. The plot shows that with the Ziegler–Nichols settings the Nyquist curve touches the circle.

The three tuning gains of the PID controller affect the Nyquist curve of the loop transfer function L in specific manners, and by varying them a designer can in principle shape the Nyquist curve of L.

Figure 4.13 is a *tuning map* showing the consequences of changing the Ziegler–Nichols settings by ±20%. Since the Ziegler–Nichols settings only give approximate values, the tuning map provides empirical knowledge about how changes affect the behaviour of the closed-loop system. It provides a visual overview, which can otherwise be difficult to achieve.

The figure shows that some changes are unfortunate (sub-figures a, c, e, g), while one change in particular makes the system more robust (sub-figure d): decreasing K_p while increasing T_i and T_d. To interpret, it is an advantage to slow down the controller (decrease K_p), reduce the integral action (increase T_i), and increase the damping (increase T_d). A test showed that this

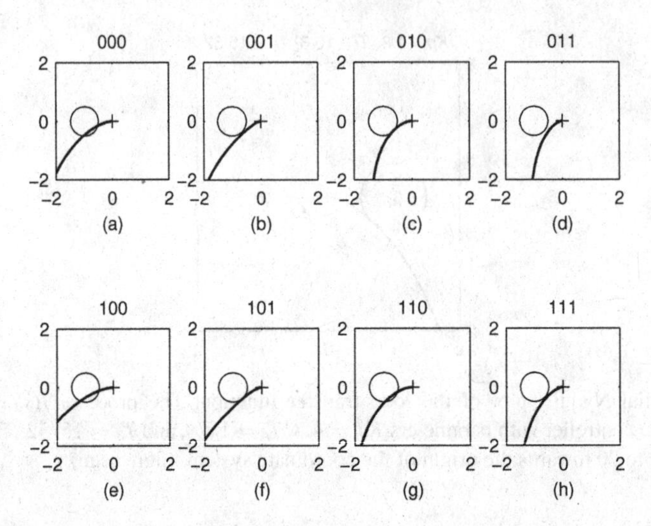

Figure 4.13 Tuning map for PID control of $G(s) = 1/(s+1)^3$. Each plot is a Nyquist curve of the transfer function round the closed loop. The three parameters K_p, T_i, and T_d are changed successively by +20%, indicated by a 1, or −20%, indicated by a 0, relative to the Ziegler–Nichols settings. For instance, the indicator 001 (above b) shows that K_p and T_i are decreased while T_d is increased. (fignyqs.m)

indeed results in a better response, reducing the initial overshoot to roughly one-third, while at the same time attenuating the oscillations in the load response.

We can explain how each PID gain affects the Nyquist curve. Having defined the loop transfer function $L = CG$, insert the expression for C from Equation (4.38) and switch from s to $i\omega$ to analyse the Nyquist curve,

$$\begin{aligned} L(i\omega) &= \left(K_p + K_i \frac{1}{i\omega} + K_d i\omega\right) G(i\omega) \\ &= KpG(i\omega) + K_i \frac{1}{i\omega} G(i\omega) + K_d i\omega G(i\omega) \\ &= KpG(i\omega) - iK_i \frac{1}{\omega} G(i\omega) + iK_d \omega G(i\omega) \end{aligned} \tag{4.39}$$

Now it is clear that $L(i\omega)$ is composed of three terms for a given ω: the first term is proportional to $G(i\omega)$, the second term is proportional to $-iG(i\omega)$, and the third term is proportional to $iG(i\omega)$. Geometrically, the last two terms lie in directions orthogonal to the vector representing $G(i\omega)$ in the complex plane; see Figure 4.14. The figure illustrates that we can map a point Q_1 on the process Nyquist curve 'left and right' by adjusting the tuning gains K_i and K_d, while K_p moves it 'in and out', loosely speaking. As ω increases and the point Q_1 moves inward towards the origin, the three component vectors rotate. Furthermore, the length of component vector $\mathbf{g}_I$ decreases while the length of $\mathbf{g}_D$ increases. It is therefore quite difficult to predict the shape of the whole L-curve for a given choice of settings K_p, K_i, and K_d, but it is possible to construct and inspect the image on a computer.

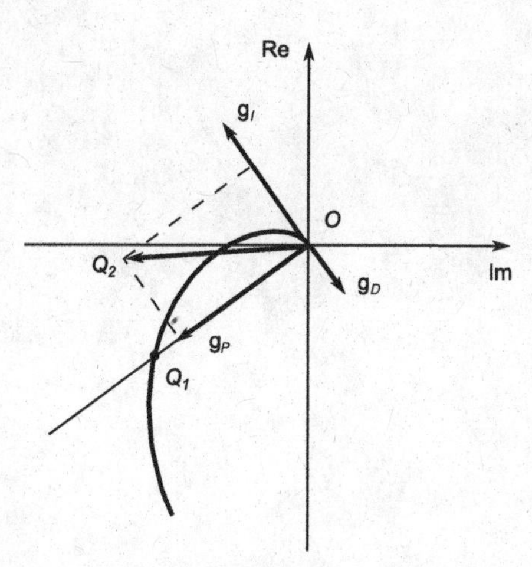

Figure 4.14 The PID controller maps point Q_1 on the process Nyquist curve to Q_2. The image vector $\overrightarrow{OQ_2}$ is the sum of three components: $\mathbf{g}_P$ related to the proportional action, $\mathbf{g}_I$ related to the integral action, and $\mathbf{g}_D$ related to the derivative action.

4.11 Notes and References*

The link between the fuzzy and the PID gains has been sought by various authors (e.g. Siler and Ying 1989, Mizumoto 1992, Qiao and Mizumoto 1996). With the specially crafted linear fuzzy control surface in this chapter, the relationship is particularly transparent, opening up for a transfer of PID tuning methods.

The authoritative reference on PID control by Åström and Hägglund (2006) presents implementation algorithms and tuning methods, and develops the so-called AMIGO tuning method. The methods are illustrated by simulation examples, one of them being the third-order process used in this chapter. Recent tuning methods include: Ziegler–Nichols, kappa-tau, pole placement, stability margins, D-partitioning, OLDP, polynomial, Nyquist, genetic algorithms, adaptive interaction, cancellation, K-B parametrization, multiple integration, and frequency loop shaping; see the overview article by Cominos and Munro (2002). The article is part of a special issue on PID control (Isaksson and Hägglund 2002). For shaping the frequency response, we would ideally like the Nyquist curve to be the vertical line passing through −0.5 on the real axis, as that would mean unity closed-loop gain at all frequencies and an infinitely fast response (Gyöngy and Clarke 2006).

Historically, Ziegler and Nichols were the first to publish optimum settings found from open-loop tests and closed-loop tests (1942, 1943). Cohen and Coon (1953) recognized, however, that alternatives were necessary for certain types of process. For a historical account of the events and the instrumentation leading up to that point, see the article by Bennett (1993) or the summary in Åström and Hägglund (2006 p. 93).

5

Nonlinear Fuzzy PID Control

Nonlinear systems are difficult to analyse, and, as an alternative, standard nonlinear elements have been developed for simulation, including: dead zone, saturation, and quantizer. By analogy, this chapter develops a set of standard control surfaces: the dead zone surface, the saturation surface, and the quantizer surface. The standard surfaces are applied in four cases: an unstable frictionless vehicle, a nonlinear valve compensator, a motor actuator with limits, and for regulating a mass load in the Autopilot simulator. The fuzzy PID controllers are analysed in the phase plane as far as possible, that is, in two dimensions. Geometrically, the crisp PD controller commands a control signal, which is proportional to the distance of a moving point in the phase-plane from a switching line.

A linear time-invariant system behaves well, at least in the sense that (1) its response has an analytical solution, and (2) a sinusoidal input causes a sinusoidal output of the same frequency. Nonlinear systems, on the contrary, behave more or less unpredictably, and designers must brace themselves for the next step into the nonlinear domain. The superposition principle does *not* hold (superposition is briefly: $y_1 + y_2 = G(u_1 + u_2)$, when $y = G(u)$ is a transfer function from input u to output y). Furthermore, the response depends not only on the frequency of the input, but also on the amplitude. The overall plan now is to approach the final two steps of the design procedure: make the fuzzy controller nonlinear (step 3), and fine-tune it (step 4).

5.1 Nonlinear Components

The local behaviour of a nonlinear system can be quite different from the global behaviour. For example, a system can be locally stable, and still be globally unstable when disturbed from its stable state. A control valve is an example of a component which is often nonlinear. The flow through the valve depends on the fraction that the valve is open (between 0 and 1) as well as its flow characteristic $f(x)$. Pneumatic air or a motor moves the valve stem to open or close the valve. The flow characteristic can be linear or nonlinear depending on the design of the valve.

We can often model and simulate nonlinear systems, using software libraries of typical nonlinearities that mimic the behaviour of physical components. Figure 5.1 shows a few examples of simplified behaviour: a rate limiter, a saturation, a dead zone, and a quantizer.

Foundations of Fuzzy Control: A Practical Approach, Second Edition. Jan Jantzen.

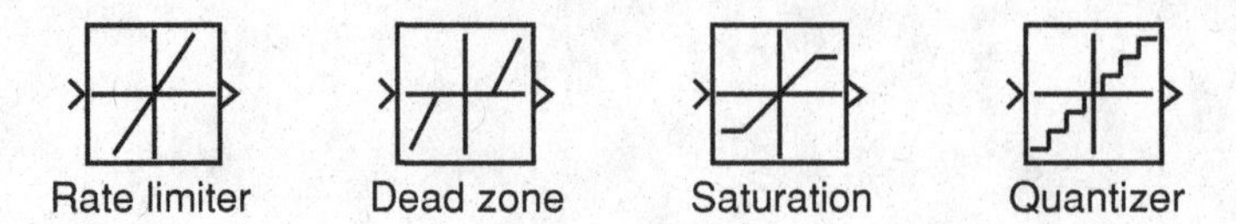

Figure 5.1 A sample collection of four nonlinearities from the Simulink standard library. (figstandardnonlins.mdl)

A fuzzy controller is a nonlinearity itself, generally, but a designer may intentionally shape it to compensate other nonlinearities or to improve the response of the closed-loop system.

Example 5.1 *Nonlinear valve (adapted after Åström and Wittenmark 1995)*
A nonlinear control valve is the actuator in a PI control loop that controls a linear process with the transfer function $P(s) = 10/(s+1)^3$. *The following is a model of the valve:*

$$f(x) = x^4, 0 \leq x \leq 1$$

Here $f(x)$ *is the flow through the valve. The variable* x *is the fraction of valve opening, such that fully closed corresponds to* $x = 0$ *and fully open to* $x = 1$. *The PI controller has the gains* $K_p = 0.10$, $T_i = 1$. *Does the nonlinearity affect the stability of the controlled system?*
► *Solution*
The slope of $f(x)$ *in a given point is the local gain of the valve, when it operates in a narrow region about the point. Clearly, the slope of* $f(x)$ *is small for small values of* x *and steep for large values of* x. *If the slope is very steep, the gain is high, and that could potentially upset the stability.*
Figure 5.2 shows the response to three consecutive steps into higher and higher regions of operation. For small reference signals the closed-loop response is well damped, but as the

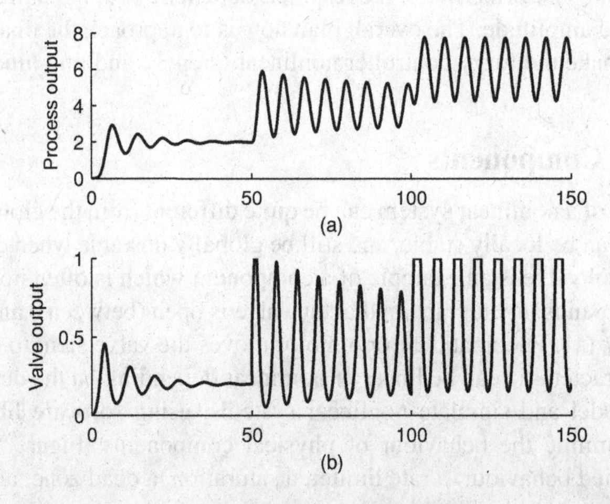

Figure 5.2 Step responses with a nonlinear valve at three different operating points. The reference steps incrementally two units every 50 seconds (a) while the controller (b) tries to keep the system stable. The process is $P(s) = 10/(s+1)^3$ and the PI controller settings are $K_p = 0.10$ and $T_i = 1$. (figvalve.m)

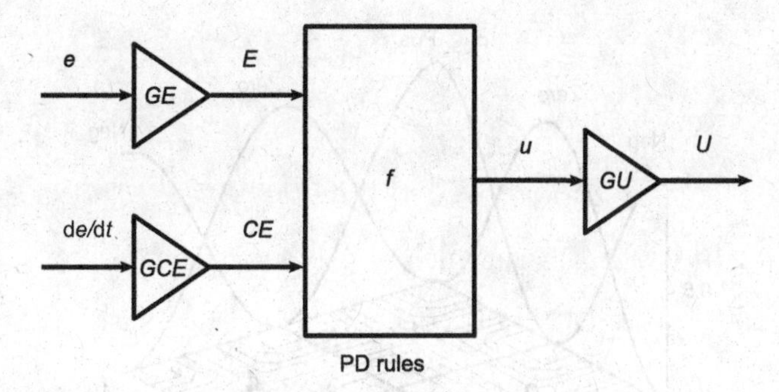

Figure 5.3 Fuzzy PD controller, FPD.

reference steps up, the response gets worse and worse. The controller is tuned to give a good response in the lower region, but the system is unstable in the upper region. There are several ways to cope with such a problem, one way being to devise a nonlinear compensator, which is the inverse of the valve characteristic.

5.2 Phase Plot

A *phase plot* is generally a plot of a time-dependent variable against its own time derivative. The fuzzy PD (FPD) controller operates on error and its time derivative, hence our interest in phase plots. Its configuration is repeated in Figure 5.3; it has only two inputs, which makes it suitable for plotting. The control surface is a plot of *change in error* against *error*, with the control signal along the third axis, and it therefore forms a surface.

Take for example the standard fuzzy PD controller with nine rules,

1. If *error* is Neg and *change in error* is Neg then *control* is NB (5.1)
2. If *error* is Neg and *change in error* is Zero then *control* is NM (5.2)
3. If *error* is Neg and *change in error* is Pos then *control* is Zero (5.3)
4. If *error* is Zero and *change in error* is Neg then *control* is NM (5.4)
5. If *error* is Zero and *change in error* is Zero then *control* is Zero (5.5)
6. If *error* is Zero and *change in error* is Pos then *control* is PM (5.6)
7. If *error* is Pos and *change in error* is Neg then *control* is Zero (5.7)
8. If *error* is Pos and *change in error* is Zero then *control* is PM (5.8)
9. If *error* is Pos and *change in error* is Pos then *control* is PB (5.9)

Figure 5.4 displays the *phase plane*, and the contour plot shows the region of influence of each rule. For example, rule 1 concerns a negative error and a negative change in error; the maximum influence of this rule is in the third quadrant. A designer can adjust the value of negative big (NB) in order to adjust the control in this region. Similarly for the other rules:

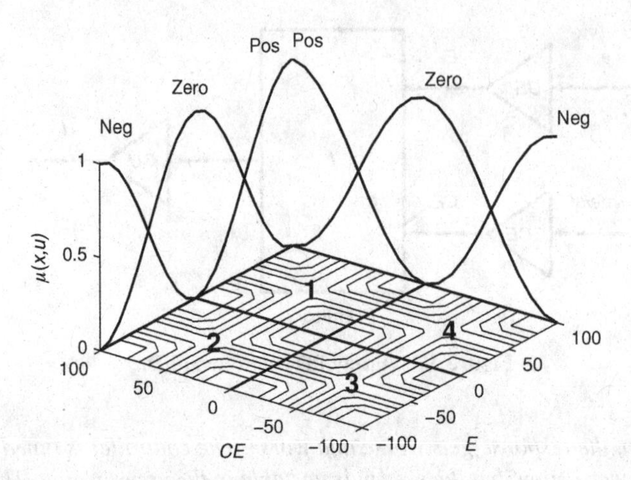

Figure 5.4 Contour plot showing the firing strengths in a rule base with nine rules. The numbers 1, 2, 3, and 4 mark the four quadrants of the phase plane. (figruleregions.m)

rule 3 concerns quadrant 2; rule 7 concerns quadrant 4; rule 9 concerns quadrant 1; and rule 5 concerns the centre. The remaining rules concern the boundary zones between quadrants.

With two inputs and one output the rule base mapping is the relationship between error and change in error on the premise side, and control action on the conclusion side, as portrayed in the control surface. To be precise, the control surface concerns the input signals E and CE, which are after the gains GE and GCE, and the output u, which is before the gain GU (Figure 5.3). A dynamic response can be plotted, $CE(t)$ against $E(t)$, to form a *phase trajectory* in the phase plane or, possibly, on the control surface. The (E, CE)-plane is bounded by the limits of the universes, and the trajectory will always stay within the boundary. In a table based controller, the phase plane trajectory points to the cells in the control table visited by the controller.

Figure 5.5 shows a typical step response with overshoot. The upper right plot shows the trajectory in the phase plane. It is the projection of the trajectory on the control surface (lower right). The trajectory traverses the four quadrants of the phase plane, as indicated by four circles in each plot, in a clockwise direction towards the origin of the coordinate system.

- *Quadrant 4* ($E > 0$, $CE < 0$). Initially the error is large and positive, and the plant output is moving towards the reference. The error $E = GE * e = GE * (r - c)$ is positive (for $GE > 0$) as long as the plant output is below the reference. Furthermore, the change in error is negative (for $GCE > 0$), since $\dot{e} = -\dot{c}$, as long as the plant output is increasing. The phase trajectory spirals in a clockwise direction.
- *Quadrant 3* ($E < 0$, $CE < 0$). The plant output has overshot the reference and is still moving away from the reference. The error is negative, since the plant output is above the reference. Furthermore, the change in error is negative, since the plant output is still increasing.
- *Quadrant 2* ($E < 0$, $CE > 0$). The plant output is returning towards the reference. The error is negative, since the plant output is above the reference. Furthermore, the change in error is positive, since the plant output is now decreasing.

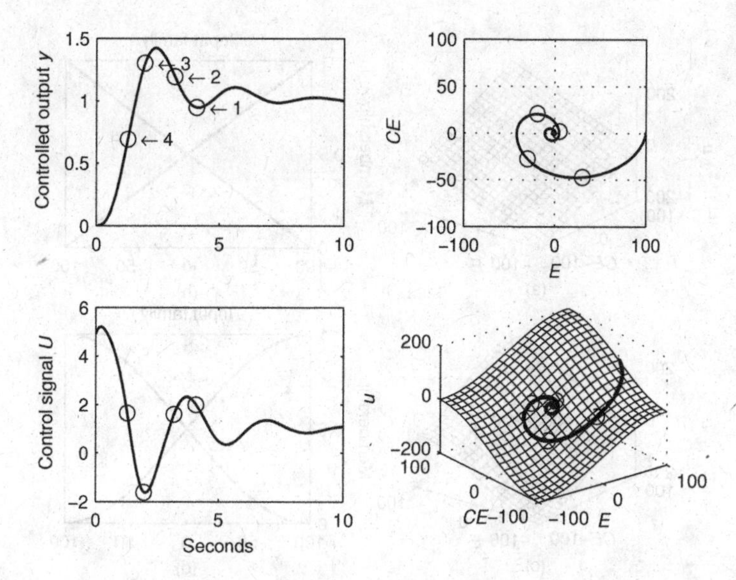

Figure 5.5 The four circles indicate four typical stages of a step response. There is one circle in each quadrant (1 to 4) of the phase plane. (figregio.m)

- *Quadrant 1* ($E > 0, CE > 0$). The plant output is moving away from the reference during an undershoot. The error is positive, the plant output is below the reference. Furthermore, the change in error is positive, and the plant output is decreasing.

The diagonal in the north-west–south-east direction of the phase plane corresponds to zero control signal, and points above the diagonal correspond to positive control signals, while those below the diagonal correspond to negative control signals. The diagonal is thus a *switching line,* that is, the control signal changes sign when the trajectory crosses the line. The designer may affect the response locally, either by changing the overlap between neighbouring rules or by adjusting one or several conclusion singletons.

If the trajectory ends in the centre (the origin), the plant is on the reference and not moving, that is, in equilibrium. But a plant does not necessarily settle in the centre: A steady-state error results in an end point on the $CE = 0$ line, but off the centre.

5.3 Four Standard Control Surfaces

The third step in the overall design procedure is to make the fuzzy controller nonlinear. The control surface affects the dynamics of the closed-loop system; in fact, it is the only component which governs the nonlinear behaviour of the controller. Inspired by the standard two-dimensional nonlinearities (Figure 5.1), this book employs four standard control surfaces: linear, saturation, dead zone, and quantizer (Figures 5.6 and 5.7). The surfaces are soft versions of the common nonlinearities, only three-dimensional.

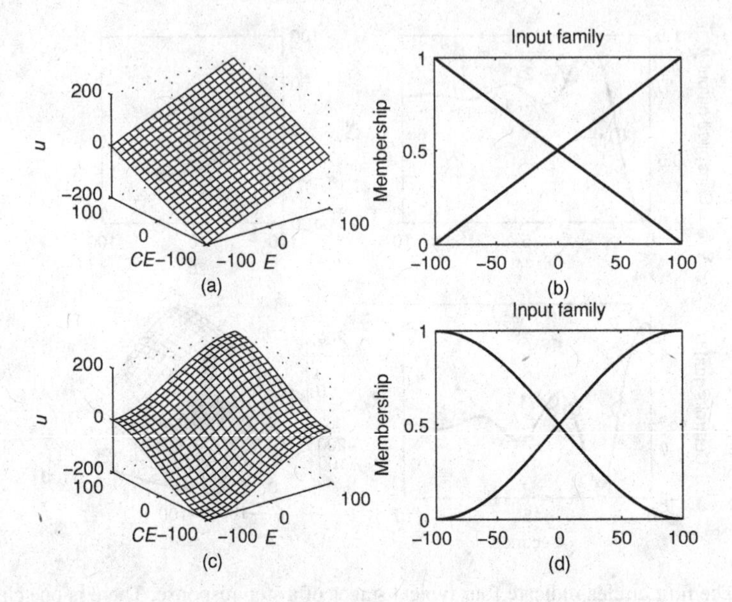

Figure 5.6 Linear surface (a) and the input family that generated it (b). Saturation surface (c) with the input family that generated it (d). (figsurfs.m)

- The *linear* surface (Figure 5.6a, b) results from the rule base (5.5) using only rules 1, 3, 7, and 9, with the triangular premise sets,

$$\mu_{Neg} = \mu_{Trapezoid}(x; -100, -100, -100, 100)$$
$$\mu_{Pos} = \mu_{Trapezoid}(x; -100, 100, 100, 100)$$

The surface in the figure is equivalent to the summation $E + CE-$; compare the values on the axes. Since the surface is equivalent to a summation, the controller is similar to a crisp PD controller.

- The *saturation* surface (Figure 5.6c, d) is built using only rules 1, 3, 7, and 9, together with the premise sets,

$$\mu_{Neg} = \mu_{STrapezoid}(x; -100, -100, -100, 100)$$
$$\mu_{Pos} = \mu_{STrapezoid}(x; -100, 100, 100, 100)$$

These are smooth trapezoids, built from segments of cosine functions (Chapter 2). Notice the absence of the central rule (rule 5) with zero *error* and zero *change in error*. When the error is near zero, a disturbance will increase the magnitude of the control signal, but when the error reaches a certain level, a further disturbance causes little or no increase of the control signal. The same can be said for the change in error. That surface has a steeper slope – higher gain – near the centre of the table than the linear surface has. It has the same values pairwise in the four corners as the linear surface.

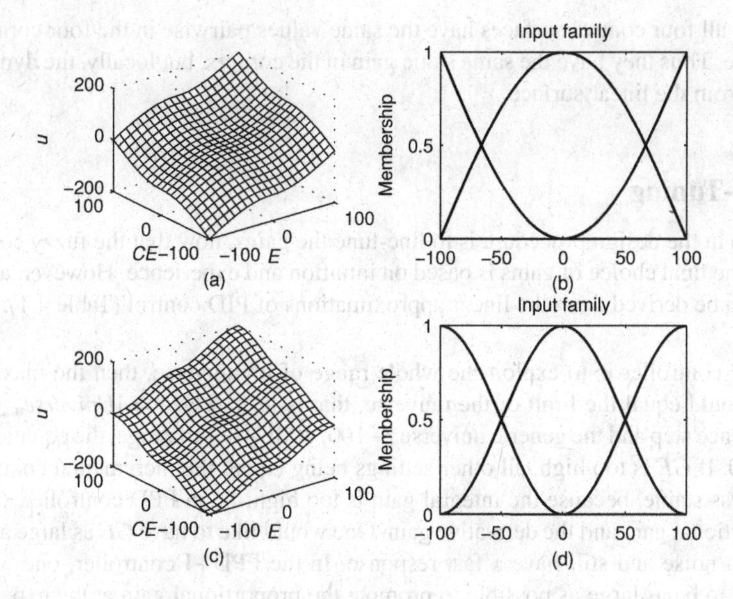

Figure 5.7 Dead zone surface (a) and the input family that generated it (b). Quantizer surface (c) with the input family that generated it (d). (figsurfs.m)

- The *dead zone* surface (Figure 5.7a, b) is built from all nine rules. The premise sets are defined as

$$\mu_{Neg} = 2\mu_{STrapezoid}(x; -200, -200, -200, 0)$$
$$\mu_{Zero} = 2\mu_{STrapezoid}(x; -200, 0, 0, 200)$$
$$\mu_{Pos} = 2\mu_{STrapezoid}(x; 0, 200, 200, 200)$$

  They are smooth trapezoids, built from segments of cosine functions with half the frequency and twice the amplitude of the previous. That surface has a more gentle slope – lower gain – near the centre of the table. When the controller operates near the centre, a disturbance only affects the control signal a little until the operating point is at some distance from the centre. It also has the same values pairwise in the four corners as the linear surface.
- The *quantizer* surface (Figure 5.7c, d) is a blend of the previous two surfaces. It is built using all nine rules, with nonlinear membership functions, defined as,

$$\mu_{Neg} = \mu_{STrapezoid}(x; -100, -100, -100, 0)$$
$$\mu_{Zero} = \mu_{STrapezoid}(x; -100, 0, 0, 100)$$
$$\mu_{Pos} = \mu_{STrapezoid}(x; 0, 100, 100, 100)$$

  They are smooth trapezoids, built from segments of cosine functions. It has a flat plateau near the centre and other plateaus in several places. Even this surface has the same values as the other surfaces in the four corners.

In summary, all four control surfaces have the same values pairwise in the four corners as the linear surface. Thus they have the same static gain in the corners, but locally, the dynamic gain is different from the linear surface.

5.4 Fine-Tuning

The last step in the design procedure is to fine-tune the gains, now that the fuzzy controller is nonlinear. The final choice of gains is based on intuition and experience. However, a few rules of thumb can be derived from the linear approximations of PID control (Table 4.1):

- *GE*. If the controller is to exploit the whole range of its universe, then the maximal $E = GE * e$ should equal the limit of the universe, that is $|e_{\max} * GE| = |Universe_{\max}|$. With a unit reference step and the generic universe $[-100, 100]\%$ of full range, the equation implies $GE = 100$. If *GE* is too high, all other settings being equal, the incremental controller will become less stable, because the integral gain is too high. In an FPD controller, *GE* affects the proportional gain and the derivative gain. One would like to have *GE* as large as possible to dampen noise and still have a fast response. In the FPD+I controller, one would also prefer *GE* to be as large as possible to promote the proportional gain at the expense of the integral gain and the derivative gain.
- *GCE*. Similarly if *change in error* is to exploit the whole range of its universe, $|\dot{e}_{\max} * GCE| = |Universe_{\max}|$. In an FPD controller, a larger *GCE* means a larger derivative gain with no side effect on the proportional gain. To dampen noise to a minimum, one will therefore prefer *GCE* as small as possible. In the FInc controller, an increase in *GCE* will decrease the integral gain and increase the proportional gain; thus one would like to keep *GCE* as large as possible to preserve stability. In the FPD+I controller, an increase in *GCE* will increase the derivative gain, so one would keep it as small as possible.
- *GCU or GU*. These affect the proportional gain, so one would like to have them as large as possible without creating too much overshoot. If too small, the system will be too slow, and if too large the system might become unstable.

Algorithm 5.1 summarizes the procedure for hand-tuning an FPD+I controller (with trivial modifications this procedure covers the FPD and FInc controllers as well).

Algorithm 5.1 Hand-tuning FPD+I.

---

1. Adjust *GE* (or *GCE*) according to step size and universe to exploit the range of the universe of *E* (or *CE*) fully.
2. Remove integral action and derivative action by setting $GIE = GCE = 0$. Tune *GU* to give the desired response, ignoring any final value offset.
3. Increase the proportional gain by means of *GU*, and adjust the derivative gain by means of *GCE* to dampen the overshoot.
4. Adjust the integral gain by means of *GIE* to remove any final value offset.
5. If dissatisfied, return to step 1.

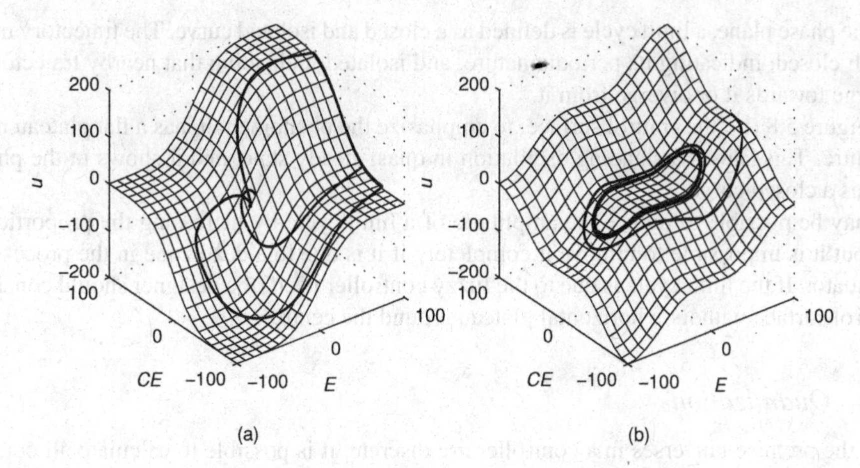

Figure 5.8 Saturation in the right limit of the E-universe (a) and a limit cycle (b). (figs2load.m)

The stability will be similar to the stability of the system's linear approximation resulting from a Taylor series expansion near an equilibrium, mathematically speaking. In simulation, it is possible to experiment with different controller surfaces to find the stability margin and the sensitivity to deadtimes. The responses are amplitude dependent, however, and thereby depend on the step size.

Up to this point, we have ignored a number of practical points for the sake of clarity. Thus we now turn to some phenomena that must be considered, and possibly avoided, during the tuning phase.

5.4.1 *Saturation in the Universes*

Saturation in the premise universes affects the dynamic response. The phase plot in Figure 5.8 (a) displays saturation in the E-universe, since the response follows the edge of the control surface for a period of time. It is obviously a limitation of the signal magnitude, and it may result in a larger overshoot, slower response, or a sluggish load response.

When the error signal is noisy, the *CE*-universe is likely to exhibit saturation, because differentiation of a noisy signal produces spikes. This can be an advantage, though, because the universe will limit the magnitude of the spikes.

If saturation is unintended, it can be removed by α-scaling.

5.4.2 *Limit Cycle*

A *limit cycle* is a periodic motion in a quasi-steady state of the system. It is a unique feature of nonlinear systems. It can occur, for example, if there is a dead zone such that the plant drifts away from the equilibrium until a certain point, where the controller takes action to move it back towards the equilibrium.

In the phase plane, a limit cycle is defined as a closed and isolated curve. The trajectory must be both closed, indicating its periodic nature, and isolated, indicating that nearby trajectories converge towards it or diverge from it.

In Figure 5.8 (b), the control surface, to emphasize the phenomenon, has a flat plateau near the centre. This causes a standing oscillation in quasi-steady state, and it shows in the phase plane as a closed curve.

It may be possible to reduce the amplitude of a limit cycle by increasing the proportional gain, but it is impossible to remove it completely if it is due to a dead zone in the process or the actuator. If the limit cycle is due to the fuzzy controller itself, the designer should consider a control surface without a horizontal plateau around the centre.

5.4.3 *Quantization*

When the premise universes in a controller are discrete, it is possible to calculate all combinations of $E(n)$ with $CE(n)$ before putting the controller into operation; that is the principle underlying the table based controller. The quantization in the table affects the performance, however.

With a quantum size of 10 in premise universes $[-100, 100]$ the resolution is 5% of full range. This causes a limit cycle, because the controller allows the plant to drift within a cell in the lookup table, until it shifts to another cell with a corrective control action. This is especially noticeable in steady state. The amplitude of the limit cycle is affected by the input gains.

There are three ways to reduce the limit cycle: (1) increase the input gains, (2) make the resolution finer, and (3) use interpolation. The first option makes the controller more sensitive to small deviations from the setpoint, and may also cause a saturation that changes the dynamics. A variant of the second option is to use a new table with a finer resolution when the plant approaches the setpoint. The third option removes the quantization effect completely.

5.4.4 *Noise*

Measurement noise causes an oscillatory behaviour. If the noise frequency is high, it will drive the phase plot to the edges of the *change in error* universe; beyond, the input universe will limit spikes in the signal.

If the noise causes instability or disturbs the control, the first option is to decrease the derivative gain. A second option is to install a filter that attenuates the noise, although a filter will always introduce some amount of time delay.

The whole design procedure is now complete, and Algorithm 5.2 summarizes its four steps. It refers to an FPD+I controller, because it is the most general controller, but it covers the other controllers as well with slight modifications. It is now time to demonstrate what fuzzy controllers can do, that linear PID controllers cannot.

Algorithm 5.2 Fuzzy controller design procedure.

1. *Design a crisp PID controller*. Use a PID tuning method to find K_p, $1/T_i$, and T_d.
2. *Replace it with a linear fuzzy controller*. Use the FPD+I configuration, transfer K_p, $1/T_i$, and T_d to *GE*, *GCE*, *GIE*, and *GU* (GCU) using Table 4.1. Run the controller, and check

for saturation in the premise universes. When it is removed, by means of α-scaling, check that the closed-loop response is exactly the same with the fuzzy rule base replaced by a pure summation.

3. *Make it nonlinear.* Use a standard nonlinear control surface to begin with. Modify the rules by trial and error to reshape the control surface.
4. *Fine-tune it.* Use hand-tuning: use GE to improve the rise time, use GCE to dampen overshoot, and use GIE to remove any steady state error.

5.5 Example: Unstable Frictionless Vehicle

Assume a frictionless vehicle – let us say a hovercraft or a spacecraft – can be modelled by Newton's second law: $F = ma$. Here F is an external force, m is the mass, and a is the acceleration. Assume further that the 'external' force is supplied by a controller, which may actually be on-board the vehicle. For the sake of clarity, let $m = 1$.

Acceleration is the derivative of velocity, which in turn is the derivative of position, and therefore the vehicle can be modelled as a double integrator $1/s^2$ in the Laplace domain. The double integrator $1/s^2$ is open-loop unstable, but it is possible to stabilize it with PD control. Can we use a fuzzy PD controller in this case?

► Solution

The short answer is: Yes, we can always construct a fuzzy PD controller that performs the same way as a crisp PD controller. Furthermore, the FPD gives some extra options, and we may be able to exploit the nonlinearities to our advantage. We go as far as the first three steps of the standard design procedure to demonstrate the point.

1. *Design a crisp PID controller.* Hand-tuning resulted in the controller parameters $K_p = 0.5$, $T_d = 1$ and we chose sampling time $T_s = 0.05$ for the simulations.
2. *Replace it with a linear fuzzy controller.* The error is less than or equal to one, and to accommodate the whole range, the gain on the error is chosen as $GE = 100$. The remaining gains are given by the gain relationships for a linear controller (Table 4.1),

$$GU = K_p/GE = 0.5/100 = 0.005$$
$$GCE = GE * T_d = 100 * 1 = 100$$

   Figure 5.9 shows the response with a linear control surface and four rules. The response was identical to that of the crisp PD controller. It is a loose tuning, with 31% overshoot and some oscillation.
3. *Make it nonlinear.* Since the process is unstable in open-loop, tight control around the settling point is necessary. We therefore choose a saturation type control surface; it has a larger dynamic gain around the centre than the linear surface. Figure 5.10 shows the new response. The overshoot is now 20%, and the response is less oscillatory.

The only change from crisp PD to fuzzy PD was the change of control surface, that is, the tuning remained the same. That alone provided a better response, and it shows that even for a

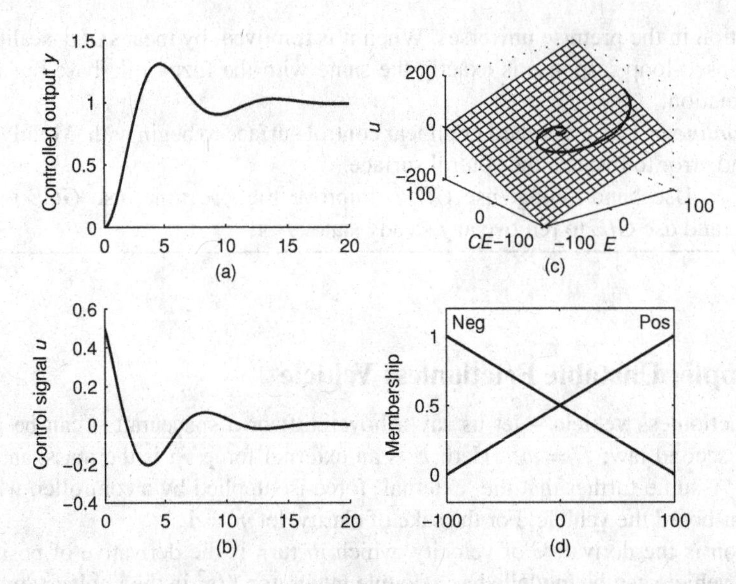

Figure 5.9 Closed-loop step response. Double integrator $1/s^2$ process with a linear FPD controller. (figfpdi4s2.m)

linear process, it may be an advantage to apply a nonlinear controller. Note that it might also be possible to achieve an even better result by retuning the crisp (linear) PD controller; the example does not find the optimal controller.

Inserting a saturation type surface diminished the overshoot for the double integrator, indicating that this was a good choice. Can we improve the response even more?

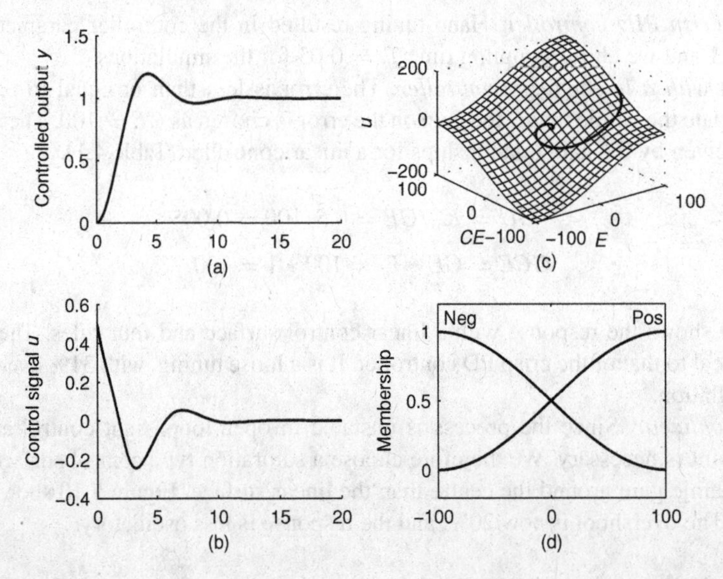

Figure 5.10 Closed-loop step response. Double integrator $1/s^2$ with a saturation surface FPD controller. (figfpdi4s2.m)

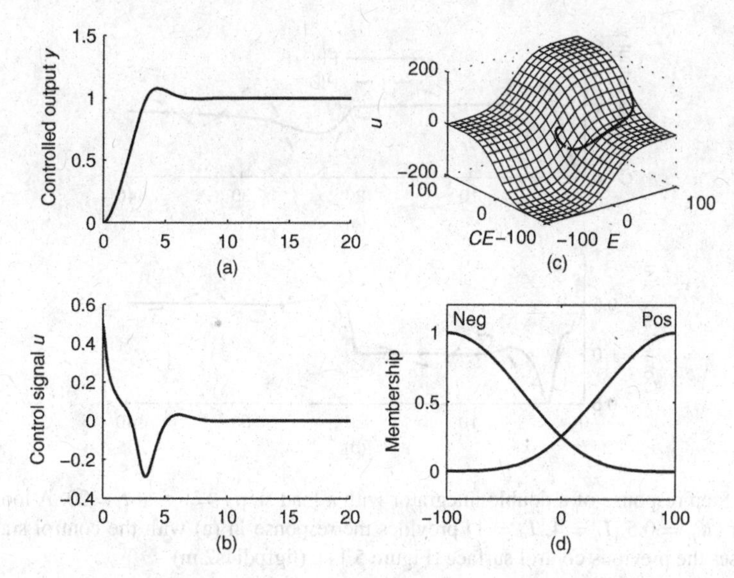

Figure 5.11 Closed-loop step response. Double integrator $1/s^2$ with a saturation surface FPD controller generated by squared membership functions. (figfpdi4s2.m)

► Solution

It was the steep slope around the centre of the surface that proved to be an advantage. The following rule base generates an even more pronounced slope:

If *error* is very Neg and *change in error* is very Neg then *control* is NB

If *error* is very Neg and *change in error* is very Pos then *control* is Zero

If *error* is very Pos and *change in error* is very Neg then *control* is Zero

If *error* is very Pos and *change in error* is very Pos then *control* is PB

The rule base includes the fuzzy logic hedge 'very', which squares the membership functions.

Figure 5.11 shows the result. The overshoot is now 8%, and the response settles in a well-damped manner. The control signal is clearly nonlinear: It slows down before crossing zero the first time, then it dips distinctly before pulling back towards zero.

Now include a load step of $l = 0.5$. What do we have to do to redesign the controller?

► Solution

A load on the system requires integral action in the controller in order to provide a sustained control signal in steady state to counterbalance the load. The design steps are this time as follows:

1. *Design a crisp PID controller.* Hand-tuning resulted previously in the controller parameters $K_p = 0.5$, $T_d = 1$. As a quick guess, set $T_i = 4T_d = 4$, which is commonly used, even by

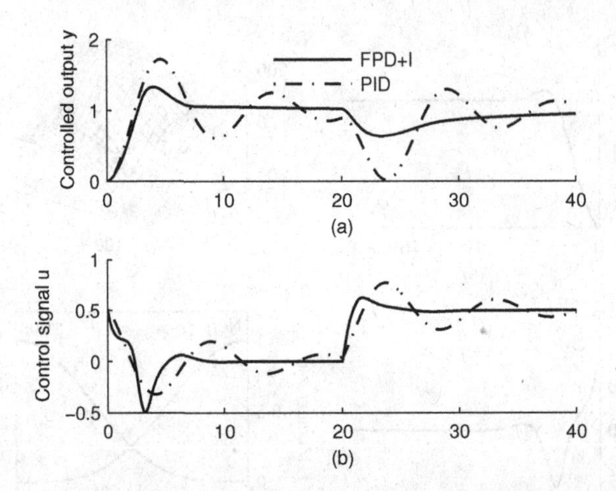

Figure 5.12 Step response of a double integrator with a load step $l = 0.5$ at $t = 20$. A loosely tuned PID controller ($K_p = 0.5, T_i = 4, T_d = 1$) provides the response in (a) with the control signal in (b). The FPD+I uses the previous control surface (Figure 5.11). (figfpdi4s2.m)

Ziegler and Nichols (if $T_i < 4T_d$ the PID controller will have complex zeros). We choose sampling time $T_s = 0.05$ as before.

2. *Replace it with a linear fuzzy controller*. We choose again $GE = 100$, $GU = 0.005$, $GCE = 100$. For GIE we have the relationship (Table 4.1) $GIE = GE * 1/Ti = 25$.
3. *Make it nonlinear*. We keep the successful control surface made from squared membership functions (Figure 5.11).

Figure 5.12 shows the new responses with a PID controller and the FPD+I controller. The response with the PID controller is very oscillatory, because of the loose tuning, but although the tuning is the same the FPD+I response is very good in comparison.

5.6 Example: Nonlinear Valve Compensator

An earlier example demonstrated how a nonlinear valve affects the stability margin in different regions of operation (Example 5.1). The process is linear with the transfer function $P(s) = 10/(s+1)^3$. The valve is modelled by the function

$$f(x) = x^4, 0 \leq x \leq 1$$

where $f(x)$ is the fraction of flow through the valve and x the fraction it is open. Can a nonlinear fuzzy rule base improve the response?

►Solution (adapted from Åström and Wittenmark 1995)

Figure 5.13 shows the valve characteristic. The flow stays below 10% as long as the valve is less than 55% open; at larger openings it increases rapidly. Notice that both x and $f(x)$ are limited to the interval [0, 1]. To compensate the nonlinearity $f(x)$, the output u of the controller is fed through a function $g(u)$, which is approximately the inverse of $f(x)$. Their

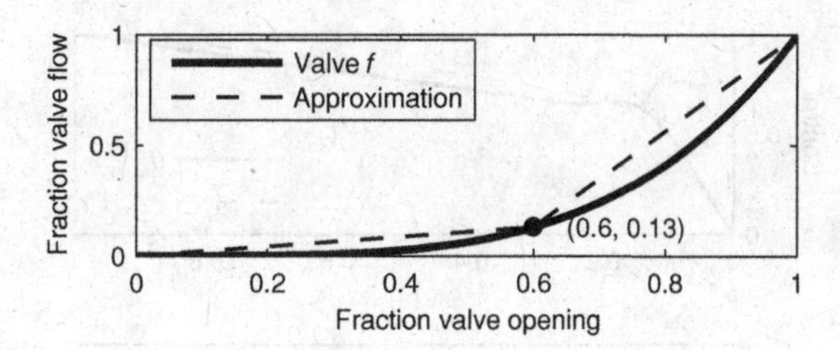

Figure 5.13 Valve characteristic $f(x)$. The flow through the valve is a nonlinear function of valve opening from fully closed ($x = 0$) to fully open ($x = 1$). The piece-wise linear approximation can be inverted. (figsugvalve.m)

composition is $f(g(u))$, which is expected to have less variation in the gain than $f(x)$ has; if g is an exact inverse then u passes straight through to the process.

We approximate $f(x)$ by a piece-wise linear function (see Figure 5.13) as follows:

$$\hat{f}(x) = \begin{cases} \dfrac{0.13}{0.6}x & 0 \leq x \leq 0.6 \\ \dfrac{0.87}{0.4}x - 1.18 & 0.6 < x \leq 1 \end{cases}$$

Its inverse, which is also a piece-wise linear function, is easy to find:

$$g(u) = \begin{cases} 4.62u & 0 \leq u \leq 0.13 \\ \dfrac{0.4}{0.87}u + 1.18\dfrac{0.4}{0.87} & 0.13 < u \leq 1 \end{cases}$$

The following Sugeno rule base (of the first order) interpolates between the two line segments, in order to achieve a smooth transition from one to the other:

If u is Low then *control* $= 4.62u$

If u is High then *control* $= 0.46u + 0.543$

Figure 5.14 shows the result of the interpolation (*interpolant*) and a suitable set of membership functions that are hand-tuned. The interpolant is smooth in the transition zone between the two line segments as required. The membership functions are smooth trapezoids with some overlap in the neighbourhood of the point where the two line segments meet. The overlap between the membership functions determine the width of the transition zone. The overlap is chosen by trial and error to give an acceptable response. The figure also shows the true inverse of the valve characteristic. The estimate is clearly better in the high end than in the low end; there is room for improvement.

Figure 5.15 shows the resulting step responses. Compared to the earlier responses (Figure 5.2) the compensator stabilized the system and provided a well-damped response at the low end. The responses at the three operating points are somewhat different from each

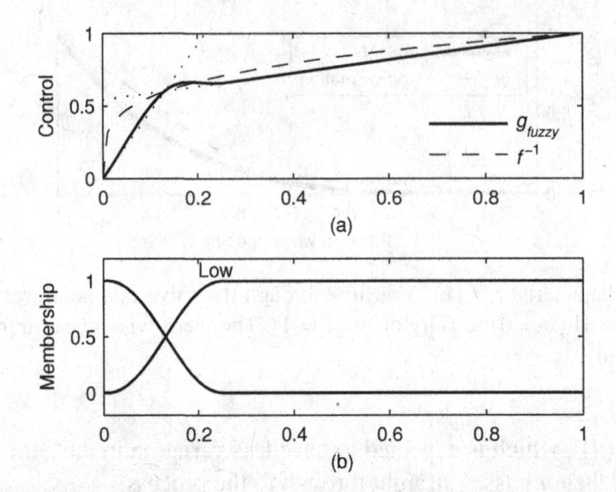

Figure 5.14 Fuzzy interpolation between two (dotted) line segments (a). The membership functions in (b) control the interpolation. The piece-wise linear function is the inverse of an approximation of the nonlinear valve characteristic. The true inverse is also shown (dashed line). (figsugvalve.m)

other: the lowest response is good, while the two following responses are oscillatory. Their shapes differ from each other, because the function $g(u)$ is only an approximate inverse of $f(x)$. Nevertheless, the responses are better than the previous result (Figure 5.2).

The controller is still the same crisp PI controller ($K_p = 0.1$, $T_i = 1$; see Example 5.1). It is followed by a fuzzy single-input rule base, which acts as a static compensator, in order to improve the performance of the crisp PI controller. This is thus an example of a hybrid controller consisting of a crisp PI controller connected in series with a fuzzy rule base.

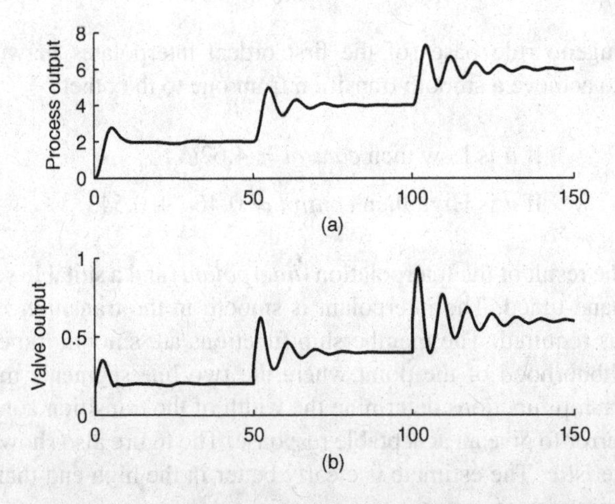

Figure 5.15 Step responses with a compensated valve at three different operating points. The compensator is a fuzzy rule base. The reference steps up two units every 50 seconds (a), while the control signal (b) tries to keep the system stable. The process is $P(s) = 10/(s+1)^3$ and the PI controller settings are $K_p = 0.10$ and $T_i = 1$. (figvalve.m)

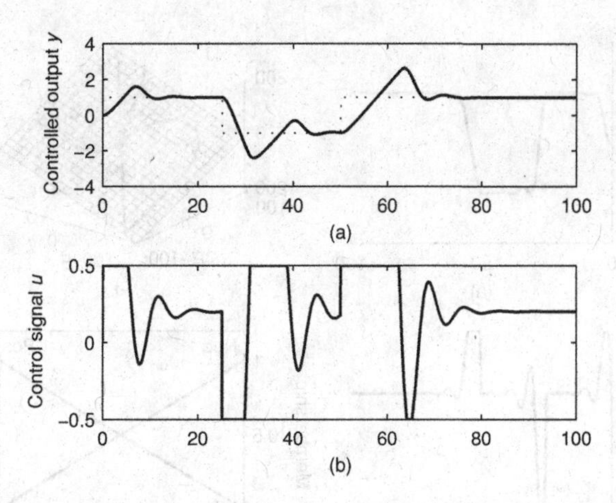

Figure 5.16 Integrator windup. The process output is supposed to follow the reference steps in (a). The control signal is saturated during the overshoots (b). The process is $G(s) = 1/s\,(s+1)$, with a load of 0.2, and it is controlled by a PI controller. (figwindup.m)

5.7 Example: Motor Actuator with Limits

Consider a simple motor with the transfer function

$$G(s) = 1/s\,(s+1)$$

In front of the motor is an actuator that saturates when the magnitude of the control signal is 0.5. There is a load on the motor, and it is controlled by a PI controller ($K_p = 1, T_i = 3, T_d = 0$, load $l = 0.2$, and sample-time $T_s = 0.05$). Unfortunately, the integrator in the controller winds up when the actuator saturates. The response in Figure 5.16 shows considerable overshoot when the system responds to reference steps.

The plot of the control signal shows that the actuator saturates in the limits several times. Can a fuzzy controller solve the integrator windup problem?

► Solution

The FInc controller can replace a PI controller. It has an integrator in the output end (Figure 4.5), which is relatively simple to limit to the limits of the actuator. This avoids windup.

Figure 5.17 shows the result. The motor output follows the reference better, although there is still a little overshoot. The rule base is linear, but α-scaling was chosen to cut off a very large spike in CE ($\alpha = 0.41$). The only other action was to limit the contents of the integrator, such that it stays within the actuator limits.

The FInc gains are chosen according to the gain relationships between PID and fuzzy PID controllers (Table 4.1), that is, $GE = 100$, $GCE = GE * T_i$, $GCU = K_p/GCE$.

5.8 Autopilot Example: Regulating a Mass Load

Again we let the Autopilot train car drive on a track shaped like a parabola. The car is initially standing still, but this time at the reference position $x = 5$. This position is on a slope, therefore

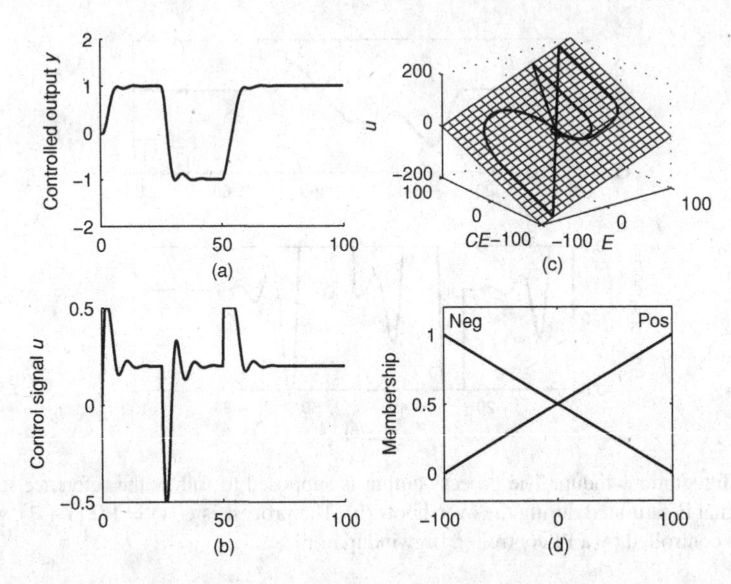

Figure 5.17 The motor response to reference steps (a) and the control signal, which saturates (b). The control surface (c) is linear, but α-scaling is tuned to cut a large spike off by means of the universe limit. The membership functions are simply triangular (d). (figwindup.m)

the controller holds it in position by means of a non-zero control effort. At the time $t = 10$ the car is loaded with an extra mass of $5 \times m$, that is, five times it own mass ($m = 1500$ kg). At $t = 30$ the load is taken off again.

The load changes will force the train car away from its reference position. Can a fuzzy controller regulate the position, such that excursions from the reference are less than 1.5 m and preferably without oscillatory responses?

► Solution

When the load comes on, the system responds with a dip (downwards in a plot of the position), and when the load goes off, the system responds with a flare (upwards in a plot of the position). Dips and flares may have different shapes in a nonlinear system, and local changes in the control strategy may improve specific local behaviour. Again we go as far as the first three steps of the standard design procedure.

1. *Design a crisp PID controller*. Integral action is necessary, because gravity exerts a sustained load on the system. Let us therefore try PID control, and an equivalent FPD+I controller. We could perhaps apply the Ziegler–Nichols frequency response method to find the controller gains, but we found earlier some settings: $K_p = 6000$ and $T_d = 1$. These work very well in this case also, except that we must add integral action. As a quick guess, set $T_i = 4T_d = 4$ as we did earlier (Section 5.5). The response is not optimal, but fair (dip 1.9 m and oscillatory, flare 1.75 m). Figure 5.18 shows the reference position of the train car on the track, and how far the car travels during the responses to the step load.
2. *Replace it with a linear fuzzy controller*. The product $GE * GU$ is equivalent to the proportional gain K_p. Let $GU = 1$, then $GE = K_p/GU = 6000$. The gain GCE is determined by $GCE/GE = T_d$, and it follows that $GCE = 6000$ also. Now that we have an integrator,

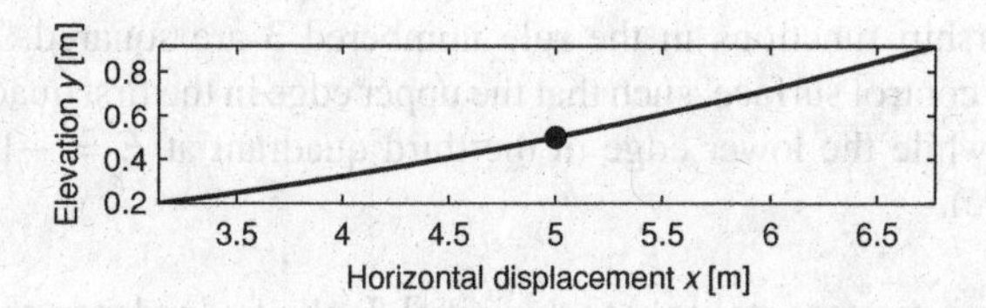

Figure 5.18 Position of the train car. The dot marks the referernce point, and the curve indicates how far the train car travels during the load step responses. The track is parabolic. (figrunautopilot4.m)

we must choose a setting for GIE also, that is, $GIE = GE/T_i = 1500$. Choose the standard rule base with four rules. With triangular membership functions, and other specific design choices (Theorem 3.1), the controller acts like a summation.

3. *Make it nonlinear*. Figure 5.19 shows the final design and the result. The PID response when the load comes on is rather oscillatory (Figure 5.19a), and the dip is larger than the flare. The response from the FPD+I is better in the sense that the magnitude of the dip is smaller (1.2 m), the flare also (1.13 m), and the response is not oscillatory (Figure 5.19a); the design thus meets the specifications. We used the standard saturation surface (Figure 5.19d), but changed one rule. The rule base is as follows:

1. If *error* is Neg and *change in error* is Neg then *control* is NB
3. If *error* is very Neg and *change in error* is very Pos then *control* is Zero
7. If *error* is Pos and *change in error* is Neg then *control* is Zero
9. If *error* is Pos and *change in error* is Pos then *control* is PB

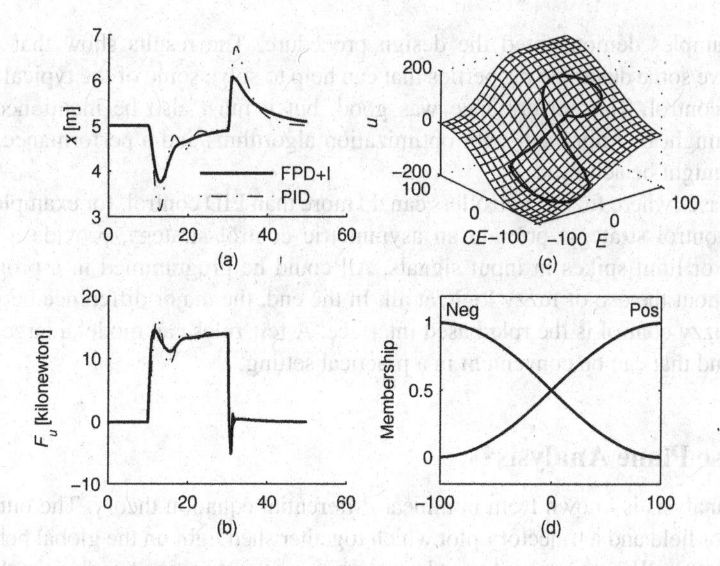

Figure 5.19 Autopilot response to load on ($t = 10$), then load off ($t = 30$). The step response with the FPD+I is together with the response with the PID controller in (a). The two control signals are in (b). The control surface (c) is a slightly warped version of the standard saturation surface due to a modification of the rule base. The membership functions are smooth (d). (figrunautopilot4.m)

Thus the membership functions in the rule numbered 3 are squared. This modification slightly warps the control surface, such that the upper edge in the first quadrant at $CE = 100$ is raised a little, while the lower edge in the third quadrant at $E = -100$ is depressed a little (Figure 5.19c).

The control surface reacts more strongly to the initial flank of a load step compare the control signal in Figure 5.19d. The response when the load comes on (the dip) starts in the centre of the surface, then it extends into the first quadrant while rotating clockwise. It returns through the fourth quadrant, almost to the centre. The flare trajectory continues in a clockwise direction through the third quadrant, then into the second quadrant. The difference between the PID control signal and the FPD+I control signal (Figure 5.19b) is barely visible at the beginning of the two steps, but the FPD+I reacts earlier, and that is enough to attenuate the dip and the flare. The oscillatory response, when pulling back from the dip, disappeared (fourth quadrant).

Apart from being nonlinear, the control strategy is asymmetric. In contrast, a crisp PID controller is always symmetric.

5.9 Summary

The gain relationships between the crisp PID controller and the fuzzy PID controllers opened for a transfer of tuning gains to the fuzzy controller. The linear fuzzy controller performs the same way as the equivalent PID controller. But the fuzzy controller has more design options.

The control surface is the single component that governs the nonlinearity of the fuzzy controller. The four surfaces presented in this chapter are just four out of many possibilities. It is an attempt to standardize the design, in order to bring down the number of possible design choices.

Several examples demonstrated the design procedure. The results show that the fuzzy controllers have some desirable properties that can help to solve some of the typical problems in nonlinear control. The performance was good, but it must also be mentioned that the performance might be improved; with optimization algorithms and a performance criterion, better results might be achieved.

There are cases where fuzzy controllers can do more than PID control, for example: provide a nonlinear control strategy, provide an asymmetric control strategy, provide a nonlinear compensator, or limit spikes in input signals. All could be programmed in a programming language, without the use of fuzzy logic at all. In the end, the major difference between PID control and fuzzy control is the rule based interface. A few rules can model a large region of state-space, and that can be convenient in a practical setting.

5.10 Phase Plane Analysis*

Phase plane analysis is known from nonlinear differential equation theory. The outcome is a map of a vector field and a trajectory plot which together shed light on the global behaviour of a nonlinear system. We can apply phase plane analysis without having to solve the differential equations analytically. Furthermore, it accommodates discontinuous (hard) nonlinearities.

The plots can provide hints to the design of a fuzzy controller, even for systems of higher order than two, but the method is here restricted to PD type controllers (without integral action). Phase plane analysis uses state-space models, and it is an advantage to have some basic knowledge of the state-space approach, some of which is provided in the following.

A *state-space model* of a linear system is a model in the form of a vector–matrix equation, that is,

$$\dot{\mathbf{x}} = \boldsymbol{A}\mathbf{x} + \boldsymbol{B}\mathbf{u} \tag{5.10}$$

Here $\mathbf{x}$ is the *state-variable vector*, $\boldsymbol{A}$ is the *system matrix*, and $\boldsymbol{B}$ the *input matrix*. The state-vector $\mathbf{x}$ holds a number of state-variables. If that number is n, then the system is an nth-order system, and the dimension of $\boldsymbol{A}$ is n-by-n. The number of rows in $\boldsymbol{B}$ is n, and the number of columns matches the number of elements in the input vector $\mathbf{u}$. The vector on the left-hand side is the *velocity vector*, because it is the time derivative of the *position vector* $\mathbf{x}$. The equation relates – in n dimensions – the position and velocity of a motion depending on the inputs in $\mathbf{u}$.

A *phase plot* is generally a plot of a variable and its time derivative plotted against each other. The fuzzy PD (FPD) controller operates on error and its time derivative. These therefore span a phase plane, and the response of the system follows a certain trajectory in that phase plane.

Phase plane analysis is a graphical method that provides a valuable visual overview. It has a drawback, as it is limited to two-dimensional plots, and so systems of order higher than two are normally excluded from phase plane analysis. Nevertheless, even when an FPD controller controls a system of higher order than two, we still wish to observe the two-dimensional phase plane spanned by the two controller inputs.

5.10.1 Trajectory in the Phase Plane

The starting point is a second-order *autonomous* system

$$\dot{\mathbf{x}} = \boldsymbol{A}\mathbf{x} \tag{5.11}$$

or written out,

$$\dot{x}_1 = a_{11}x_1 + a_{12}x_2 \tag{5.12}$$

$$\dot{x}_2 = a_{21}x_1 + a_{22}x_2 \tag{5.13}$$

where x_1 and x_2 are the state-variables of the system, and $\boldsymbol{A}$ the system matrix. An autonomous system is a system without inputs, thus the usual input vector $\mathbf{u}$ is missing.

Geometrically, the coordinate axes of x_1 and x_2 span a two-dimensional space, which is a phase plane. The solution in time $\mathbf{x}(t)$ to Equation (5.11) – given an initial value $\mathbf{x}(0)$ of the state-vector – is a curve in the phase plane with x_2 along the vertical (ordinate) axis and x_1 along the horizontal (abscissa) axis. The curve is called a *phase trajectory*.

5.10.2 Equilibrium Point

If the system settles down to an equilibrium state, as the result of some initial state away from the equilibrium, the equilibrium is characterized by zero motion in the phase plane. Such an *equilibrium point* must satisfy the equation

$$\mathbf{0} = \boldsymbol{A}\mathbf{x}$$

for all time. If the $\boldsymbol{A}$ matrix is non-singular (its determinant is non-zero), then the only solution is the origin $\mathbf{x} = \mathbf{0}$, which is then an equilibrium point. Otherwise (the determinant is zero) there may be many solutions, that is, a collection of equilibrium points lying on a line through the origin. Nonlinear systems can have several equilibrium points, one, or none at all.

The ratio of Equation (5.13) to Equation (5.12) is the slope of the phase trajectory, since we have

$$\frac{\dot{x}_2}{\dot{x}_1} = \frac{\frac{dx_2}{dt}}{\frac{dx_1}{dt}} = \frac{dx_2}{dx_1}$$

The slope S at any point (x_1, x_2) is therefore

$$S = \frac{dx_2}{dx_1} = \frac{a_{21}x_1 + a_{22}x_2}{a_{11}x_1 + a_{12}x_2} \tag{5.14}$$

Conversely, given a particular slope $S = S^*$, and solving the equation for x_2, the result is an equation for a line along which the slope, or inclination, of all trajectories crossing it, is the same. More generally, the line is a curve, and such a curve is called an *isocline*. A collection of isoclines can be used to graphically construct phase plane trajectories (see for example Slotine and Li 1991).

Drawing isoclines is the classical route to take, but we shall deviate slightly and apply a view from kinematic geometry instead. Geometrically, the vector differential Equation (5.11) expresses the rate of change $\dot{\mathbf{x}}(t)$ at the time t of the state-vector $\mathbf{x}(t)$ of the system. As $\mathbf{x}(t)$ is the locus vector of a point in the phase plane travelling along a trajectory, $\dot{\mathbf{x}}(t)$ is its *velocity vector*. The velocity vector is tangential to the phase trajectory. Geometrically, the velocity vector indicates the speed (its length) and the direction (its angle) of the motion.

The matrix $\boldsymbol{A}$ can be viewed as an operator which maps $\mathbf{x}$ to $\dot{\mathbf{x}}$. For almost all vectors in the plane, the operation is the composition of a rotation and a scaling, such that $\dot{\mathbf{x}}$ is at an angle to $\mathbf{x}$, and it has a different length. For an eigenvector $\mathbf{v}$, however, $\boldsymbol{A}\mathbf{v} = \lambda\mathbf{v}$ where λ is the corresponding eigenvalue; thus the mapping is just a scaling, and the rotation is absent. Eigenvectors are said to be *fixed lines* under the mapping.

The geometrical view allows us to visually inspect the phase plane.

5.10.3 Stability

For any particular instance (x_1^*, x_2^*) of the state-vector, Equations (5.12–5.13) determine the speed and direction of motion $(\dot{x}_1^*, \dot{x}_2^*)$. Taking a number of such points (x_1^*, x_2^*) – possibly distributed in the phase plane in an even manner – the velocity vectors provide an overview

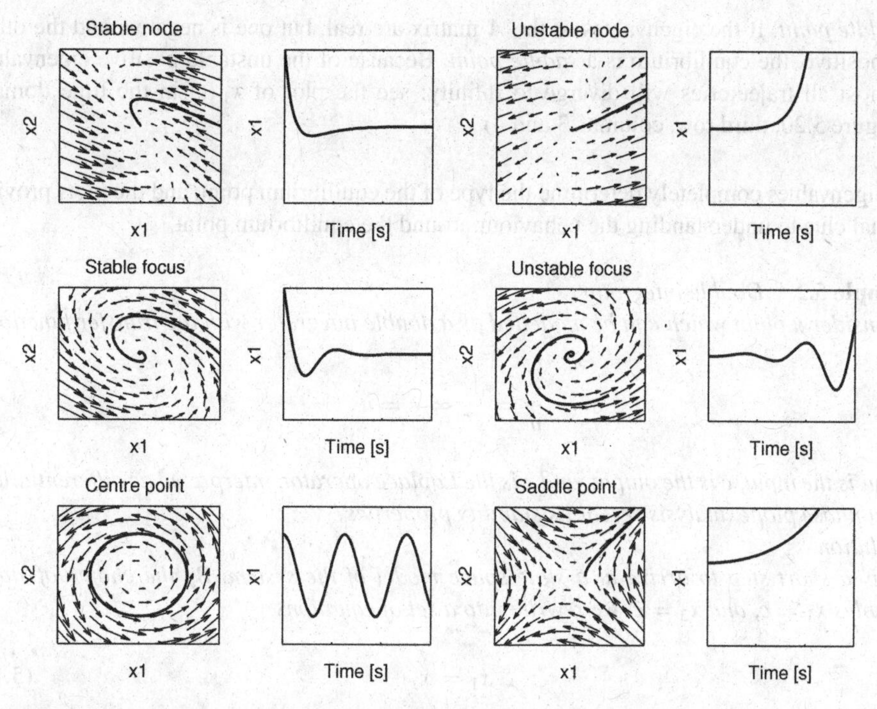

Figure 5.20 Equilibrium points. Read column-wise, columns 1 and 2 are stable or marginally stable, while columns 3 and 4 are unstable. The trajectory in an (x_1, x_2)-plane is equivalent to the time response on its right resulting from an individual initial state. (figequil.m)

of the 'flow' of the system. Figure 5.20 shows six different types of behaviour around the equilibrium point of various $\boldsymbol{A}$ matrices:

- *Node*. If the eigenvalues of the $\boldsymbol{A}$ matrix are real and negative, the equilibrium is a *stable node*, because $\mathbf{x}(t)$ converges to $\mathbf{0}$ in an exponential decay. If both eigenvalues are positive, the equilibrium is an *unstable node*, because $\mathbf{x}(t)$ diverges exponentially. Since the eigenvalues are real, there are no oscillations in the motion; compare the plots of $x_1(t)$ in the time domain (Figure 5.20, first row).
- *Focus*. If the eigenvalues of the $\boldsymbol{A}$ matrix are complex conjugates, with negative real parts, the equilibrium is a *stable focus*. The state-vector $\mathbf{x}(t)$ converges to $\mathbf{0}$, but in an oscillatory manner, whereby the trajectory in the phase plane encircles the origin one or more times, unlike the stable node. If both eigenvalues have positive real parts, the equilibrium is an *unstable focus*, because $\mathbf{x}(t)$ diverges in an oscillatory manner; compare the plots of $x_1(t)$ in the time domain (Figure 5.20, second row).
- *Centre point*. If the eigenvalues of the $\boldsymbol{A}$ matrix are complex conjugates, with real parts equal to zero, the equilibrium is a *centre point (vortex)*, because $\mathbf{x}(t)$ encircles the equilibrium point, without converging or diverging. The plot of $x_1(t)$ in the time domain shows a sustained oscillation, which indicates a marginally stable system (Figure 5.20, third row, columns 1 and 2).

- *Saddle point*. If the eigenvalues of the A matrix are real, but one is negative and the other is positive, the equilibrium is a *saddle point*. Because of the unstable positive eigenvalue, almost all trajectories will diverge to infinity; see the plot of $x_1(t)$ in the time domain (Figure 5.20, third row, columns 3 and 4).

The eigenvalues completely determine the type of the equilibrium point, and the plots provide a visual clue to understanding the behaviour around the equilibrium point.

Example 5.2 *Double integrator*
Consider a plant which can be modelled as a double integrator with the transfer function

$$\frac{c}{u} = \frac{1}{s^2} \Leftrightarrow \ddot{c} = u$$

Here u is the input, c is the output, and s is the Laplace operator, interpreted as differentiation. Can a phase plane analysis reveal its stability properties?
► *Solution*
It is a short step to arrive at a state-space model of the system. By the choice of state-variables $x_1 = c$, and $x_2 = \dot{c}$, we rewrite into a set of equations

$$\begin{aligned}\dot{x}_1 &= x_2 \\ \dot{x}_2 &= u\end{aligned} \tag{5.15}$$

Comparing with the state-space form, Equation (5.10), the matrices are:

$$\mathbf{A} = \begin{bmatrix} 0 & 1 \\ 0 & 0 \end{bmatrix}, \mathbf{b} = \begin{bmatrix} 0 \\ 1 \end{bmatrix}$$

By setting $\dot{x}_1 = \dot{x}_2 = 0$ in Equation (5.15), and $u = 0$ as well, because we are considering the autonomous system, we find that at equilibrium $x_2 = 0$, while x_1 can be anything. In other words, all points on the whole x_1-axis are equilibria. If $x_2 \neq 0$ – the moving point is above or below the x_1-axis – the velocity is constant and parallel to the x_1-axis, which will move the point towards infinity.

*Since the **A** matrix is upper triangular, the eigenvalues are displayed on the diagonal: $\lambda_1 = \lambda_2 = 0$. Therefore each equilibrium point is an unstable node.*

Even though the method is restricted to autonomous systems, it can be applied to closed-loop systems, and more easily so if the reference is constant. If the local behaviour of a nonlinear system can be approximated by that of a linear system, the equilibrium type is the same.

5.11 Geometric Interpretation of the PD Controller*

Assume for now that both the process and the controller are linear. Furthermore, assume that the process is of second order, the reference r is constant, and, for simplicity, that the process has only a single input u. The process is then characterized by a set of state-space equations,

$$\dot{\mathbf{x}} = \mathbf{A}\mathbf{x} + \mathbf{b}u \tag{5.16}$$

The controller delivers a control signal u, related to the error $e = r - c$ where c is the controlled output, by a transfer function C,

$$u = Ce \tag{5.17}$$

Substituting Equation (5.17) into Equation (5.16), the closed-loop system equations appear,

$$\dot{\mathbf{x}} = \mathbf{A}\mathbf{x} + \mathbf{b}Ce \tag{5.18}$$

Thus the velocity vector $\dot{\mathbf{x}}$ is the sum of two vector components: the open loop velocity vector $\mathbf{A}\mathbf{x}$ plus a contribution from the controller $\mathbf{b}Ce$. A plot of the two components shows the influence of the controller at various stages of the closed-loop response and thus provides a visual tuning aid (Figure 5.21).

Example 5.3 *Proportional control of a double integrator*

Take the previous double integrator (Example 5.2), and assume that it is controlled by a P-controller, $u = K_p e$, where K_p is the proportional gain, and the error $e = r - x_1$. Can the controller stabilize the system?

► *Solution*

Insert into Equation (5.16) to obtain the system equations

$$\dot{\mathbf{x}} = \mathbf{A}\mathbf{x} + \mathbf{b}u = \mathbf{A}\mathbf{x} + \mathbf{b}K_p e \tag{5.19}$$

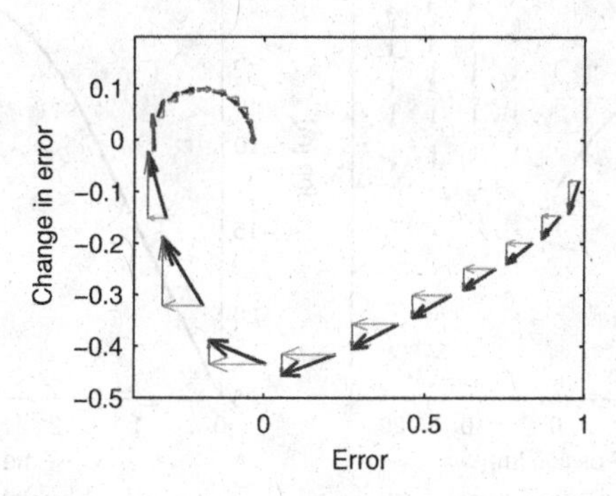

Figure 5.21 Velocity vectors along the trajectory. The velocity vector at each point is the sum of two vectors: a horizontal open loop velocity plus a vertical controller contribution. Compare Equation (5.18). (figs2load.m)

It is the same **A** *matrix and* **b** *vector as in the open-loop system. Writing the equations out, the closed-loop system is:*

$$\dot{x}_1 = x_2$$
$$\dot{x}_2 = K_p(r - x_1) = -K_p x_1 + K_p r$$

The closed-loop system matrix and input vector are thus

$$\mathbf{A}_{cl} = \begin{bmatrix} 0 & 1 \\ -K_p & 0 \end{bmatrix}, \mathbf{b}_{cl} = \begin{bmatrix} 0 \\ K_p \end{bmatrix} \tag{5.20}$$

The eigenvalues of $\mathbf{A}_{cl}$ $\left(\lambda = \pm\sqrt{-K_p}\right)$ *are either complex conjugates on the imaginary axis* ($K_p > 0$)*, indicating a marginally stable centre point in equilibrium, or the eigenvalues are real with one negative eigenvalue and one positive* ($K_p < 0$)*, indicating an unstable saddle point.*

We can therefore conclude that it is impossible to stabilize the system with proportional control.

Example 5.4 *Stopping a car, PD control*

Consider again the problem of stopping a car in front of a red stop light (Chapter 1). We saw that a PD controller resulted in a good response. Can phase plane analysis explain the motion?

► *Solution*

The car model is also a double integrator. Figure 5.22 shows that the controller twists the vector field, especially in the upper half of the phase plane, while the rest of the plane remains unchanged due to the constraints on the control signal.

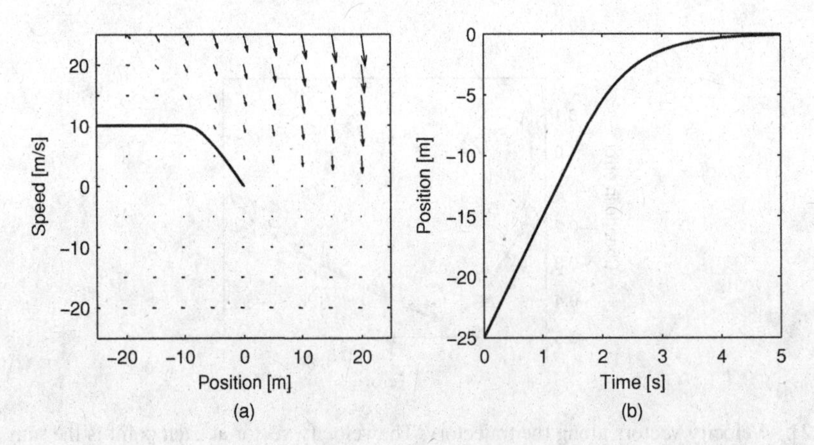

Figure 5.22 Stopping a car. The controller twists the vector field (a) such that the trajectory ends in the centre. Braking starts at a position about 10 metres before the stop line (b). (figcar.m)

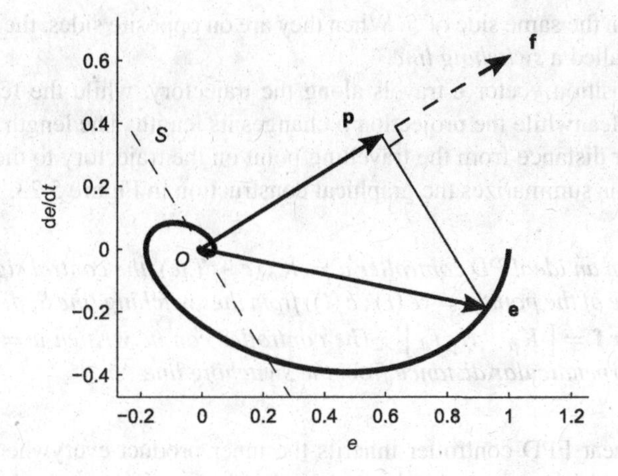

Figure 5.23 Trajectory in the $(e, \dot{e})$-plane. The PD feedback vector $\mathbf{f} = K_p \begin{bmatrix} 1 & T_d \end{bmatrix}^T$ (dashed) defines a fixed switching line S (dashed). The control signal $u = K_p (e + T_d \dot{e})$ is proportional to the length of the projection of the position vector $\mathbf{e}$ on $\mathbf{f}$, that is $u = \|\mathbf{p}\| \, \|\mathbf{f}\|$. As the position vector rotates clockwise, the magnitude of u changes with $\|\mathbf{p}\|$. Consequently, $u = 0$ on S, and as the trajectory crosses S, the sign of u changes. (figphase3rd.m)

The example shows first of all that the phase plane provides valuable overview. Secondly, the phase plane method could also accommodate the nonlinear constraints on the control signal.

5.11.1 The Switching Line

The ideal PD controller $u = K_p (e + T_d \dot{e})$ is equivalent to a vector product in the $(e, \dot{e})$ phase plane. The control signal u depends on the position vector $\mathbf{e} = \begin{bmatrix} e & \dot{e} \end{bmatrix}^T$ as defined in Figure 5.23. This is useful to know for a fuzzy controller designer, because the $(e, \dot{e})$-plane is related to the control surface of an FPD controller.

The controller can be written as an inner product of two vectors, that is,

$$u = \begin{bmatrix} K_p & K_p T_d \end{bmatrix} \begin{bmatrix} e \\ \dot{e} \end{bmatrix} = \mathbf{f}^T \mathbf{e} = \mathbf{f} \cdot \mathbf{e}$$

Here, $\mathbf{e} = \begin{bmatrix} e & \dot{e} \end{bmatrix}^T$ is a time-dependent position vector, which describes a trajectory in the $(e, \dot{e})$-plane, which is a phase plane. The *feedback vector* $\mathbf{f} = \begin{bmatrix} K_p & K_p T_d \end{bmatrix}^T$, which is constant but adjustable, defines a line S perpendicular to the vector and passing through the origin O. For a stable system, the trajectory will end at the origin, which is the point of zero error and zero change in error. The inner (dot) product, $\mathbf{f} \cdot \mathbf{e} = \|\mathbf{f}\| \, \|\mathbf{e}\| \cos a$, where a is the angle between the two vectors, is the same as the length of $\mathbf{f}$ times the length of the projection of $\mathbf{e}$ on $\mathbf{f}$.

In other words, a particular setting of K_p and T_d defines a particular line S and a scaling in the phase plane. The moment the trajectory cuts S, $\mathbf{e}$ is perpendicular to $\mathbf{f}$, and the inner product is therefore zero (Figure 5.23). Furthermore, the inner product has a positive sign as long as $\mathbf{e}$ and $\mathbf{f}$ are on the same side of S. When they are on opposite sides, the sign is negative, and S is therefore called a *switching line*.

The tip of the position vector $\mathbf{e}$ travels along the trajectory, while the feedback vector $\mathbf{f}$ remains constant. Meanwhile the projection $\mathbf{p}$ changes its length. The length of $\mathbf{p}$ is the same as the perpendicular distance from the travelling point on the trajectory to the switching line. The following lemma summarizes the graphical construction in Figure 5.23.

Lemma 5.1 *Given an ideal PD controller $u = K_p(e + T_d\dot{e})$, the control signal u is proportional to the distance of the point $\mathbf{e} = (e(t), \dot{e}(t))$ from the switching line S, defined by the normal feedback vector $\mathbf{f} = \begin{bmatrix} K_p & K_pT_d \end{bmatrix}^T$. The controller can be written $u = \mathbf{f} \cdot \mathbf{e} = \|\mathbf{f}\| \, \|\mathbf{p}\|$, where $\|\mathbf{p}\|$ is the perpendicular distance from the switching line.*

A table based linear FPD controller inherits the inner product everywhere in the control table. The control table therefore contains numbers of increasing magnitude with increasing distance from the switching line. This is confirmed by the control surface $u = E + CE$ valid for the linear FPD acting as a summation.

The e-axis is an isocline, even in the nonlinear domain, because the trajectory always crosses the e-axis at a right angle. This can be seen as follows. The derivative of the position vector is the velocity vector $\dot{\mathbf{e}} = \begin{bmatrix} \dot{e} & \ddot{e} \end{bmatrix}^T$. The velocity vector is tangent to the trajectory, and it points in the direction of the motion. Now form the inner product

$$\mathbf{e} \cdot \dot{\mathbf{e}} = \begin{bmatrix} e & \dot{e} \end{bmatrix} \begin{bmatrix} \dot{e} \\ \ddot{e} \end{bmatrix} = e\dot{e} + \dot{e}\ddot{e} = \dot{e}(e + \ddot{e})$$

On the e-axis $\dot{e} = 0$, and the inner product is zero. The velocity vector is thus perpendicular to the position vector when crossing the e-axis. There is no assumption of linearity, and the tangent being parallel to the second axis is therefore a geometric property of the trajectory – valid in any coordinate transformation of the phase plane.

Example 5.5 *Controller contribution, third-order process*

Let the process be a third-order system with the transfer function

$$\frac{c}{u} = \frac{1}{s(s+1)^2} \tag{5.21}$$

Here u is the input, c is the output, and s is the Laplace operator. What is the geometric contribution from a PD controller in the controller phase plane?

► *Solution*

It is necessary to obtain a state-space model of the system in order to get a picture of the geometry. The process is of third order, however, and we must therefore expect a three-dimensional state-space, which is slightly complicated. The controller is only of second order, though, and it is therefore possible to perform a phase plane analysis in the controller plane.

To obtain a state-space model of the system, rearrange Equation (5.21) as follows,

$$s\,(s+1)^2\,c = u \Leftrightarrow$$
$$\left(s^3 + 2s^2 + s\right)c = u \Leftrightarrow$$
$$\dot{c} + 2\ddot{c} + \dot{c} = u$$

Choose state-variables $x_1 = c$, $x_2 = \dot{c}$, $x_3 = \ddot{c}$, *and rewrite to obtain a model on state-space form*

$$\begin{aligned}\dot{x}_1 &= x_2\\ \dot{x}_2 &= x_3\\ \dot{x}_3 &= -2x_3 - x_2 + u \end{aligned}\tag{5.22}$$

Here u *is the control signal coming from the ideal PD controller, that is* $u = K_p(e + T_d\dot{e})$. *The equation thus contains a mixture of* x*-variables and* e*-variables. The following variable substitutions change the viewpoint from* x*-coordinates to controller coordinates:*

$$\begin{aligned} e &= r - x_1\\ \dot{e} &= -\dot{x}_1\\ \ddot{e} &= -\dot{x}_2\\ \dot{e} &= -\dot{x}_3 \end{aligned}$$

To form the derivatives above we assumed for simplicity that r *is constant, such that its derivative is zero. The phase trajectory of* $(e\,(t)\,, \dot{e}\,(t))$ *is thus a simple linear mapping of the phase trajectory of* $(x_1\,(t)\,, \dot{x}_1\,(t))$. *That is the same as* $(x_1\,(t)\,, x_2\,(t))$, *according to the state-space model in Equation (5.22), and the controller thus acts as a model (an observer) of the first two state-variables. When the controller – with a proper tuning – tries to stabilize the system within some reasonable time, it is in effect acting counter to the motion of the partial position vector* $(x_1\,(t)\,, x_2\,(t))$.

After substituting the variables and rearranging, we obtain the following closed-loop state-space model in controller coordinates:

$$\begin{aligned}\dot{e}_1 &= e_2\\ \dot{e}_2 &= e_3\\ \dot{e}_3 &= -2e_3 - e_2 - u \end{aligned}\tag{5.23}$$

This state-space model shows first of all that the controller contributes a linear change $(-u)$ *to the open-loop* $\dot{e}_3$. *The change depends on the two other state-variables, namely* e_1 *and* e_2, *according to the PD control law. Secondly, the controller directly affects the velocity of only the third state-variable* e_3, *not the other two.*

Taking $(e\,(t)\,, \dot{e}\,(t))$ *as a two-dimensional position vector, its velocity vector is the derivative* $(\dot{e}\,(t)\,, \ddot{e}\,(t))$. *Its acceleration vector is the derivative of the velocity vector, that is* $(\ddot{e}\,(t)\,, \ddot{e}\,(t))$.

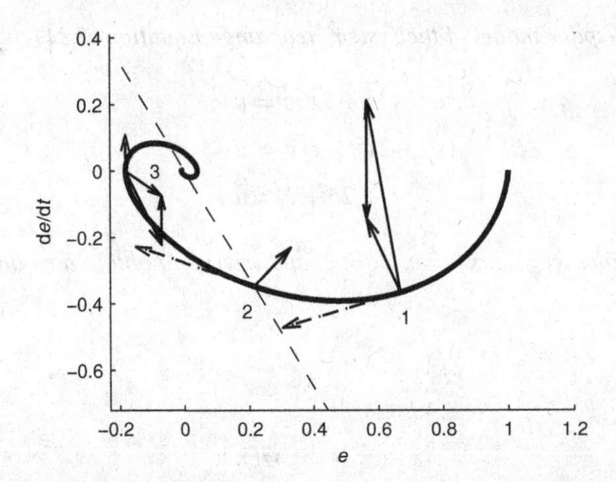

Figure 5.24 Step response in the $(e, \dot{e})$ plane. Solid arrows are acceleration vectors and dash-dot arrows are velocity vectors. In each of the three points (1, 2, 3), the closed loop acceleration is the vector sum of the open loop acceleration and the controller contribution (always vertical). The controller contribution is in a downward direction before reaching the dashed switching line (1), zero on the line (2), and in an upward direction after the line (3). The process is $P(s) = 1/s(s+1)^2$, and the controller is a PD controller with $K_p = 0.8$, $T_d = 0.9$. (figphase3rd.m)

The controller thus affects the last coordinate of the acceleration vector. In other words, it adds only upwards or downwards contributions to the acceleration vector, not sideways. Everything is in two dimensions, and it is possible to draw the effect of the controller along the trajectory of $(e(t), \dot{e}(t))$.

Figure 5.24 illustrates how the controller contributes at three subsequent points on the trajectory. The switching line plays an important role, in the sense that contributions are downward above the switching line, zero on the line, and upward below the line. We can say that the controller tries to pull the moving point towards the switching line, although it is restricted to act only on acceleration and then only in upward and downward directions.

The example shows that phase plane analysis gives insight even in the case of a third-order system. It shows that the PD controller is restricted to act only in particular directions, coupled to the switching line. Clearly, the picture becomes more difficult for higher orders of the process, and also if we introduce integral action into the controller.

5.11.2 A Rule Base for Switching

Going back to the rules of thumb for tuning a PD controller (Table 4.3), an increase of T_d will slow the rise time down and decrease any overshoot. In terms of the phase plane, the feedback vector **f** will turn more upwards, the switching line S will rotate with it, and its slope will become less negative. The slope will always be negative, because T_d is positive. An increase of the proportional gain K_p will scale the magnitude of the control signal everywhere in the phase plane.

The tuning factor T_d thus determines the slope of the switching line. For a fast response, T_d should be as small as possible; that is, the slope of the switching line should be steep (large negative). There is a limit, though, to what is acceptable, because the overshoot will grow as the slope becomes steeper.

The switching line acts as an *attractor* in the sense that the control action tries to pull the trajectory towards the switching line. Let us, for a moment, perform the thought experiment that the system initially starts on the switching line and follows it all the way to the origin of the phase plane. In that case, the control signal of the PD controller will be zero all along, that is

$$0 = K_p (e + T_d \dot{e}) \Rightarrow$$
$$\dot{e} = -\frac{1}{T_d} e$$

This is a first-order differential equation with the solution $e(t) = e(0) \exp^{-t/T_d}$. In words, it is an exponential decay with the time constant T_d. It would be convenient to make the system follow the switching line, because the designer decides the time constant (the slope).

A control strategy for a fuzzy controller could be to try and follow the switching line, or at least get as close as possible. The definition of the linear FPD controller $U = (GE * e + GCE * \dot{e}) * GU$ defines also the switching line. Its equation is:

$$0 = (GE * e + GCE * \dot{e}) * GU \Rightarrow$$
$$GCE * \dot{e} = -GE * e \Rightarrow$$
$$CE = -E$$

The last step just uses the naming convention for the input signals after gains. The last line shows that the switching line always has a slope of −45 degrees in the (E, CE)-plane. In the table based controller, the switching line is thus the diagonal of the table. Since the control signal U is proportional to $E + CE$, the following rule base is a simple nonlinear version of the summation:

If $E + CE$ is Pos then *control* is 200

If $E + CE$ is Neg then *control* is −200 (5.24)

Figure 5.25 shows the control surface with a particular choice of the membership functions Pos and Neg. The surface is a saturation type of surface with a soft proportional band in the transition zone around the switching line. The breakpoints of the membership functions control the width of the proportional band, which could be anything between 1 and 200, because the range of $E + CE$ is the interval $[-200, 200]$, assuming that E and CE are defined on the standard universes $[-100, 100]$.

Example 5.6 *Sliding mode control*

Given the previously mentioned process $P(s) = 1/s(s+1)^2$. To what extent is it possible to keep the closed-loop system on or near the switching line with an FPD controller?

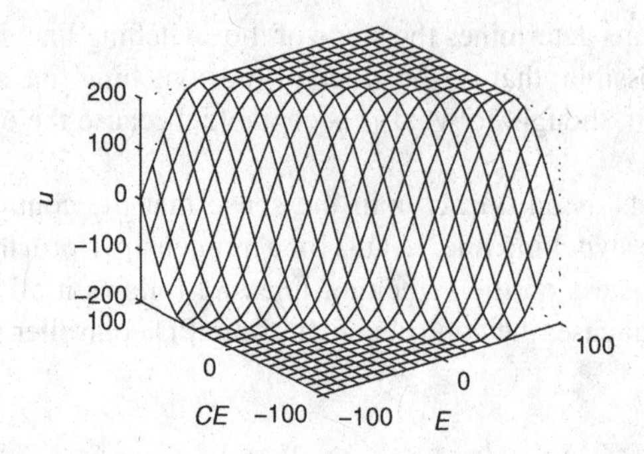

Figure 5.25 Saturation surface around the switching line. The two rules in Equation (5.24) define the surface with a proportional band of 100. (plotsurf.m)

▶ *Solution*

With a control strategy based on the distance from the switching line, it might be possible to keep the system trajectory in the neighbourhood of the switching line. This strategy is called sliding mode *control.*

The controller in Figure 5.26 uses a control surface with a rather steep transition zone. The surface works almost like a relay. The process output response is faster than the corresponding linear PD controller, but it also contains some ripple due to the relay nature of the

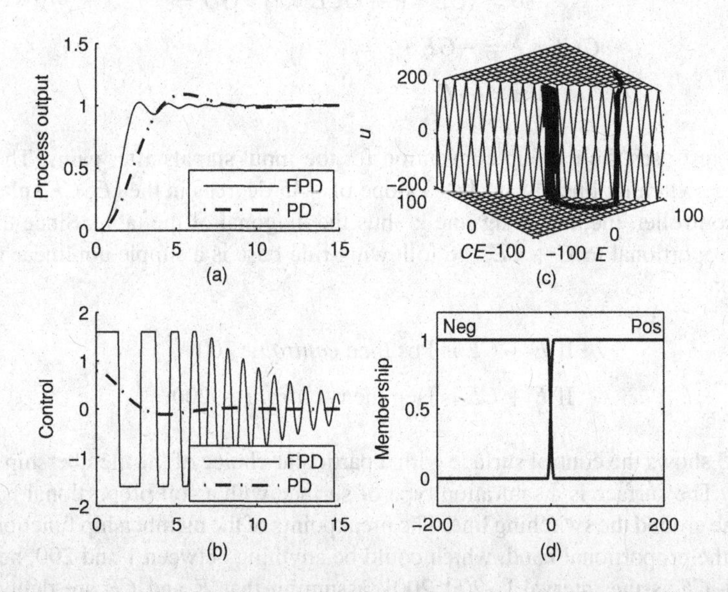

Figure 5.26 Step response with a saturation surface around the switching line. Compared to a PD controller, the response is faster, but more oscillatory (a). The control signal chatters with a high amplitude (b). The control surface is a steep saturation surface built along the switching line (c). The overlap between the two fuzzy sets is narrow compared to the whole operating range (d); the proportional band is 10. The process is $P(s) = 1/s(s+1)^2$, and the controller is an FPD controller with settings corresponding to $K_p = 0.8$, $T_d = 0.9$. (figphase3rd.m)

controller. The control signal switches from one extreme to the other to begin with. After a while the amplitude decreases as the response settles, because the transition zone is sloping. The trajectory is sometimes outside of the sloping transition zone, but all in all fairly close to the switching line.

The example shows that a rather crude sliding mode strategy provides a different kind of step response. The control signal oscillates, and if this can be tolerated, the sliding mode kind of strategy might be an option. Some feedback systems are born with a switching controller, for instance switched mode power supplies, and in that case a sliding mode strategy is natural. It is obviously straightforward to implement a basic sliding mode controller, which is a well-developed control paradigm (see for example Slotine and Li 1991, Edwards and Spurgeon 1998).

In the nonlinear domain, the switching line may be a curve, as opposed to a straight line. This option opens up new possibilities of shaping the rule base to get a desired response. For example, if a second-order type of response is desired, then the switching line should be a soft z-shaped curve passing through the origin. If we include integral action, loosely denoted by $\int e$, the $\left(\int e, e\right)$ plane will be the phase plane, and the vector $(e, \dot{e})$ will be the velocity vector.

5.12 Notes and References*

Aracil, García-Cerezo, and Ollero introduced in 1988 the idea of studying the vector field, with the process and controller represented by vectors (Aracil *et al.* in Driankov *et al.* 1996). Being geometrical, the approach lends itself to graphical illustration. Otherwise, a vector interpretation of the PID controller in state-space is uncommon in the literature, but for a fuzzy control engineer it is natural to think in geometric terms (the control table and the control surface).

The book by Slotine and Li (1991) contains an easy introduction to phase plane analysis. Atherton (1975) gives the topic a deeper treatment with many interesting interpretations. The traditional focus is on isoclines, in order to draw the phase trajectories, while today this is easily done by computer.

An early attempt to apply the phase plane to estimate stability is based on the linguistic phase plane (Braae and Rutherford 1979a, 1979b). To overcome time delays, attempts have been made to shift the phase plane (Li and Gatland 1995) or rotate it (Tanaka *et al.* 1991).

6

The Self-Organizing Controller

The self-organizing controller is an adaptive controller that adjusts its control surface according to a performance measure. Formally, it belongs to the class of model reference adaptive systems. The chapter gives an example with a long deadtime that the SOC is able to control after some adaptation runs. The chapter simplifies the performance measure to the transfer function of a first-order model, and it shows how the performance measure is equivalent to the deviation of the actual state from the model.

Mamdani and his PhD students developed the self-organizing controller (SOC) as an extension to the original fuzzy controller. They called it *self-organizing* because it was able to adjust the control table of a fuzzy controller without human intervention. It was developed specifically to cope with time delays in a process.

At that time, the distinction between the terms *self-organizing* and *adaptive* was unclear. Today, self-organization refers to a changing structure, such as the connections in a network. In daily conversation, *to adapt* means to modify according to changing circumstances, for example, 'they adapted themselves to the warmer climate'. Today, the SOC is regarded as an adaptive controller.

6.1 Model Reference Adaptive Systems

An adaptive controller is intuitively a controller that can modify its behaviour if the process varies nonlinearly, if the disturbances vary nonlinearly, or if the same controller is to be used under different conditions. Lacking a formal definition, a pragmatic definition is as follows: An adaptive controller is a controller with adjustable parameters and a mechanism for adjusting the parameters (Åström and Wittenmark 1995).

An adaptive controller has a distinct architecture consisting of two loops: (1) an inner control loop, which is the basic feedback loop; and (2) an outer loop, which adjusts the parameters of the controller, see the block diagram in Figure 6.1 of a *model reference adaptive system*

Foundations of Fuzzy Control: A Practical Approach, Second Edition. Jan Jantzen.

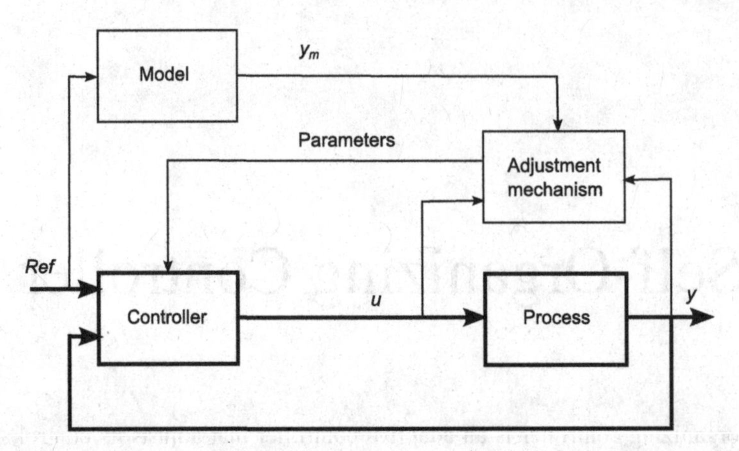

Figure 6.1 Model reference adaptive system, MRAS. The outer loop (thin lines) adjusts the controller parameters such that the error $\varepsilon = y - y_m$ becomes close to zero.

(MRAS). The idea is to have a fixed reference model specify the desired output of the inner loop. There are some general assumptions concerning the blocks in the diagram:

- *Process*. We assume that we know the structure of the process, although the parameters are unknown. For linear plants, this means knowing the number of zeros and the number of poles, but not the exact locations. For nonlinear plants, this means knowing the structure of the equations, but some parameters are unknown.
- *Model*. The reference model provides a *performance specification*. The choice of reference model is part of the design. It should reflect the desired performance, such as rise time, settling time, and overshoot. This ideal behaviour should be achievable, which means that there are some constraints on the structure of the reference model (order, type, relative degree) given the assumed structure of the process. The model returns the *desired response* y_m to a reference signal *Ref*.
- *Controller*. The controller should have *perfect model-following* capacity; that is, the closed loop transfer function of the inner loop should be identical to the model, when the process is known. This imposes some constraints on the structure of the controller. When the process parameters are unknown, the controller parameters will achieve perfect model-following asymptotically.
- *Adjustment mechanism*. The adjustment mechanism adjusts the parameters of the controller. The *model error* ε, which is the deviation of the process response from the desired response, governs the adjustment of the parameters. A strategy, the *adaptation law*, adjusts the parameters such that the error ε – or a function of ε – is minimized.

It is a challenge to design the adjustment mechanism such that the inner loop system remains stable. If the performance criterion can be optimized analytically, it may also be possible to analytically guarantee stability. Noise may cause problems, however, for the adaptation mechanism.

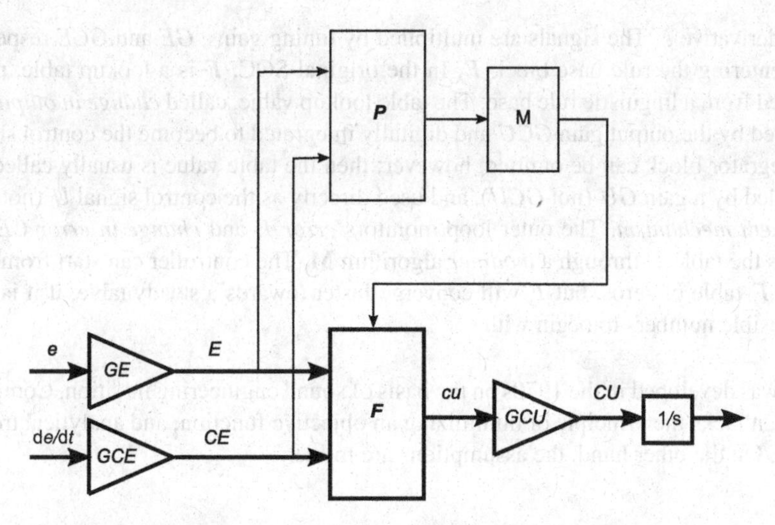

Figure 6.2 Self-organizing fuzzy controller, SOC. The outer loop (thin lines) adjusts the controller lookup table $\boldsymbol{F}$ according to the performance measure in $\boldsymbol{P}$. A modifier algorithm M finds which cell to adjust and decides the amount of adjustment.

6.2 The Original SOC

The SOC has a hierarchical structure in which the inner loop is a table based controller and the outer loop is the adjustment mechanism (Figure 6.2). The idea behind the adaptation is to let the adjustment mechanism update the values in the control table $\boldsymbol{F}$, on the basis of the current performance of the controller. If the performance is undesirable, the cell in $\boldsymbol{F}$ that is believed to be responsible receives a penalty, such that next time that cell is visited, the performance measure will be closer to or even equal to zero. The following items describe the SOC in accordance with the MRAS framework.

- *Process*. The developers of SOC made no assumptions about the structure of the process, but they required that the step response be monotonic, in other words, a minimum-phase system. The process can be nonlinear, and the process may contain a large deadtime. The dominant time constant and any deadtime must be known approximately, for example as a result of a step response experiment.
- *Model*. A *performance specification* p governs the magnitude of each change to $\boldsymbol{F}$. The performance specification depends on the current values of error and the change in error. The performance specification is a set of preset numbers organized in a *performance table* $\boldsymbol{P}$ of the same size as $\boldsymbol{F}$. Table $\boldsymbol{P}$ expresses what is desirable – or undesirable rather – in a transient response. Table $\boldsymbol{P}$ can be built using linguistic rules, but is often built by hand; see the two examples in Figures 6.3 and 6.4. The same performance table $\boldsymbol{P}$ may be used with a different process, without prior knowledge of the process, since it just expresses the *desired* transient response.
- *Controller*. The inner loop is an incremental, digital controller. The change in output CU_n at the current time n is added to the control signal U_{n-1} from the previous time instant, modelled as a summation in Figure 6.2. The two inputs to the controller are the error e

and its derivative $\dot{e}$. The signals are multiplied by tuning gains, *GE* and *GCE* respectively, before entering the rule base block $\boldsymbol{F}$. In the original SOC, $\boldsymbol{F}$ is a lookup table, possibly generated from a linguistic rule base. The table lookup value, called *change in output, cu*, is multiplied by the output gain *GCU* and digitally integrated to become the control signal U. The integrator block can be omitted, however; then the table value is usually called u (not cu), scaled by a gain *GU* (not *GCU*), and used directly as the control signal U (not CU).
- *Adjustment mechanism.* The outer loop monitors *error E* and *change in error CE*, and it modifies the table $\boldsymbol{F}$ through a *modifier* algorithm **M**. The controller can start from scratch with an $\boldsymbol{F}$-table of zeros, but $\boldsymbol{F}$ will converge faster towards a steady table, if it is primed with sensible numbers to begin with.

The SOC was developed in the 1970s on the basis of sound engineering intuition. Compared to an MRAS, it lacks the principle of minimizing an objective function, and analytical treatment is difficult. On the other hand, the assumptions are milder.

6.2.1 Adaptation Law

The SOC adapts the system in accordance with the desired response. At the sampling instant n,

1. it records the deviation from the desired state; and
2. it corrects table $\boldsymbol{F}$ accordingly.

The performance table $\boldsymbol{P}$ evaluates the current state and returns a performance specification $\boldsymbol{P}(i_n, j_n)$. Index i_n corresponds to E_n, such that $E_n = \mathcal{U}_e(i_n)$, where $\mathcal{U}_e$ is the input universe. Index j_n corresponds to CE_n, such that $CE_n = \mathcal{U}_{ce}(j_n)$, where $\mathcal{U}_{ce}$ is the other input universe.

Figures 6.3 and 6.4 are examples of early performance tables. Intuitively, a zero performance specification $\boldsymbol{P}(i_n, j_n) = 0$ implies that the state (E_n, CE_n) is satisfactory. If the performance specification is non-zero, that state is unsatisfactory to a degree indicated by the magnitude of the number p_n. When p_n is non-zero, the modifier **M** assumes that the control signal must be adjusted by the amount p_n. The current control signal cannot be held responsible, however, because it takes some time before a control action propagates to the process output.

The simple strategy is to go back a number of samplings d in time to correct an earlier control signal. The modifier must therefore know the time lag in the process. The integer d is comparable to the process time lag; here d is called the *delay-in-penalty* (in the literature it is called delay-in-*reward,* but that seems slightly misleading).

It is required that an increase in process output calls for an adjustment of the control signal always in the same direction, whether it be an increase or a decrease. The modifier thus assumes that the process output is a monotonic function of the input.

The precise adjustment mechanism is simply $u_{n-d} = u_{n-d} + p_n$. In terms of the tables $\boldsymbol{F}$ and $\boldsymbol{P}$, the adjustment rule is

$$\boldsymbol{F}(i, j)_{n-d} = \boldsymbol{F}(i, j)_{n-d} + \boldsymbol{P}(i, j)_n$$

Notice the subscripts: the time subscript n denotes the current sampling instant, and subscript $n - d$ denotes the sampling instant d samples earlier.

E	CE −6	−5	−4	−3	−2	−1	0	1	2	3	4	5	6
−6	−6	−6	−6	−6	−6	−6	−6	0	0	0	0	0	0
−5	−6	−6	−6	−6	−6	−6	−6	−3	−2	−2	0	0	0
−4	−6	−6	−6	−6	−6	−6	−6	−5	−4	−2	0	0	0
−3	−6	−5	−5	−4	−4	−4	−4	−3	−2	0	0	0	0
−2	−6	−5	−4	−3	−2	−2	−2	0	0	0	0	0	0
−1	−5	−4	−3	−2	−1	−1	−1	0	0	0	0	0	0
0	−4	−3	−2	−1	0	0	0	0	0	1	2	3	4
1	0	0	0	0	0	0	1	1	1	2	3	4	5
2	0	0	0	0	0	0	2	2	2	3	4	5	6
3	0	0	0	0	2	3	4	4	4	4	5	5	6
4	0	0	0	2	4	5	6	6	6	6	6	6	6
5	0	0	0	2	2	3	6	6	6	6	6	6	6
6	0	0	0	0	0	0	6	6	6	6	6	6	6

Figure 6.3 Performance table adapted from Procyk and Mamdani (1979). Note the universes.

Example 6.1 *Adjustment mechanism*

Illustrate how the modifier applies the updates to the control table.

► *Solution*

Assume $d = 2$ and variables as in Table 6.1 after a step in the setpoint. From $t = 1$ to $t = 4$ the process moves up towards the setpoint. Apparently, it follows the desired trajectory, because the performance specification p is 0. At $t = 5$, error changes sign, indicating an overshoot, and the performance table reacts by dictating $p_5 = -1$. Since d is 2, the entry to be adjusted in $\boldsymbol{F}$ will be at the position corresponding to $t = n - d = 5 - 2 = 3$. At that

E	CE −6	−5	−4	−3	−2	−1	0	1	2	3	4	5	6
−6	−6	−6	−6	−6	−6	−6	−6	−5	−4	−3	−2	−1	0
−5	−6	−6	−6	−6	−5	−4	−4	−4	−3	−2	−1	0	0
−4	−6	−6	−6	−5	−4	−3	−3	−3	−2	−1	0	0	1
−3	−6	−6	−5	−4	−3	−2	−2	−2	−1	0	0	1	2
−2	−6	−5	−4	−3	−2	−1	−1	−1	0	0	1	2	3
−1	−5	−4	−3	−2	−1	−1	0	0	0	1	2	3	4
0	−5	−4	−3	−2	−1	0	0	0	1	2	3	4	5
1	−3	−2	−1	0	0	0	0	1	1	2	3	4	5
2	−2	−1	0	0	0	1	1	1	2	3	4	5	6
3	−1	0	0	0	1	2	2	2	3	4	5	6	6
4	0	0	0	1	2	3	3	3	4	5	6	6	6
5	0	0	1	2	3	4	4	4	5	6	6	6	6
6	0	1	2	3	4	5	6	6	6	6	6	6	6

Figure 6.4 Performance table adapted from Yamazaki (1982).

Table 6.1 Performance data for Example 6.1.

	Time instance				
Variable	$t = 1$	$t = 2$	$t = 3$	$t = 4$	$t = 5$
E	6	3	1	0	−1
CE		−3	−2	−1	−1
u		0	−1	−1	−2
p		0	0	0	−1

sampling instant, the state was $(E_3, CE_3) = (1, -2)$ *and the element* $\mathbf{F}(i, j)_3$ *was* $u_3 = -1$. *The adjusted entry is* $u_3 = u_3 + p_5 = -1 - 1 = -2$, *which is to be inserted into* $\mathbf{F}(i, j)_3$.

6.3 A Modified SOC

The original performance tables in Figures 6.3 and 6.4 were built by hand using a trial and error approach. Presumably, if the numbers in the table $\boldsymbol{P}$ are small in magnitude, many updates are required before $\boldsymbol{F}$ converges to a steady table. On the other hand, if the numbers in $\boldsymbol{P}$ are large in magnitude, the convergence should be faster, but it may also be unstable. The following analysis leads to a simplified adaptation mechanism.

Procyk and Mamdani (Figure 6.3) preferred to keep the zeros in a z-shaped region, while Yamazaki (Figure 6.4) kept the zeros in a more or less diagonal band. As the zeros indicate no penalty, those states are desired. Assuming that the system stays within a zero band along the diagonal, what does this imply?

Noticing the numbers on the axes, a zero diagonal is equivalent to keeping the sum of the rule base inputs at zero in that region. It is a discrete version of the continuous relation,

$$GE * e + GCE * \dot{e} = 0 \tag{6.1}$$

The right-hand side corresponds to the entries '0' in the table, the term $GE * e$ is the rule base input E (error), and $GCE * \dot{e}$ is the rule base input CE (change in error). This is an ordinary differential equation, which determines the ratio between the variable e and its own time derivative. Its solution is,

$$e(t) = e(0) \exp\left(-\frac{t}{GCE/GE}\right)$$

That is a first-order exponential decay with time constant GCE/GE; the error e will gradually decrease by a fixed ratio determined by the time constant, and after $t = GCE/GE$ seconds it has decreased to 37% of the initial value $e(0)$. Note that the equation concerns the error $e = Ref - y$, such that the process output y after a step input will reach 63% of its final value in GCE/GE seconds.

To interpret, the modifier $\mathbf{M}$ tries to adapt the system to a first-order behaviour of the error.

The z-shaped table (Figure 6.3) is more generous. It allows a zero slope to begin with, for instance $\boldsymbol{P}(100, 0) = 0$, and some overshoot near the final value, around the centre of the table. This behaviour is similar to a second-order transient response.

Apparently, a simpler way to build a performance table is to use Equation (6.1) replacing the zero on the right-hand side by p. The time constant of the desired response would thus be GCE/GE.

We would perhaps like to tune GCE and GE independently, without affecting the desired time constant directly. Therefore, we introduce the *desired time constant* τ. A simple approximation to the table $\boldsymbol{P}$ is the performance specification,

$$p(n) = \gamma\,(e(n) + \tau * \dot{e}(n)) * T_s \tag{6.2}$$

The *adaptation gain* γ affects the convergence rate. The index n is the time instant in discrete time, and T_s is the sample period. In fact, Equation (6.2) is an incremental update – because the output is a change to an existing value – and therefore the multiplication by the sample period T_s. The longer the sample period, the fewer the updates within a given time span and the larger the penalty per update in order to keep the convergence rate independent of the choice of sample period.

6.4 Example with a Long Deadtime

Consider the process

$$G(s) = \exp^{-9s} \frac{1}{s\,(s+1)^2}$$

It is difficult to control, because the deadtime of 9 seconds is large (approximately 1/3) compared to the apparent time constant of the process, and the process contains an integrator, which has a destabilizing effect. The strategy is the following:

1. tune a fuzzy controller with a linear control surface;
2. start adaptation without changing the tuning; and
3. measure the performance of the resulting response.

At time $t = 0$, the reference changes abruptly from 0 to 1 and after 500 seconds, a load of 0.05 units is forced on the system. We are using the sampling time $T_s = 1$ second.

6.4.1 *Tuning*

Since the test system includes a load, it will be necessary to maintain a non-zero control signal in the steady state while the load is on; therefore, the controller must contain integral action. The choice is between a fuzzy PD+I controller and a fuzzy incremental controller, Finc. The latter is chosen, because it has one gain less to tune, and derivative action is of little help when there is a large deadtime. The control table is a 21 × 21 lookup table, which is an arbitrary choice of resolution. The initial table is linear and the lookup is with interpolation.

A loose hand-tuning of an incremental PD controller, equivalent to a PI controller, resulted in the gains

$$K_p = 0.002$$
$$T_d = 20$$

The gain K_p is the proportional gain and T_d is the differential gain, but since the controller output is integrated, T_d is in effect the gain on the error term and K_p is the gain on the integral term. The step response in Figure 6.5 shows that the response is rather oscillatory, that is, its stability margin is rather poor. It is possible to detune the system to oscillate less, but it is fairly difficult to achieve damping of both the reference step response and the load response at the same time. It is hoped that the SOC can improve the response by attenuating the oscillations while maintaining the overall settling time. The equivalent fuzzy gains are

$$GE = 100$$
$$GCE = GE * Td = 2000$$
$$GCU = \frac{Kp}{GE} = 2 \times 10^{-5}$$

The Finc response is the same as the incremental PD controller response, which is omitted to save space. During the initial seconds of the response (Figure 6.5), the deadtime appears as a horizontal section. When the load comes on at $t = 500$, there is a large dip in the response followed by some oscillation.

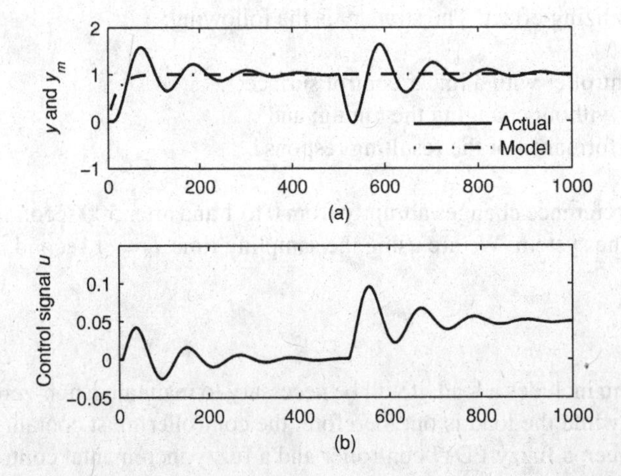

Figure 6.5 Incremental PD control of process $e^{-9s}/s\,(s+1)^2$. The output response in (a) compares the process output y and the desired response y_m. The lower plot (b) is the control signal u before self-organization. (figsocdelay.m)

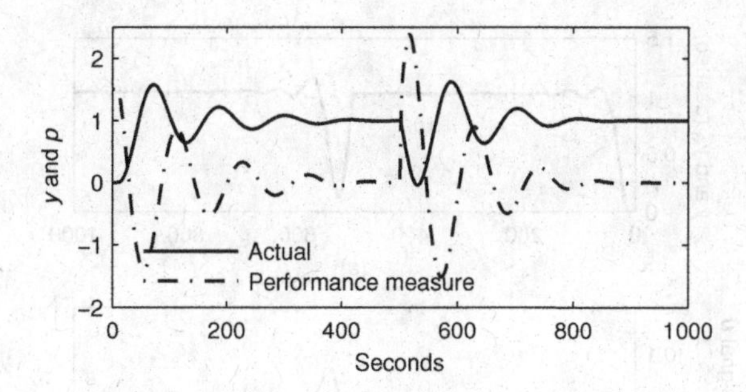

Figure 6.6 SOC performance. Step response after the first run of self-organization (solid) in comparison with the performance measure (dashed). (figsocdelay.m)

6.4.2 *Adaptation*

Three parameters control the adaptation: the desired time constant τ in seconds, the delay-in-penalty T_{dip} in seconds, and the adaptation gain γ. The dashed line in Figure 6.5 shows a step response corresponding to a desired time constant of $\tau = 20$ seconds. It is chosen equal to T_d in the hope that this will be achievable. If the desired time constant is chosen too fast, the response will begin to oscillate during adaptation.

A plot (Figure 6.6) of the performance specification, Equation (6.2), with an arbitrary adaptation gain to be determined in a moment, shows the response of the process with the performance specification as a dashed line. A closer study shows that the performance specification peaks about 14–17 s earlier than the process output, due to the differential term in Equation (6.2).

By inspection, it is seen that the time constant of the closed-loop system is about 27 s. This is the time it takes to reach 63% of the steady-state value, after the deadtime of 9 s, during the initial seconds of the response. We may thus reason that it takes approximately $9 + 27 = 36$ s for a control signal to show on the output of the process. We can now make a guess at the delay-in-penalty T_{dip}. We observed that the performance specification is *ahead* of the process output; thus the T_{dip} may be less than 36 s. Indeed, $T_{dip} = 20$ s is chosen; the difference is 16 s, which is within the interval 14–17 s. With a sampling period of 1 s, the delay-in-penalty is $d = T_{dip}/T_s = 20/1 = 20$ samples.

The adaptation gain is, by trial and error, set to $\gamma = 1.5$. The strategy is to choose it as high as possible, but still keep the response steady. With these choices, the adaptation can be switched on.

6.4.3 *Performance*

The SOC is able to dampen the oscillations after a few runs, and the performance improves after each run. Figure 6.7 shows the step response after 29 iterative runs. The large oscillations have disappeared and the load dip at $t = 500$ is also a little smaller. The process follows more or less the desired first-order response, but owing to the deadtime and the higher order of the process, it is unable to follow the desired response perfectly. Notice that the steady-state value is a little different from the reference value. The modifier has modified the control table, such

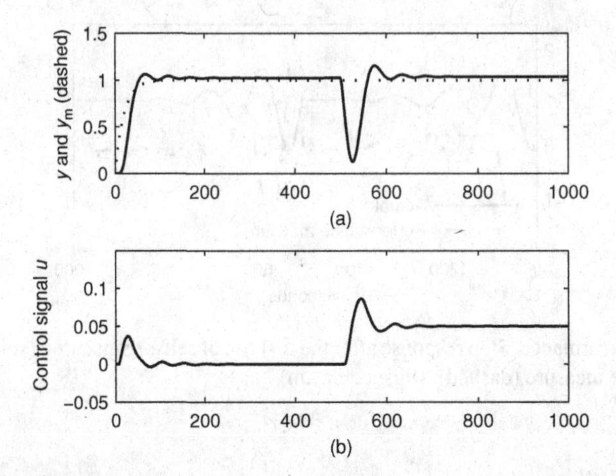

Figure 6.7 After 29 runs. The process step response y is close to the desired step response y_m (a) and the settling time is thus greatly improved. The control signal u (b) shows less control effort than before adaptation. (figsocdelay.m)

that one or more cells close to the centre of the control table now contain a numerically small control signal. The result is a small dead zone.

The modified control surface contains several jagged peaks (see Figure 6.8). During each run the controller visited a number of cells in the lookup table, sometimes the same cell several times; the accumulated changes result in the sharp peaks. One might expect jumps in the control signal as a consequence, but in fact it is rather smooth thanks to the integrator in the output end of the incremental controller.

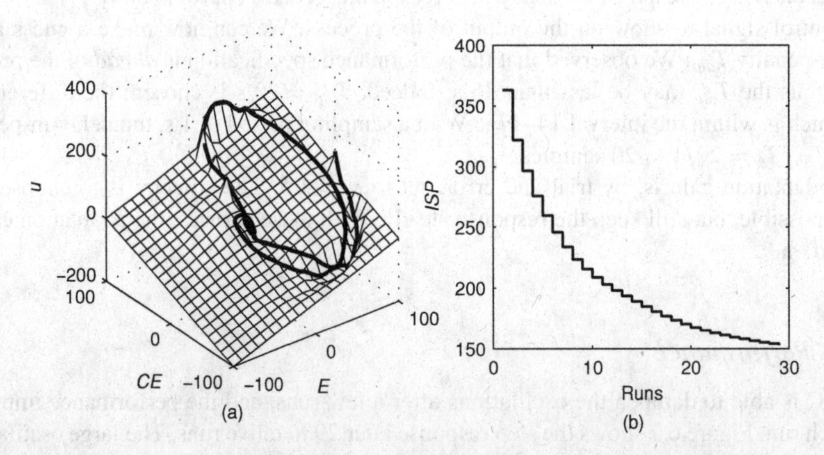

Figure 6.8 Convergence. After 29 runs the control surface (a) has been updated to an irregular shape. The trajectory passes through regions with large changes. Only visited regions are changed. The decreasing performance measure ISP (b) levels out indicating convergence to a steady control surface. (figsocdelay.m)

When the response is close to the desired response, the corrections to the control table will be small in magnitude. The SOC seeks to minimize the performance specification p_n at each time instant n, and the discretized integral squared performance (ISP) index,

$$ISP = \sum_n p_n^2 * T_s$$

expresses the modifier's effort during each run. The smaller the ISP the better.

Experiments on laboratory models have shown that various plants can be stabilized, but the rule modifier is sensitive to noise on the process output – it cannot tell if a poor performance specification is due to noise or an undesired state.

6.5 Tuning and Time Lock

In practice, additional questions arise, for example, how to tune the gains, how to choose the design parameters, how to stop the adaptation if it behaves unintentionally, and how to cope with noise. Such problems may seem unimportant, but they do deserve attention, since they may finally stand in the way for a successful implementation.

6.5.1 *Tuning of the SOC Parameters*

We associate the term *tuning* with the adjustment of the gains and design parameters, while the term *self-organization* is associated with adjusting the parameters of the control table.

- *GE, GCE, G(C)U*. The controller gains must be set near some sensible settings, with the exact choice less important compared to a conventional fuzzy controller. Imagine, for example, that the output gain is lowered between two training sessions. The adjustment mechanism will then compensate by producing an $\boldsymbol{F}$-table with numbers of larger magnitude. Even if the input gains were changed, it would still manage to adapt. It is therefore possible to start with a linear $\boldsymbol{F}$-table, and set the gains loosely according to a PID tuning rule or hand-tuning. That is a good starting point for the self-organization.
- *Desired time constant* τ. The smaller the value of τ, the faster the desired response. If it is too small, however, the inner loop cannot possibly follow the desired trajectory, but the modifier will try anyway. As a result, the $\boldsymbol{F}$-table winds up, and the step response will overshoot more and more. The lower bound for τ is when this overshoot starts to occur. A process with a time constant τ_p and a deadtime T_p requires that

$$T_p \leq \tau \leq T_p + \tau_p$$

A value somewhat smaller than the right-hand side of the inequality is often achievable, because the closed loop system is usually faster than the open loop system. A good choice is $\tau = T_d$, which is the derivative gain of the chosen PID settings.

- *Delay-in-penalty* d. Parameter d, measured in samplings, should be chosen with due respect to the sample period. The delay should in principle be the desired time constant divided by the sample period and rounded to the nearest integer,

$$d = round(\tau / T_s)$$

The results are often better, however, with a somewhat smaller value.
- *Adaptation gain* γ. The larger the value of γ, the faster the F-table winds up, but if it is too large, the training becomes unstable. A reasonable magnitude of a large modification p is less than one-fifth of the maximum value in the output universe. This rule results in the upper bound:

$$\gamma \leq \frac{0.2 * |F(i, j)|_{\max}}{|(e_n + \tau * \dot{e}_n)|_{\max} * T_s}$$

Compared to conventional fuzzy control, the tuning task is shifted from having to accurately tune $\{GE, GCE, G(C)U\}$ to tuning $\{\tau, d, \gamma\}$ loosely.

6.5.2 Time Lock

The delay-in-penalty d causes a problem when the reference or the load changes abruptly.

Consider this case. If the error and the change in error, for a time period longer than d samples, have been zero, then the controller is in a desired state (the steady state). Suddenly there is a disturbance from the outside: the performance specification p becomes non-zero, and the modifier will modify the F-table d samples back in time. It should not do so, however, because the state was acceptable there. The next time the controller visits that state, the control signal will fluctuate. The problem appears after step changes in the reference or the load.

A solution is to implement a *time-lock* (Jespersen 1981). The time lock stops the self-organization for the next d samples; if it is activated at the sampling instant T_n, self-organization stops until sampling instant $T_n + d + 1$. In order to trigger the time-lock, it is necessary to detect disturbances, abrupt changes in the load, and abrupt changes in the reference. If these events cannot be measured directly and fed forward into the SOC, it is necessary to try to detect it in the process output. If it changes more than a predefined threshold, or if the combination of *error* and *change in error* indicates an abrupt change, then activate the time-lock.

The time-lock is conveniently implemented by means of a *queue* of length $d + 1$ samples.

$$\mathbf{q} = \left[(i, j)_{n-d} \quad \cdots \quad (i, j)_{n-1} \quad (i, j)_n \right]$$

The element to update in $\boldsymbol{F}$ is indicated by the front (leftmost) element in $\mathbf{q}$. At the next sampling instant, the current index pair is appended to the (rightmost) end of the queue, while the first index pair is removed. If an event triggers the time-lock, flush the queue; that is, reset all cells to zero. New updates to $\boldsymbol{F}$ will only be possible when the queue is full again, that is, when the first element in the queue is non-zero, $d + 1$ samplings after the flush.

An extra precaution is to protect the centre of $\boldsymbol{F}$ from updates; this is the steady state $(E, CE) = (0, 0)$ in which the control action must always be zero.

If necessary, the time-lock can be activated each time the modifier makes a change in $\boldsymbol{F}$. In that case, the modifier waits to see the influence of the change before making a new change.

A practical strategy is to flush the time-lock queue in connection with a manual switch to *adaptation mode*. When the controller is switched to adaptation mode, a series of steps on the reference (or the load) will occur at known times. The time-lock queue can then be flushed in connection with the step changes.

6.6 Summary

The adjustment mechanism is simple, but in practice the design is complicated by the time lock and noise precautions. Also, if the resolution of the control table is too low, the adaptive system may become unsteady or even unstable. Although the tuning of the traditional input and output gains (GE, GCE, $G(C)U$) is made less cumbersome by the adjustment mechanism, it introduces other parameters that have to be tuned. The tuning task is nevertheless easier, because the response is less sensitive to these parameters (delay-in-penalty, adaptation gain, desired time constant).

The SOC is based on the assumption that the process response is monotonic (is minimum-phase), and an example showed that it works for a process with a large deadtime and an integrator. There is no guarantee, however, that the adaptive system will be stable. To prove stability, further theoretical research is necessary, but it will be a difficult task, because the updates to the control table are local updates rather than global with regard to the whole state-space.

One may ask where fuzzy logic enters the picture, since the adjustment law is an equation and the control table is automatically adjusted. Fuzzy logic is absent, but historically the design sprang from the fuzzy controller. Any control table, derived from fuzzy rules, can be used as the initial table to be adapted. It is conceivable, also, to use a performance specification based on fuzzy rules.

6.7 Example: Adaptive Control of a First-Order Process*

We shall use an example (Åström and Wittenmark 1995) with a first-order process to illustrate how an adaptive control system can be designed, and to introduce the concepts in model reference adaptive control. The first-order system is chosen not only owing to its simplicity but also because we shall use the results later in connection with the fuzzy SOC.

The room temperature, the level of liquid in a tank being emptied, or the discharge of an electronic flash, are examples of systems that may be approximated by a first-order differential equation

$$\dot{y}(t) = -ay(t) + bu(t)$$

where $u(t)$ is the control signal, $y(t)$ is the measured output, and a and b are constant process parameters that are unknown. The negative sign of a emphasizes that the process is stable when a is positive.

Let a first-order reference model specify the desired performance of the inner-loop system,

$$\dot{y}_m(t) = -a_m y_m(t) + b_m Ref(t)$$

where a_m and b_m are constant parameters and $Ref(t)$ is a bounded reference signal. The parameter a_m must be positive in order to ensure that the model is stable, and b_m is chosen positive without loss of generality. Using Laplace notation, where s is the Laplace variable, and ignoring initial conditions, the model is represented by

$$\begin{aligned}
sy_m(s) &= -a_m y_m(s) + b_m Ref(s) \Leftrightarrow \\
(s + a_m)y_m(s) &= b_m Ref(s) \Leftrightarrow \\
y_m(s) &= \frac{b_m}{s + a_m} Ref(s)
\end{aligned}$$

In other words, the model has the transfer function

$$G_m(s) = \frac{y_m(s)}{Ref(s)} = \frac{b_m}{(s + a_m)} \tag{6.3}$$

In order to simplify the notation, we shall omit the argument (s) in the Laplace domain and (t) in the time domain, except when we wish to distinguish parameters from signals.

The objective is to form a control law and an adaptation law, such that the model error $\varepsilon = y - y_m$ converges to zero.

6.7.1 The MIT Rule

One strategy is to define a positive objective function to be minimized according to the MIT rule (invented at the Massachusetts Institute of Technology, MIT). Define an objective function

$$J(\theta) = \frac{1}{2}\varepsilon^2$$

that depends on the adjustable parameter θ. This objective function is always non-negative, and minimizing $J(\theta)$ will also minimize ε. The MIT rule suggests changing the parameter θ in the direction of the negative gradient of J, that is,

$$\frac{\mathrm{d}\theta}{\mathrm{d}t} = -\gamma \frac{\partial J}{\partial \theta} = -\gamma \varepsilon \frac{\partial \varepsilon}{\partial \theta}$$

The *adaptation gain* γ is a parameter that the designer chooses; it determines the step length during the iterative search for the minimum. The *sensitivity derivative* $\partial\varepsilon/\partial\theta$ is the sensitivity of the error to changes in the parameter θ; if the sensitivity is large, then the parameter change will be large – all other parameters being equal. The parameter change also depends directly on the magnitude of the error ε. Assuming that the parameter θ changes much slower than the other dynamics, derivatives can be calculated assuming θ as a constant. For example $\mathrm{d}(\theta u)/\mathrm{d}t = (\mathrm{d}\theta/\mathrm{d}t)u + \theta(\mathrm{d}u/\mathrm{d}t)$ using the product rule for differentiation, but if θ is assumed as a constant, an approximation is $\theta(\mathrm{d}u/\mathrm{d}t)$. In general, approximation is necessary to evaluate the sensitivity derivative.

The objective function is arbitrary. For example, the alternative function $J(\theta) = |\varepsilon|$ results in the parameter rate of change $d\theta/dt = -\gamma(\partial\varepsilon/\partial\theta)sign(\varepsilon)$. An even simpler adaptation law is $d\theta/dt = -\gamma\, sign(\partial\varepsilon/\partial\theta)sign(\varepsilon)$ which avoids the evaluation of the sensitivity derivative.

6.7.2 *Choice of Control Law*

To continue the design, choose the control law

$$u(t) = \theta_1 Ref(t) - \theta_2 y(t) \tag{6.4}$$

This choice of structure allows for perfect model-following. The closed loop dynamics are

$$\dot{y}(t) = -ay(t) + bu(t) \tag{6.5}$$

$$= -ay(t) + b(\theta_1 Ref(t) - \theta_2 y(t)) \tag{6.6}$$

$$= (-a - \theta_2)y(t) + b\theta_1 Ref(t) \tag{6.7}$$

Indeed, if the process parameters were known, the parameter values

$$\theta_1 = \frac{b_m}{b}$$

$$\theta_2 = \frac{a_m - a}{b}$$

would make the dynamics of the inner loop and the model identical, and provide a zero model-following error. Since a and b are unknown, the controller must achieve this objective adaptively.

6.7.3 *Choice of Adaptation Law*

To apply the MIT rule, introduce the model error

$$\varepsilon = y - y_m$$

The sensitivity derivative is

$$\frac{\partial\varepsilon}{\partial\theta} = \frac{\partial(y - y_m)}{\partial\theta} = \frac{\partial y}{\partial\theta}$$

since the model output y_m does not depend on the controller parameter θ. It follows from Equation (6.7) that the closed-loop process output y is determined by

$$y(s) = \frac{b\theta_1}{s + a + b\theta_2} Ref(s) \tag{6.8}$$

Now differentiate with respect to the controller parameters,

$$\frac{\partial y}{\partial \theta_1} = \frac{b}{s + a + b\theta_2} Ref(s)$$

$$\frac{\partial y}{\partial \theta_2} = -\frac{b^2\theta_1}{(s + a + b\theta_2)^2} Ref(s) = -\frac{b}{s + a + b\theta_2} y(s)$$

The switch from signal $Ref(s)$ to signal $y(s)$ in the last equation used Equation (6.8). But the equations still contain the process parameters a and b, which are unknown; we would like to dispose of these parameters. An approximation is therefore required. We observe that the denominator $s + a + b\theta_2$ is the closed loop denominator, which will be the same as the model denominator under perfect model-following. Therefore, we apply the approximation

$$s + a + b\theta_2 \approx s + a_m$$

The approximation will be reasonable when the parameters are close to their correct values. We thus achieve the following adaptation laws for adjusting the parameters

$$\frac{d\theta_1}{dt} = -\gamma\varepsilon\frac{\partial \varepsilon}{\partial \theta_1} = -\gamma\varepsilon\left[\frac{b}{s + a + b\theta_2} Ref(s)\right] \approx -\gamma\varepsilon\left[\frac{b}{s + a_m} Ref(s)\right]$$

$$\frac{d\theta_2}{dt} = -\gamma\varepsilon\frac{\partial \varepsilon}{\partial \theta_2} = -\gamma\varepsilon\left[-\frac{b}{s + a + b\theta_2} y(s)\right] \approx \gamma\varepsilon\left[\frac{b}{s + a_m} y(s)\right]$$

The final step is to dispose of the unknown parameter b. Since it is constant, it can be absorbed in the adaptation gain, but we have to know the sign of b in order to use the correct sign for the adaptation gain. This is a mild condition in practice, since it can be determined experimentally: if a positive step on the reference increases $y(t)$, then the sign is positive, and if it decreases, the sign is negative. By the substitution $\gamma b = \gamma' sign(b) a_m$, we achieve the adaptation laws,

$$\frac{d\theta_1}{dt} \approx -\gamma' sign(b)\varepsilon\left[\frac{a_m}{s + a_m} Ref(s)\right]$$

$$\frac{d\theta_2}{dt} \approx \gamma' sign(b)\varepsilon\left[\frac{a_m}{s + a_m} y(s)\right]$$

The expressions in square brackets are filtered versions of the signals $Ref(s)$ and $y(s)$, such that the steady-state gain of the filter is 1. Thus the change in θ_1 depends on Ref, which is the signal that θ_1 multiplies, and the change in θ_2 depends on y, which is the signal that θ_2 multiplies; this is a consequence of the sensitivity derivative.

6.7.4 Convergence

The behaviour of the system is now illustrated in simulation. The parameters are chosen to be $a = 1, b = 0.5, a_m = b_m = 2, \gamma = 1$. The input signal is a square wave with amplitude 1, and Figure 6.9 shows the response. The control is quite good as early as the second or third step in

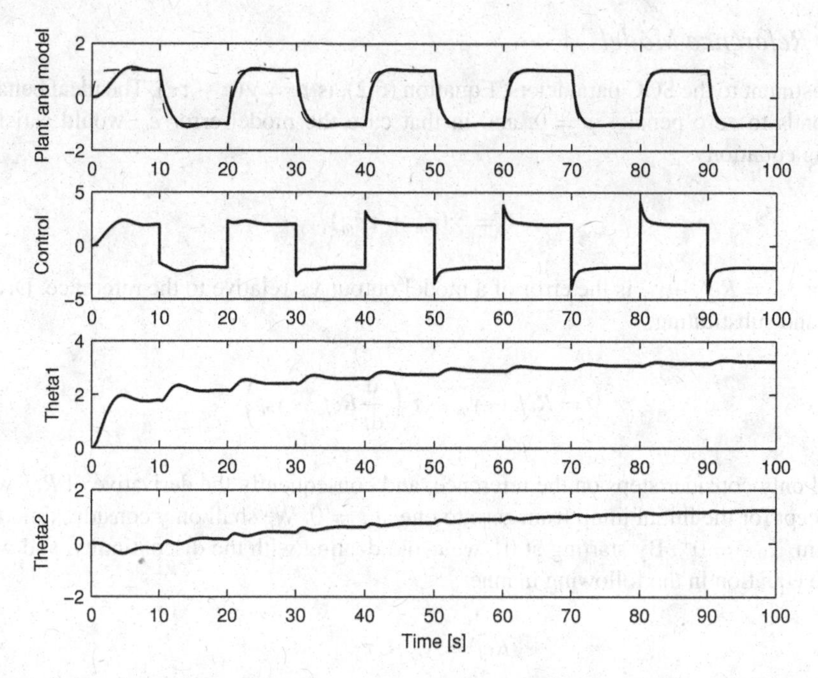

Figure 6.9 First-order MRAS with the parameters $a_m = b_m = 2$, $a = 1$, $b = 0.5$, $\gamma = 1$. The system is stable, and the process output and the system output converges towards the model (top, dashed). (figadap1.m)

the reference. The process output (row 1 in the figure, solid line) follows the model (dashed) with increasing accuracy. The control signal (row 2) becomes more and more articulated as the parameters θ_1 and θ_2 increase (rows 3 and 4). They are far from the correct values $\theta_1 = 4$ and $\theta_2 = 2$, but they will converge given a longer simulation period.

The system is stable and it converges towards perfect model-following, which can be shown by means of Lyapunov theory (see for example Åström and Wittenmark 1995 or Slotine and Li 1991).

In general, the model error can converge even though the parameters do not converge to their correct values; the controller may work for a subset of frequencies if only the ratio of θ_1 to θ_2 is correct.

The parameters change more when the control signal switches, which illustrates the requirement for an input signal sufficiently rich in frequencies – so-called *persistently exciting*. It is particularly apparent in the fuzzy SOC, which makes local changes in the control surface; the system must in principle visit the whole phase plane in order to complete the adaptation.

6.8 Analytical Derivation of the SOC Adaptation Law*

The SOC adjustment mechanism can be interpreted from the viewpoint of model reference adaptive control, which provides a deeper insight into the mechanism.

6.8.1 Reference Model

The adjustment to the SOC parameters, Equation (6.2), is $p = \gamma(e + \tau\dot{e})$. The ideal behaviour corresponds to zero penalty $p = 0$, and in that case the model error e_m would satisfy the following equation

$$0 = \gamma(e_m + \tau\dot{e}_m) \tag{6.9}$$

The error $e_m = Ref - y_m$ is the error of a model output y_m relative to the reference. Dividing γ away and substituting,

$$0 = Ref - y_m + \tau\left(\frac{\mathrm{d}}{\mathrm{d}t}Ref - \dot{y}_m\right)$$

We shall only consider steps on the reference, and consequently the derivative of *Ref* will be zero, except for the initial jump from zero to one at $t = 0$. We shall only consider time $t > \epsilon$, where $\lim_{\epsilon\to 0}\epsilon = 0^+$. By starting at 0^+ we avoid dealing with the discontinuity, and we can write the equation in the following manner:

$$Ref = y_m + \tau\dot{y}_m$$

Switching to Laplace notation, the equation is equivalent to the following transfer function model

$$y_m = \frac{1}{1+\tau s}Ref \tag{6.10}$$

This is a first-order reference model (Figure 6.1). Summarizing, an ideal behaviour, without any adjustments, is equivalent to saying that the system output perfectly follows a reference model, which returns a first-order desired response y_m to a step in the reference in accordance with Equation (6.10).

6.8.2 Adjustment Mechanism

The question now is, what is the interpretation of p_n in Equation (6.2) when it is *non*-zero? In continuous time, Equation (6.2) is

$$p = \gamma(e + \tau s e) \tag{6.11}$$

Subtracting Equation (6.9) from (6.11) we have

$$\begin{aligned} p &= \gamma(e + \tau s e) - \gamma(e_m + \tau s e_m) \\ &= \gamma(Ref - y + \tau s\,(Ref - y)) - \gamma(Ref - y_m + \tau s(Ref - y_m)) \\ &= \gamma((y_m - y) + \tau s(y_m - y)) \end{aligned}$$

By insertion of the model error $\varepsilon = y - y_m$,

$$p = \gamma((-\varepsilon) + \tau s(-\varepsilon)) \Leftrightarrow$$
$$p = -\gamma(\varepsilon + \tau s \varepsilon)$$

We have thus transformed the performance specification p into an expression in the model error ε. The expression in parentheses is the model error ε plus a term proportional to the change in error $s\varepsilon$; this can be interpreted as the first-order prediction $\hat{\varepsilon}$ of the future model error, τ seconds ahead of current time. Thus

$$p = -\gamma \hat{\varepsilon}, \qquad \hat{\varepsilon} = (\varepsilon + \tau s \varepsilon) \tag{6.12}$$

To recapitulate, the performance specification p is equivalent to a parameter adjustment proportional to the predicted model error $\hat{\varepsilon}$, with adaptivity gain γ. Since the adjustment p is an incremental adjustment $\mathrm{d}\theta/\mathrm{d}t$, the parameter θ itself is the integral of Equation (6.12) plus the initial value. The adjustment mechanism is thus a proportional-integral type, rather than just integral.

What happens if we use $\hat{\varepsilon}$ (instead of ε) from Equation (6.12) in the objective function? For comparison, we will take the previously worked out first-order example, and proceed in the same manner, step by step.

Error function The model error function is

$$\hat{\varepsilon} = \varepsilon + \tau s \varepsilon = y - y_m + \tau s(y - y_m) \tag{6.13}$$

Objective function We choose the quadratic objective function

$$J(\theta) = \frac{1}{2}\hat{\varepsilon}^2$$

Minimizing $J(\theta)$ will minimize $\hat{\varepsilon}$. It may happen, that $\hat{\varepsilon}$ is zero while ε is non-zero. Nevertheless, $\varepsilon + \tau s \varepsilon = 0$ is a differential equation that will drive ε to zero, given that τ is positive. The adaptation stops, but the model error tends to zero in an equilibrium.

MIT rule The MIT rule suggests the parameter changes

$$\frac{\mathrm{d}\theta}{\mathrm{d}t} = -\gamma \frac{\partial J}{\partial \theta} = -\gamma \hat{\varepsilon} \frac{\partial \hat{\varepsilon}}{\partial \theta}$$

Model The model is, as before,

$$\dot{y}_m(t) = -a_m y_m(t) + b_m Ref(t) \tag{6.14}$$

Control law We choose the same control law,

$$u(t) = \theta_1 Ref(t) - \theta_2 y(t)$$

Closed loop dynamics With that control law, the closed loop dynamics of the inner loop is

$$\dot{y}(t) = (-a - \theta_2)y(t) + b\theta_1 Ref(t) \tag{6.15}$$

In transfer function form,

$$y(s) = \frac{b\theta_1}{s + a + b\theta_2} Ref(s) \tag{6.16}$$

Perfect model-following We know from the previous investigation that the control law allows perfect model-following.

Sensitivity derivative The sensitivity derivative is

$$\frac{\partial \hat{\varepsilon}}{\partial \theta} = \frac{\partial (y - y_m + \tau s(y - y_m))}{\partial \theta} = \frac{\partial (y + \tau s y)}{\partial \theta} = (1 + \tau s)\frac{\partial y}{\partial \theta}$$

since the model output y_m is independent of the controller parameter θ. Differentiating y in Equation (6.16) with respect to the controller parameters,

$$\frac{\partial y}{\partial \theta_1} = \frac{b}{s + a + b\theta_2} Ref(s)$$

$$\frac{\partial y}{\partial \theta_2} = -\frac{b^2\theta_1}{(s + a + b\theta_2)^2} Ref(s)$$

Closed loop substitution In the equation for $\partial y/\partial \theta_2$ we make use of the inner-loop transfer function from Equation (6.15),

$$\frac{\partial y}{\partial \theta_2} = -\frac{b^2\theta_1}{(s + a + b\theta_2)^2} Ref(s) = -\frac{b}{s + a + b\theta_2} y(s)$$

Adaptation law We arrive at the following adaptation laws for adjusting the parameters:

$$\frac{d\theta_1}{dt} = -\gamma\hat{\varepsilon}\frac{\partial \hat{\varepsilon}}{\partial \theta_1} = -\gamma\hat{\varepsilon}(1 + \tau s)\left[\frac{b}{s + a + b\theta_2} Ref(s)\right]$$

$$\frac{d\theta_2}{dt} = -\gamma\hat{\varepsilon}\frac{\partial \hat{\varepsilon}}{\partial \theta_2} = -\gamma\hat{\varepsilon}(1 + \tau s)\left[-\frac{b}{s + a + b\theta_2} y(s)\right]$$

Approximation Again, we approximate the inner-loop characteristic polynomial with the model characteristic polynomial,

$$s + a + b\theta_2 \approx s + a_m$$

After approximation, the adaptation laws are therefore,

$$\frac{d\theta_1}{dt} \approx -\gamma\hat{\varepsilon}\,(1+\tau s)\left[\frac{b}{s+a_m}Ref(s)\right]$$

$$\frac{d\theta_2}{dt} \approx \gamma\hat{\varepsilon}\,(1+\tau s)\left[\frac{b}{s+a_m}y(s)\right]$$

Dispose of process parameters Since the unknown parameter b is constant, it can be absorbed in the adaptation gain, but we have to know the sign of b in order to use the correct sign for the adaptation gain. By the substitution $\gamma b = \gamma' sign(b) a_m$,

$$\frac{d\theta_1}{dt} \approx -\gamma' sign(b)\hat{\varepsilon}\,(1+\tau s)\left[\frac{a_m}{s+a_m}Ref(s)\right] \tag{6.17}$$

$$\frac{d\theta_2}{dt} \approx \gamma' sign(b)\hat{\varepsilon}\,(1+\tau s)\left[\frac{a_m}{s+a_m}y(s)\right] \tag{6.18}$$

The square brackets are filtered versions of the signals Ref and y, and the steady state gain of the filter is 1.

Cancellation Since τ is the prediction horizon for the predicted model error, it is a design parameter. If we choose

$$\tau = \frac{1}{a_m}$$

the adaptation laws are simply,

$$\frac{d\theta_1}{dt} \approx -\gamma' sign(b)\hat{\varepsilon}(t)Ref(t) \tag{6.19}$$

$$\frac{d\theta_2}{dt} \approx \gamma' sign(b)\hat{\varepsilon}(t)y(t) \tag{6.20}$$

We notice that the adaptation law Equation (6.19) is similar to Equation (6.12), except for the factor *sign*(b), which is just a convenience, and the factor Ref. We have earlier assumed that Ref is constant, and therefore it can be absorbed in γ'.

We can conclude that our choice of error function $\hat{\varepsilon}$, in Equation (6.13), caused a cancellation of the denominators in Equations (6.17) and (6.18), which simplified the adaptation law. This is possible whenever the polynomial of the model error function Equation (6.13) is chosen to be proportional to the denominator polynomial of the model transfer function G_m.

6.8.3 The Fuzzy Controller

Does this work for a fuzzy controller, and how does it compare with the SOC? So far we have chosen a control law which has a different structure from the one used in fuzzy PD control.

Let us now try a PD structure, and model the SOC adjustment by an increment δ to the control table.

Error function The model error function is

$$\hat{\varepsilon} = \varepsilon + \tau s \varepsilon = y - y_m + \tau s(y - y_m) \tag{6.21}$$

Objective function We choose the quadratic objective function

$$J(\theta) = \frac{1}{2}\hat{\varepsilon}^2$$

Minimizing $J(\theta)$ will minimize $\hat{\varepsilon}$ and ε.

MIT rule The MIT rule suggests the parameter changes

$$\frac{\mathrm{d}\theta}{\mathrm{d}t} = -\gamma \frac{\partial J}{\partial \theta} = -\gamma \hat{\varepsilon} \frac{\partial \hat{\varepsilon}}{\partial \theta}$$

Model The model is, as before,

$$\dot{y}_m(t) = -a_m y_m(t) + b_m Ref(t) \tag{6.22}$$

with the transfer function

$$\frac{y_m(s)}{Ref(s)} = \frac{b_m}{s + a_m}$$

Control law We choose the control law,

$$u = (GE * e + GCE * \dot{e} + \delta) * GU$$

Notice the adjustable parameter δ. We assume that the SOC starts from a linear controller with $\delta = 0$. For each cell in the lookup table, δ models the adjustment that the SOC makes. The control law is a local model of each cell in the lookup table. There are thus a number of δ's, one for each cell, but we will only consider one such cell in order to keep the derivations simple and transparent.

Closed loop dynamics With that control law, the closed loop dynamics of the inner loop is

$$\begin{aligned}
\dot{y}(t) &= -ay(t) + bu(t) && (6.23)\\
&= -ay(t) + b * (GE * e + GCE * \dot{e}(t) + \delta * u_0(t)) * GU && (6.24)\\
&= -b * GCE * GU * \dot{y}(t) - (a + b * GE * GU)y(t) && (6.25)\\
&\quad + b * \delta * u_0(t) * GU + b * GE * GU * Ref(t) && (6.26)
\end{aligned}$$

Here we have assumed that $Ref(t)$ is constant, so that its time-derivative vanishes. We have modelled δ as a gain on an external input signal u_0, which is always equal to 1. Solving with respect to y, and using the Laplace variable,

$$y(s) = \frac{b * \delta * GU}{(1 + b * GCE * GU)s + (a + b * GE * GU)} u_0(s) \tag{6.27}$$

$$+ \frac{b * GE * GU}{(1 + b * GCE * GU)s + (a + b * GE * GU)} Ref(s) \tag{6.28}$$

Perfect model-following We hope to achieve perfect model-following. Assume for a moment that $\delta = 0$. By comparison of Equations (6.25) and (6.22),

$$\frac{a + b * GE * GU}{1 + b * GCE * GU} = a_m$$

$$\frac{b * GE * GU}{1 + b * GCE * GU} = b_m$$

Notice that GU multiplies gains GE and GCE, so we can regard it as a scaling factor. Reorganize into two equations in two unknowns,

$$b * GE * GU - a_m * b * GCE * GU = a_m - a$$

$$b * GE * GU - b_m * b * GCE * GU = b_m$$

The determinant $-b^2 b_m + b^2 a_m$ is non-zero if $b_m \neq a_m$ and then the system has a unique solution for $GE * GU$ and $GCE * GU$. Therefore perfect model-following is possible, as long as the model is such that $b_m \neq a_m$. The solution requires certain settings of especially GE and GCE, and since these are tuned by other means, it is unlikely that we will actually achieve perfect model-following.

Sensitivity derivative The sensitivity derivative is

$$\frac{\partial \hat{\varepsilon}}{\partial \theta} = \frac{\partial (y - y_m + \tau s(y - y_m))}{\partial \theta} = \frac{\partial (y + \tau s y)}{\partial \theta} = (1 + \tau s) \frac{\partial y}{\partial \theta}$$

since the model output y_m is independent of the parameter. Differentiating y in Equation (6.27) with respect to the adjustable parameter,

$$\frac{\partial y}{\partial \delta} = \frac{b * GU}{(1 + b * GCE * GU)s + (a + b * GE * GU)} u_0(s)$$

Closed loop substitution We do not make use of the inner-loop transfer function.

Adaptation law We arrive at the following adaptation law for adjusting the parameter:

$$\frac{d\delta}{dt} = -\gamma \hat{\varepsilon} \frac{\partial \hat{\varepsilon}}{\partial \delta} = -\gamma \hat{\varepsilon} (1 + \tau s) \left[\frac{b * GU}{(1 + b * GCE * GU)s + (a + b * GE * GU)} u_0(s) \right]$$

Approximation We approximate the inner-loop characteristic polynomial, the denominator in square brackets, with the model characteristic polynomial,

$$s + a_m$$

After insertion, the adaptation law is,

$$\frac{\mathrm{d}\delta}{\mathrm{d}t} = -\gamma\hat{\varepsilon}\,(1+\tau s)\left[\frac{b * GU}{s + a_m}u_0(s)\right]$$

Dispose of process parameters Since the unknown parameter b is constant, it can be absorbed in the adaptation gain, but we have to know the sign of b in order to use the correct sign for the adaptation gain. By the substitution $\gamma b = \gamma' sign(b) a_m$,

$$\frac{\mathrm{d}\delta}{\mathrm{d}t} \approx -\gamma' sign(b)\hat{\varepsilon}\,(1+\tau s)\left[\frac{a_m}{s + a_m}u_0(s)\right] * GU \tag{6.29}$$

The square bracket above is a filtered version of the signal u_0, and the steady state gain of the filter is 1.

Cancellation Since τ is the prediction horizon for the predicted model error, it is a design parameter. If we choose

$$\tau = \frac{1}{a_m}$$

the adaptation law is simply ($u_0 = 1$),

$$\frac{\mathrm{d}\delta}{\mathrm{d}t} \approx -\gamma' sign(b)\hat{\varepsilon}(t) * GU$$

This is similar to the adaptation law in the original SOC, apart from the factors GU and $sign(b)$. These could be absorbed in the adaptation gain, to make the two expressions identical, but they are left out as a convenience and an emphasis.

To summarize, the derivations are based on three observations. First, the performance specification $p = 0$, Equation (6.11), implicitly expresses the desired behaviour, which can be interpreted as a first-order reference model. Second, the performance specification $p \neq 0$ is interpreted as the predicted model error τ seconds ahead of current time. Third, choosing τ equal to the model time constant $1/a_m$ simplifies the adaptation law.

The theoretical considerations indicate that the SOC adjustment mechanism minimizes an objective function. Since we are adjusting only parameter δ, and not the gains, we cannot guarantee perfect model-following or stability.

6.9 Notes and References*

It is possible, and this has been done in practice, to adjust the consequence singletons in a Sugeno controller. That will make the adjustments less local, because each singleton affects a region of the phase plane determined by the premise membership functions of that rule. The corresponding adjustments to the control table cover several cells at the same time, and the modified control surface will thus be smoother. If there is sufficient computing time to execute a rule base, the control table does not have to be calculated at all.

Adaptive fuzzy controllers can be classified according to what they adapt. Systems that adapt the gains are called self-tuning, and systems that adapt the rule base are self-organizing. For an overview of the techniques, see Driankov *et al.* (1996). A third class adjusts the parameters of the membership functions, both on the premise side and on the conclusion side (e.g. Jang *et al.* 1997).

Research in adaptive control started in the early 1950s. Control engineers have tried to agree on a formal definition of adaptive control, and for example in 1973 a committee under the Institute of Electrical and Electronics Engineers (IEEE) proposed a vocabulary including the terms 'self-organizing control (SOC) system', 'parameter adaptive SOC', 'performance adaptive SOC', and 'learning control system' (Åström and Wittenmark 1995). These terms were never widely accepted, however, but the story does explain why the adaptive fuzzy controller is called the self-organizing fuzzy controller (SOC).

The SOC (Mamdani and Baaklini 1975, Procyk and Mamdani 1979, Yamazaki and Mamdani 1982) was developed specifically to cope with deadtime. To the inventors it was a further development of the original fuzzy controller (Assilian and Mamdani, 1974a, 1974b).

7

Performance and Relative Stability

The goal of the design phase is to meet performance specifications, that is, constraints associated with the closed-loop system response. The constraints specify the speed of response and the relative stability of the closed-loop system. This chapter investigates how to meet specifications on overshoot, dominant time constant, and settling time. Most of the performance of the fuzzy controllers treated in this book depends on the tuning, and the chapter develops a tuning method in order to meet specifications of performance. The method is based on a reference model, and the controller will attempt to follow the model, which expresses the desired performance. In order to identify the merits of each of the four standard control surfaces, they are investigated with respect to their relative stability in four cases of difficult processes. The relative stability is roughly measured by means of gain margin and delay margin, which is comparatively easy to do in practice. Frequency analysis and Nyquist plots complete the investigation. The chapter finds the describing function of a fuzzy controller in order to perform the frequency analysis. As a result, the chapter provides recommendations on how to match the standard control surfaces to a selection of different types of processes.

It is convenient to work with four standard control surfaces (linear, deadzone, saturation, and quantizer). It limits the selection of control surfaces, and the linear control surface can be a reference for the other, nonlinear control surfaces. But, as a result, the designer is faced with the question of which control surface to apply, and when. One would expect that for a given process, one surface would perform better than other surfaces, especially with respect to stability.

Linear systems have theory and methods for assessing the performance of a set of tuning factors, and also for calculating whether a closed-loop system is stable – there are even measures of how far a system is from instability. Robustness is a matter of keeping the system stable even under the influence of changes, such as varying temperature, and disturbances, such as a change of load. The more robust, the better. On the other hand, a very robust system will perform sluggishly. The controller parameters must therefore be tuned as a trade-off between robustness and *performance*. By performance, we mean how the closed-loop system satisfies specifications of performance, such as overshoot, dominant time constant, and settling time.

Foundations of Fuzzy Control: A Practical Approach, Second Edition. Jan Jantzen.

With nonlinear systems this is more difficult. Nevertheless, we would like some guidance regarding the choice of control surface – or at least an experimental analysis that demonstrates their different behaviour, together with different processes.

Processes that lend themselves to conventional PID control are probably best controlled this way. We are more interested in processes that are more or less difficult to control. Åström and Hägglund (2006) discuss and analyse such processes. They are processes where a more sophisticated kind of control could perform better than conventional PID control, and they mention higher-order processes, systems with a long time delay, and systems with oscillatory modes. In order to test the performance of the four standard control surfaces, we will use a test batch of representative models of such processes, and also include a fourth case: the double integrator.

If a system is stable, we usually want to know how close it is to being unstable; that is, its *relative stability*. There is a strong correlation between relative stability and performance. While stability of a linear system rests on a single concept, namely, the eigenvalues, the stability of a nonlinear system is more complex. The describing function of a nonlinearity is a simplification, being a linear approximation to a transfer function. It enables a Nyquist analysis, which is convenient, because one single plot shows whether the closed loop system is stable, and it shows the stability margin also. This chapter assesses the relative stability of the four standard control surfaces: the linear, the deadzone, the saturation, and the quantization surfaces.

7.1 Reference Model

Evidently, the ideal PID controller produces a positive control signal in certain situations and a negative control signal in other situations, depending on the error terms. States where the control signal u is zero are particularly interesting. In that case

$$0 = K_p \left(e + \frac{1}{T_i} \int e dt + T_d \dot{e} \right) \tag{7.1}$$

Given a set of controller gains K_p, T_i, T_d there will be some combinations of $e, \int e dt, \dot{e}$ that result in a zero control signal. The three terms can be viewed as state-variables in a three-dimensional state-space, that is,

$$\begin{aligned} e_1 &= \int e dt \\ e_2 &= e \\ e_3 &= \dot{e} \end{aligned} \tag{7.2}$$

Equation (7.1) defines a plane in the three-dimensional space since one of the variables, say e_3, depends linearly on the two others. As the trajectory crosses the plane, the control signal u changes sign, and we therefore call the plane a *switching plane*. Geometrically, the control action depends on the distance of the moving point from the switching surface. The following theorem establishes the exact relationship between control action and switching surface.

Theorem 7.1 *Given an ideal PID controller $u = K_p(e + \frac{1}{T_i}\int edt + T_d\dot{e})$, the control signal u is proportional to the distance of the moving point $\mathbf{e} = (\int edt, e(t), \dot{e}(t))$ from the fixed switching plane S. The switching plane is defined by the feedback vector $\mathbf{f} = [\,K_p/T_i \quad K_p \quad K_pT_d\,]^T$, which is normal to S.*

Proof. The controller can be written $u = \mathbf{f}\cdot\mathbf{e} = \|\mathbf{f}\|\,\|\mathbf{e}\|\cos\alpha$, where α is the angle between $\mathbf{f}$ and $\mathbf{e}$. But $\|\mathbf{e}\|\cos\alpha$ is the length of the projection of $\mathbf{e}$ on $\mathbf{f}$, and since $\mathbf{f}$ is normal to S, this is the same as the vertical distance of the tip of the vector $\mathbf{e}$ from S. The length of the feedback vector $\|\mathbf{f}\|$ is a proportionality constant – once the controller parameters have been selected – and the control signal is therefore proportional to the vertical distance. □

As a corollary, the control signal is positive when the moving point is on the same side of S as $\mathbf{f}$, and negative on the opposite side of S. Intuitively, we can say that the controller tries to bring the trajectory to the switching surface, or as close as possible. More precisely, a motion on the switching surface is stable, and if the control law always shortens the distance to the switching surface, then the closed-loop system is stable.

We therefore view the switching surface as a set of ideal states, or more accurately, we can take it as a reference model. It is a reference model in the sense that it expresses a behaviour that is desirable. It may be difficult, or even impossible, to achieve perfect model-following in reality. Since the model concerns the reference error, we can view it as a *deviation variable* model, that is, a model for perturbations from the steady state.

Using Laplace notation, and making the variable substitution $e_1 = \int edt$, a preliminary reference model is the following:

$$\left(s^2 + \frac{1}{T_d}s + \frac{1}{T_dT_i}\right)e_1 = 0 \tag{7.3}$$

Notice that K_p disappeared, as it was divided away. Note also that $e_1 = \int edt$ is the focus variable in the equation, not the usual error e. Clearly, this is a second-order differential equation. It is even homogeneous, since there is nothing on the right-hand side. Compare this with the standard second-order model

$$\left(s^2 + 2\zeta\omega_n s + \omega_n^2\right)e_1 = 0 \tag{7.4}$$

We can immediately reap some benefits from this comparison. The constant ζ is the *damping ratio* and the constant ω_n is the *undamped natural frequency* of the system. By comparison, there are two relationships:

$$T_d = \frac{1}{2\zeta\omega_n} \tag{7.5}$$

$$T_i = \frac{2\zeta}{\omega_n} = 4\zeta^2 T_d \tag{7.6}$$

Once ζ and ω_n are decided, then we can find the equivalent tuning gains T_d and T_i. Conversely, we can tell that the tuning of T_d affects the damping ratio while T_i affects the undamped natural frequency. Furthermore, the roots of Equation (7.3) are

$$s_{1,2} = -\frac{1}{2T_d} \pm \sqrt{\frac{1}{4T_d^2} - \frac{1}{T_i T_d}} \tag{7.7}$$

These are also the eigenvalues of the model. If $T_i = 4T_d$ the square root will be zero, and the two roots coincide in a double real root. This is a particularly interesting case, because it indicates the system is *critically damped*, that is, the *step* response is the fastest possible without overshoot. It is important to remember that the system is homogeneous, and its response depends only on the *initial* conditions. It is also important to realize that the model concerns e_1, which is the integrated error, not the error itself. It is therefore the integrated error that will have zero overshoot, in response to a step.

When $T_i < 4T_d$ the radicand under the root sign will be negative, and the eigenvalues will be two complex conjugate numbers. In that case, the step response will oscillate. However, the real part $-1/(2T_d)$ will attenuate the response in an exponential manner with a time constant. The constant $\tau = 2T_d$ is the *dominant time constant*, which is the time constant of the envelope of the oscillating response. Therefore, the tuning of T_d directly affects the dominant time constant, which is a convenient performance specification in practice.

With the state-variables defined earlier $e_1 = \int edt$ and $e_2 = e = \dot{e}_1$, we can formulate a state-space model on the form $\dot{\mathbf{e}} = \boldsymbol{A}\mathbf{e} + \boldsymbol{B}u$. To allow for a possible input u to the model, we include an input matrix $\boldsymbol{B}$, that is,

$$\begin{bmatrix} \dot{e}_1 \\ \dot{e}_2 \end{bmatrix} = \begin{bmatrix} 0 & 1 \\ -\dfrac{1}{T_i T_d} & -\dfrac{1}{T_d} \end{bmatrix} \begin{bmatrix} e_1 \\ e_2 \end{bmatrix} + \begin{bmatrix} 0 \\ \dfrac{1}{K_p T_d} \end{bmatrix} u \tag{7.8}$$

The model contains Equation (7.3) in the last row of the state-space model. Equation (7.8) is our final reference model, which accommodates two types of response: *the free response* and *the forced response*. The free response is the solution of Equation (7.8) when the input $u = 0$; the result depends only on the initial condition $\mathbf{e}(0)$ of the state-vector. The forced response is the solution of Equation (7.8) when all initial conditions are zero; in that case, the solution depends only on the input $u(t) \neq 0$.

When there is a unit step on the reference of the closed-loop system that includes a PID controller, the error jumps from zero to one. After the step, the model is in the initial state $\mathbf{e} = [0 \quad 1]^T$ and the response is a free response. When there is a step on the load of the closed-loop system, the input u jumps from zero to one, and the response type is a forced response. The differential equation is no longer homogeneous, and the combined gain $K_p T_d$ reappears in the B matrix instead of being divided out.

Referring to the usual closed-loop system (Figure 4.1), the transfer function from the load to the error is the following:

$$T = \frac{P}{1 + PC}$$

Here P is the process transfer function and C is the controller transfer function. If the magnitude of C tends to infinity, PC will dominate the denominator, and the transfer function T will tend to $-1/C$ as P cancels out. Thus, in the limit, the load step response is determined by the inverse of the PID controller transfer function. Consequently, the load step response will tend to our model step response as K_p is increased. In practice, however, noise may corrupt the feedback signal such that K_p must be limited.

Equation (7.8) is a dynamic model that uses available measurements. The second state variable e_2 is the reference model's error, which we can compare with the actual error of the closed-loop process response. The following three examples illustrate how the response follows the model.

Example 7.1 *Model-following when adjusting K_p*

Given the third-order process $P(s) = 1/(s+1)^3$ and a tuned PID controller, how well does the closed-loop system follow the reference model when the proportional gain is adjusted?

▶ *Solution*

As a starting point, we choose gains that are round numbers, that is $K_p = 2.5$, $T_i = 2$, $T_d = 5$. Then we adjust only K_p to study the effect. There is a unit step on the reference at time $t = 0$ and a unit load step at $t = 50$. Equation (7.8) shows that K_p does not appear in the **A** *matrix and therefore it should not affect the stability of the model. It appears in the* **B** *matrix, however, so K_p should affect the load response.*

Figure 7.1 shows the response with proportional gain $K_p = 2.5$ and another response with a much lower $K_p = 0.25$. The high gain controller follows the reference model well, after both the reference step and the load step. The control signal shows some oscillation. With

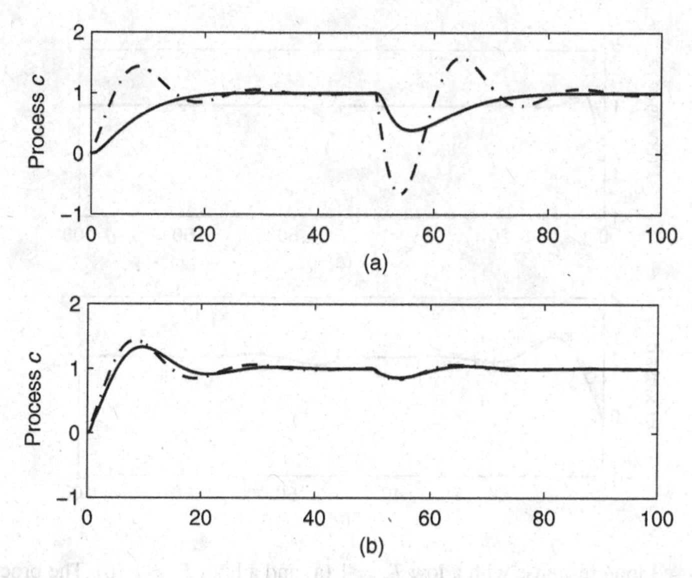

Figure 7.1 Closed-loop response with a low $K_p = 0.25$ (a) and a high $K_p = 2.5$ (b). The process is $P(s) = 1/(s+1)^3$. The response is closer to the model (dash-dot) when K_p is high. (figperfh0.m)

the low gain, the process response to the reference step is without overshoot, but far from the model response. The model load response oscillates with a high amplitude, and the closed loop system is less able to follow that. The frequencies of the oscillations match the model rather well. A further increase of the proportional gain reveals (not shown) that there is an upper limit where the control signal starts to oscillate heavily.

Overall, the high gain controller follows the model well – much better than the low gain controller does.

Example 7.2 *Model-following when adjusting T_d*

Again, given the third-order process $P(s) = 1/(s+1)^3$ and a tuned PID controller, how well does the closed-loop system follow the reference model when the derivative gain T_d is adjusted?

► *Solution*

We take the same starting point as in the previous example, that is, $K_p = 2.5, T_i = 2, T_d = 5$. Then we adjust only T_d to study the effect. The constant $\tau = 2T_d$ is the dominant time constant, and thus we expect to affect the time constant of the envelope of the response. A larger T_d should result in a slower response. Furthermore, from Equation (7.3), a larger T_d should result in a lower frequency of oscillation.

Figure 7.2 shows the response with derivative gain $T_d = 5$ and another response with a much lower $T_d = 1$. In both cases the response follows the model rather well. The dominant time constant and the frequency do indeed slow down when increasing T_d, as predicted. At the same time, however, the overshoot in the reference step response increases. If T_d is increased even further the controller will be too sensitive to noise.

Overall, the response follows the model rather well, and the dominant time constant can be controlled to some extent by adjusting T_d.

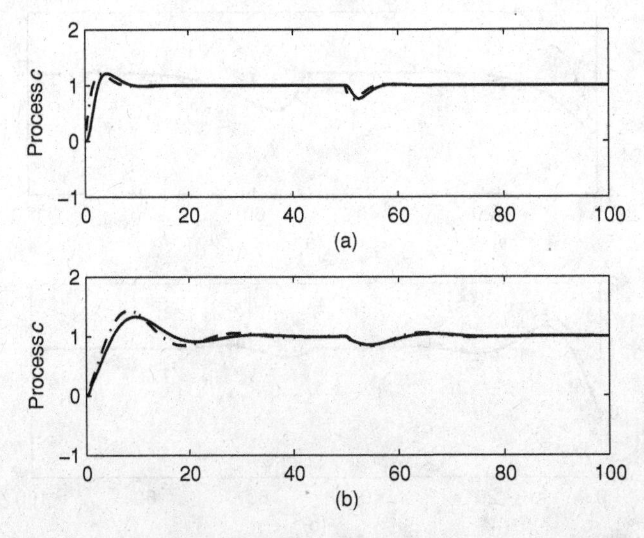

Figure 7.2 Closed-loop response with a low $T_d = 1$ (a) and a high $T_d = 5$ (b). The process is $P(s) = 1/(s+1)^3$. The model response (dash-dot) is slower when T_d is high, but the process still follows the model fairly well. (figperfh1.m)

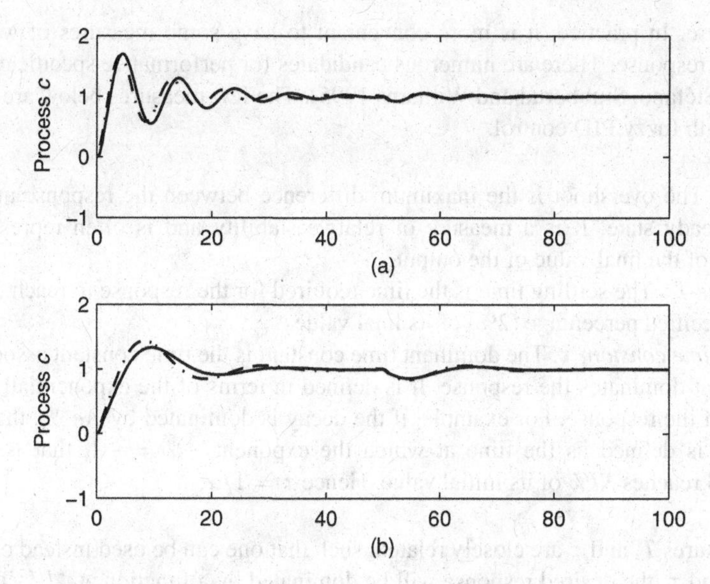

Figure 7.3 Closed-loop response with a low $T_i = 0.4$ (a) and a high $T_i = 2.0$ (b). The process is $P(s) = 1/(s+1)^3$. The model response (dash-dot) is less oscillatory when T_i is high, but the process still follows the model fairly well. (figperfh2.m)

Example 7.3 *Model-following when adjusting T_i*

Again, given the third-order process $P(s) = 1/(s+1)^3$ and a tuned PID controller, how well does the closed-loop system follow the reference model when the integral gain T_i is adjusted?

► *Solution*

We take the same starting point as in the previous example, that is, $K_p = 2.5$, $T_i = 2$, $T_d = 5$. Then we adjust only T_i to study the effect. Equation (7.7) shows that T_i affects the imaginary part of the eigenvalues without affecting the real part. We may therefore expect that a decrease in T_i results in more oscillation and more overshoot without affecting the envelope of the response.

Figure 7.3 shows the response with integral gain $T_i = 2$ and another response with a much lower $T_i = 0.4$. The higher value of T_i results in less integral action, since T_i is in the denominator. The high value simulation is therefore less oscillatory. Both responses follow the model rather well, but the oscillatory response is best at following its model. Clearly, T_i affects both overshoot and frequency in a manner as predicted.

As T_i affects both overshoot and frequency at the same time, it is a little difficult to predict the final effect. Overall, the response follows the model rather well.

7.2 Performance Measures

Specifications for an underdamped second-order system are often given in terms of ζ and ω_n, which is convenient from a theoretical point of view, that is, from the viewpoint of eigenvalues. Eigenvalues determine stability, but they do not give full information about the forced response. In practice, it is more convenient to have some measures of what can be observed in a response. There are numerous candidates for performance specifications in the literature (DiStefano, Stubberud and Williams 1995). The few measures below are feasible in connection with fuzzy PID control.

- *Overshoot.* The overshoot is the maximum difference between the response and its final value in steady state. It is a measure of relative stability and is often represented as a percentage of the final value of the output.
- *Settling time T_s.* The settling time is the time required for the response to reach and remain within a specified percentage (2%) of its final value.
- *Dominant time constant τ.* The dominant time constant is the time constant associated with the term that dominates the response. It is defined in terms of the exponentially decaying character of the response. For example, if the decay is dominated by $Ae^{-\alpha t}$, then the time constant τ is defined as the time at which the exponent $-\alpha t = -1$, that is, when the exponential reaches 37% of its initial value. Hence $\tau = 1/\alpha$.

The two measures T_s and τ are closely related, such that one can be used instead of the other. Given a desired τ, the desired response will be dominated by a function $Ae^{-t/\tau}$. The settling time is the point in time when the exponential reaches δ% of its final value, that is,

$$e^{-T_s/\tau} = \frac{\delta}{100} \Rightarrow T_s = -\ln\frac{\delta}{100}\tau$$

With $\delta = 2\%$, the solution is $T_s = 3.9\tau$ and with $\delta = 5\%$ the solution is $T_s = 3.0\tau$. The two measures τ and T_s are thus proportional, and they can be interchanged depending on which one is the more practical. Figure 7.4 illustrates the three measures in connection with an oscillatory response with an exponential decay.

The next step is to relate the model response with those performance specifications. Given values for ζ and ω_n we can easily find equivalent PID settings, but our task is to eliminate the pair ζ, ω_n and use the performance specifications instead. For this purpose, we would like to

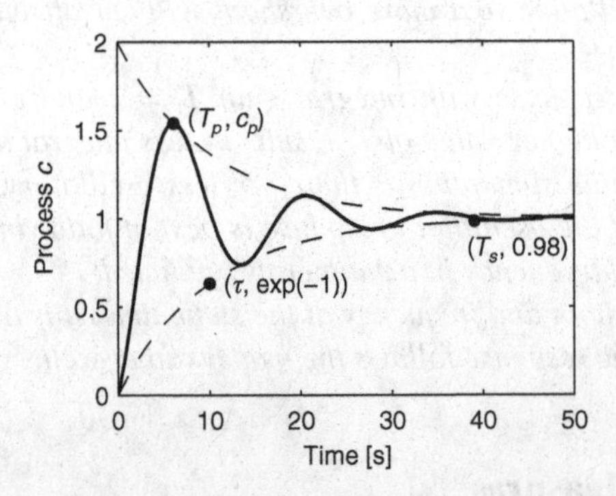

Figure 7.4 Performance of an underdamped response. The first peak $c_p = 1.54$ occurs at peak time $T_p = 6.17$ s. The envelope has the dominant time constant $\tau = 10$ and settling time $T_s = 39$ s, beyond which the response remains within a band of ±2% of the final value. Furthermore, the response has a damped natural frequency $\omega_d = 0.4$ s^{-1} and damping ratio $\zeta = 0.22$. (figperf.m)

use a solution to the free response on symbolic form in order to uncover how the PID gains affect the performance measures, and vice versa. It is fairly complicated, but possible, to find a closed form solution of the free response equation (we cannot use a textbook solution to a step response, because we are looking for an initial condition response). We defer the mathematical details until later (Example 7.6), and we just present the results at this point.

We have already presented a useful expression that relates T_d to the dominant time constant, that is,

$$T_d = \frac{1}{2}\tau$$

Therefore, given a desired dominant time constant τ, we choose T_d as half of that. The model response to a reference step will thus remain within an envelope with the desired time constant.

The upper envelope decays exponentially to the final value, which is 1. The first peak c_p of the oscillation occurs at peak time T_p, that is,

$$c_p = 1 + \exp\left(-\frac{t}{\tau}\right)|_{t=T_p} = 1 + \exp\left(-2\frac{\zeta}{\sqrt{1-\zeta^2}}\arctan\frac{\sqrt{1-\zeta^2}}{\zeta}\right) \tag{7.9}$$

The expression depends on ζ only. Since the final value of the response is 1, then $(c_p - 1) \times 100$ is the overshoot percentage. We would like to solve the equation with respect to ζ, but this seems difficult. Instead we choose to plot the function $(c_p - 1) \times 100 = f(\zeta)$ in Figure 7.5.

The figure shows that the overshoot percentage is a monotonic function of ζ, and we can thus graphically determine ζ given a desired overshoot percentage. Notice that as $\zeta \to 1$ the overshoot tends to a non-zero value of $e^{-2} \times 100 = 13.5\%$. It is possible to remove the overshoot in the reference step response, if that is preferred, by choosing larger damping ($\zeta > 1$). Our objective here, however, is to tune the controller according to specifications on overshoot, and thus $\zeta \leq 1$.

Having determined T_d and ζ, then, finally, by Equation (7.6),

$$T_i = 4\zeta^2 T_d \tag{7.10}$$

If necessary, ω_n could be found from, for example, Equation (7.5).

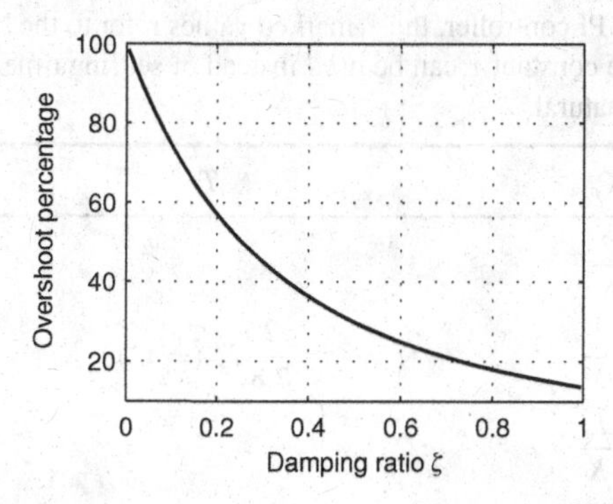

Figure 7.5 Overshoot percentage as a function of ζ. The curve allows finding ζ from a given overshoot percentage larger than 13.5%. (figzeta.m)

In summary, the desired time constant τ (or the equivalent settling time T_s) determines T_d. The desired overshoot percentage determines ζ, which in turn determines T_i.

7.3 PID Tuning from Performance Specifications

The previous equations determine the pair T_d, T_i from the pair τ, c_p. The latter two are physically meaningful, and it is possible to establish a tuning procedure based on specifications on the reference step response. In case the model-following is unsatisfactory, we hope to improve it by means of a nonlinear fuzzy controller. The procedure for tuning a PID controller is the following:

1. Start with the differential gain, and set it to $T_d = \tau/2$; that is, half of the desired time constant of the envelope.
2. Decide on a desired overshoot percentage, and use Figure 7.5 to find the equivalent damping ratio ζ.
3. Determine T_i by Equation (7.10).
4. The proportional gain is $K_p = \gamma T_i T_d / K$ where K is the static gain of the process and γ is an adaptation gain. The higher γ, the closer is the response to the desired time constant; it is in the range 2–20, and a typical value is $\gamma = 10$. The role of K_p is thus to make the closed-loop system follow the model as closely as possible. However, there is only a certain gain margin K_{GM} available, that is, $K_u = K_p K_{GM}$. Here K_u is the ultimate gain which brings the closed loop system to the verge of stability, and K_{GM} is in the range 2–5 (Åström and Hägglund 2006 p. 105).

Table 7.1 summarizes the gain settings. The table also has an entry for a PD controller. In this case, the reference model is the first-order model

$$\left(s + \frac{1}{T_d}\right) e_2 = 0$$

Table 7.1 PID settings derived from the desired settling time T_s (or τ) and overshoot percentage $(c_p - 1) \times 100$. Note: for the PI controller, the *-marked values refer to the behaviour of $\int e dt$, rather than e itself. The desired time constant τ can be used instead of settling time T_s as a performance specification, if that is more natural.

Controller	K_p	T_i	T_d
P	$\frac{\gamma}{K}$	–	–
PI	$\gamma \frac{T_i}{K}$	$\frac{2}{7.8} T_s^* (= \tau^*)$	–
PD	$\gamma \frac{T_d}{K}$	–	$\frac{2}{7.8} T_s (= \tau)$
PID	$\gamma \frac{T_i T_d}{K}$	$4\zeta^2 T_d$, $[\zeta = f^{-1}(c_p)]$	$\frac{1}{7.8} T_s (= \tau/2)$

Notice that the variable in focus is e_2, which is the reference error itself. Since the model is a first-order model, ζ and ω_n are undefined. Switching occurs around a line, and the desired time constant is directly $\tau = T_d$. Integral gain is irrelevant for a PD controller, and K_p should again be as large as possible within the available gain margin.

In case of a PI controller, the reference model is the first-order model

$$\left(s + \frac{1}{T_i}\right) e_1 = 0$$

Here the variable in focus is $e_1 = \int e dt$. Switching occurs around a line, and the desired time constant is directly $\tau = T_i$. Notice that the time constant is related to the behaviour of $\int e dt$ rather than e itself. The solution to the equation is $e_1(t) = e_1(0) \exp(-t/T_i)$. The behaviour of $e = \dot{e}_1$ is also an exponential decay. In fact, $e(t) = e_1(0)(-1/T_i) \exp(-t/T_i)$, which has the same time constant, but another final value.

Example 7.4 *Tuning according to settling time*

Given the third-order process $P(s) = 1/(s+1)^3$, as previously, how well can we tune a PID controller to obtain an overshoot of 30% and specified settling time T_s?

► *Solution*

As a starting point we choose a slow settling time of $T_s = 50$ s. Settling time is equivalent to a desired time constant $\tau = T_s/3.9$, and given τ, Table 7.1 determines T_d. An overshoot of 30% corresponds roughly to $\zeta = 0.5$ according to the graph in Figure 7.5. Knowing T_d and ζ, Table 7.1 provides T_i. For the proportional gain we use the same as in previous examples, that is, $K_p = 2.5$. In a second simulation we choose a fast settling time of $T_s = 10$ s.

Figure 7.6 shows the result of the simulation. Regarding the reference step, the two models exhibit the same overshoot, which is the specified overshoot. The two models also comply with the specifications of their settling times.

In the case of the fast response, the closed-loop response follows the model rather well. The good model-following is at the expense of an oscillating control signal, however. In the case of the slow response, the closed-loop response has less overshoot than specified.

Both load responses follow the model well.

The example shows that it is easy to tune the system based on the performance specifications. The closed-loop response may not follow the model perfectly if K_p is kept at a fixed value.

Since the tuning is based on performance measures, several different processes can be tuned to follow the same model. If they all follow closely enough, it can be an advantage in practice, because then all the closed-loop systems could be treated the same, namely the second-order reference model. The following example illustrates the performance of three different processes.

Example 7.5 *Three processes, same performance specification*

Given three process $P_1(s) = 1/(s+1)^3$, $P_2(s) = 1/(s+1)^4$, $P_3(s) = 1/s^2$, can we tune a single PID controller such that all closed loop systems obtain the same overshoot of 30% and settling time $T_s = 30$ s?

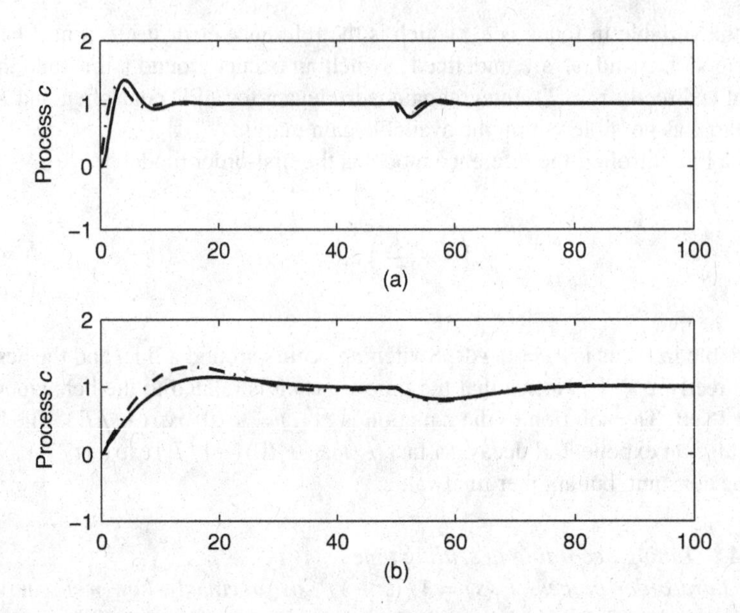

Figure 7.6 Closed-loop response with (a) a fast desired settling time $T_s = 100$ and (b) a slow desired settling time $T_s = 50$. The process is $P(s) = 1/(s + 1)^3$. The model response is slower when T_s is high, but the process still follows the model fairly well. (figperfh3.m)

► *Solution*

The specifications on overshoot and settling time determine a common model for all three processes. The question is how well each process follows the model.

The settling time $T_s = 30$ s is equivalent to a desired time constant $\tau = T_s/3.9$, and given τ, Table 7.1 determines T_d. An overshoot of 30% corresponds roughly to $\zeta = 0.5$ according to the graph in Figure 7.5. Knowing T_d and ζ, Table 7.1 provides T_i. For the proportional gain we use the same as in previous examples, that is, $K_p = 2.5$. The resulting gains are $K_p = 2.5$, $T_i = 3.85$, $T_d = 3.85$.

Figure 7.7 shows the result. The double integrator follows the model almost perfectly, while the two others achieve less overshoot (16–20%). All responses settle within the specified settling time. The fourth-order process is somewhat oscillatory, as the result of an oscillatory control signal.

A higher K_p will bring the responses of the third-order process and the fourth-order process closer to the model response, but it will also increase the oscillatory control signal of the fourth order process.

Although the three responses are similar, the specifications are not quite satisfied for all of them.

The example shows that similar responses can be obtained for different processes, but it may be difficult, or even impossible, for a given closed-loop system to comply with the performance specifications; they may be feasible for some processes, but not for others.

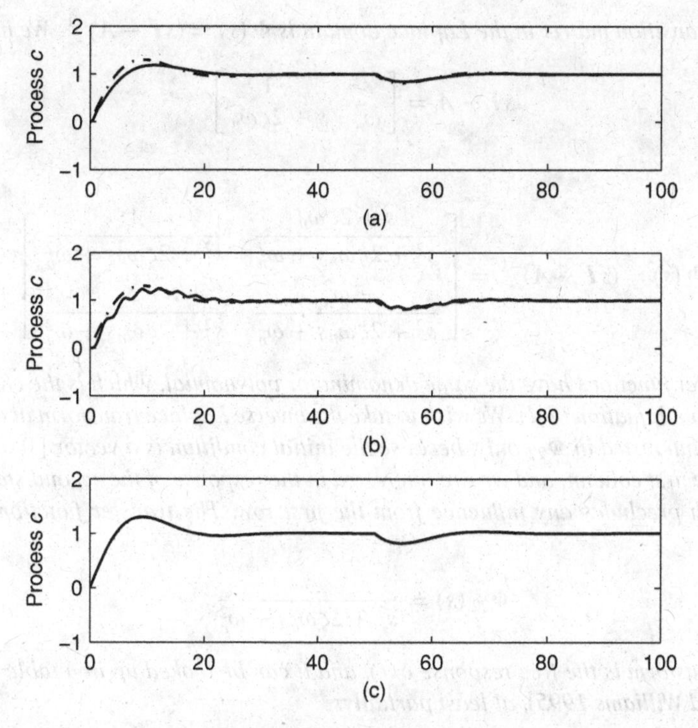

Figure 7.7 Closed-loop response of three processes with the same tuning. The processes are $P_1(s) = 1/(s+1)^3$ (a), $P_2(s) = 1/(s+1)^4$ (b) and $P_3(s) = 1/s^2$ (c). The processes follow the same model response (dash-dotted line) fairly well. (figperfmix.m)

Example 7.6 *Free response*

Given the state-space model in Equation (7.8), what is the free response as a result of the initial condition $\begin{bmatrix} e1(0) \\ e2(0) \end{bmatrix} = \begin{bmatrix} 0 \\ 1 \end{bmatrix}$? *What is the peak time and the overshoot?*

► *Solution*

In general, the solution to a free response is $\mathbf{e}(t) = \Phi(t)\,\mathbf{e}(0)$. *Here* $\mathbf{e}(0)$ *is the initial condition related to the state-vector, and* Φ *is the so-called* state-transition matrix. *It is a time-varying matrix that transforms the initial state-vector to the state-vector at time t in the future. In order to find* Φ, *we shall make use of Laplace transforms and inverse Laplace transforms. An example of this procedure can be found in the textbook by Nise (1995). Once we have the solution, we can find the maximum where the response peaks as well as the peak time.*

We are concerned with the free response, and consequently the system matrix from Equation (7.8). In terms of ζ *and* ω_n, *Equation (7.4), the system matrix is*

$$\mathbf{A} = \begin{bmatrix} 0 & 1 \\ -\omega_n^2 & -2\zeta\omega_n \end{bmatrix}$$

The state-transition matrix in the Laplace domain is $\mathbf{\Phi}(s) = (s\mathbf{I} - \mathbf{A})^{-1}$. *We have,*

$$s\mathbf{I} - \mathbf{A} = \begin{bmatrix} s & -1 \\ \omega_n^2 & s + 2\zeta\omega_n \end{bmatrix}$$

Its inverse is

$$\mathbf{\Phi}(s) = (s\mathbf{I} - \mathbf{A})^{-1} = \begin{bmatrix} \dfrac{s + 2\zeta\omega_n}{s^2 + 2\zeta\omega_n s + \omega_n^2} & \dfrac{1}{s^2 + 2\zeta\omega_n s + \omega_n^2} \\ \dfrac{-\omega_n^2}{s^2 + 2\zeta\omega_n s + \omega_n^2} & \dfrac{s}{s^2 + 2\zeta\omega_n s + \omega_n^2} \end{bmatrix}$$

All four transfer functions have the same denominator polynomial, which is the characteristic polynomial from Equation (7.4). We wish to take the inverse Laplace transform in order to find $\mathbf{\Phi}(t)$. *We are interested in* $\mathbf{\Phi}_{22}$ *only, because the initial condition is a vector* $[0 \quad 1]^T$, *which suppresses the first column, and we are interested in the response of the second state-variable* e_2 *only, which precludes any influence from the first row. The transfer function of interest is thus*

$$\mathbf{\Phi}_{22}(s) = \frac{s}{s^2 + 2\zeta\omega_n s + \omega_n^2}$$

Its inverse transform is the free response $c(t)$, *and it can be looked up in a table (DiStefano, Stubberud and Williams 1995), at least partially:*

$$\begin{aligned} c(t) &= \frac{d}{dt}\left[\frac{1}{\omega_d}\exp(-\zeta\omega_n t)\sin\omega_d t\right] \\ &= \exp(-\zeta\omega_n t)\left(-\zeta\frac{\omega_n}{\omega_d}\sin\omega_d t + \cos\omega_d t\right) \end{aligned} \tag{7.11}$$

The damped natural frequency ω_d *appears in a denominator, but it cannot be zero since we have assumed that the system is underdamped* ($\zeta < 1, \omega_d = \omega_n\sqrt{1-\zeta^2}$).

In order to find the first maximum, we could proceed the normal way, which is to differentiate and find the point in time where the derivative is zero. It is more interesting, however, to interpret the free response $c(t)$. *It contains an exponential multiplier, which is responsible for the envelope of the response; see Figure 7.4. The term in the second set of brackets of* $c(t)$ *is responsible for the oscillation inside the envelope. The first maximum* c_p *occurs when the oscillation cuts the upper exponential curve of the envelope; that is, when*

$$1 + \exp(-\zeta\omega_n t) = 1 - \exp(-\zeta\omega_n t)\left(-\zeta\frac{\omega_n}{\omega_d}\sin\omega_d t + \cos\omega_d t\right)$$

Thus, there is a maximum when $-\zeta\frac{\omega_n}{\omega_d}\sin\omega_d t + \cos\omega_d t = -1$, *and the solution to this equation is the peak time:*

$$T_p = \frac{2}{\omega_n\sqrt{1-\zeta^2}}\arctan\left(\frac{\sqrt{1-\zeta^2}}{\zeta}\right)$$

By insertion into the equation for the upper envelope, we find the peak, that is,

$$c_p = 1 + \exp\left(-\frac{2\zeta}{\sqrt{1-\zeta^2}} \arctan \frac{\sqrt{1-\zeta^2}}{\zeta}\right)$$

This is the result which was quoted earlier in Equation (7.9), and the exponential term corresponds to the overshoot. The expression depends on ζ only. Relative to the location of the eigenvalues, $\zeta = \cos\theta$, where θ is the angle between the real axis and the ray passing through the complex eigenvalue ($\zeta < 1$). Therefore, all cases with eigenvalues on the same ray, emanating from the origin, will have the same overshoot percentage.

The general idea behind model reference control is to design a controller C, such that the closed-loop transfer function equals the transfer function M of the model. More specifically, $CP/(1 + CP) = M$, where P represents the process. The solution

$$C = \frac{M}{P(1-M)}$$

accomplishes the objective seen from a mathematical point of view. In practice, however, there are possibly some difficulties: (1) the proposed solution contains the inverse P^{-1} of the process, and the solution therefore requires that it exists; (2) P may change over time, and the inverse might be highly sensitive to changes; (3) P may even be unknown.

If the controller is realizable in practice, there are still limits to its performance; if M is chosen too ambitiously then C will drive the process into saturation. The key to the design is to choose a less ambitious model M.

7.4 Gain Margin and Delay Margin

For now, we aim to achieve a first idea of the relative stability of the four standard control surfaces. We can say something about the relative behaviour, that is, relative to the linear controller. We let the linear fuzzy controller – equivalent to a conventional PID controller – act as a reference, and the relative stability of the nonlinear control surfaces can be compared with that in order to judge whether they are better or worse.

We are concerned with a feedback loop with a structure as in Figure 7.8. The controller is based on the error signal, and it is placed in the forward path. The controller is an FPD+I controller in order to accommodate all three PID actions.

The adjustable gain G_m in the control signal path is initially 1, but it can be adjusted until the closed-loop system becomes unstable. Ideally, the gain is increased until the verge of stability, and that extra gain is the *gain margin*. Similarly, the time delay D_m in the signal path is initially zero, but it can be increased gradually, until the system is on the verge of stability. The added time delay is then the *delay margin*. The two measures express the relative stability of the closed-loop system towards changes in the control signal, or more specifically: gain and phase. We can thus compare the gain margin and the delay margin with that of the linear controller, while considering various processes, to see which control surface is the more robust. An exhaustive test sequence requires 16 test set-ups, as there are four processes and

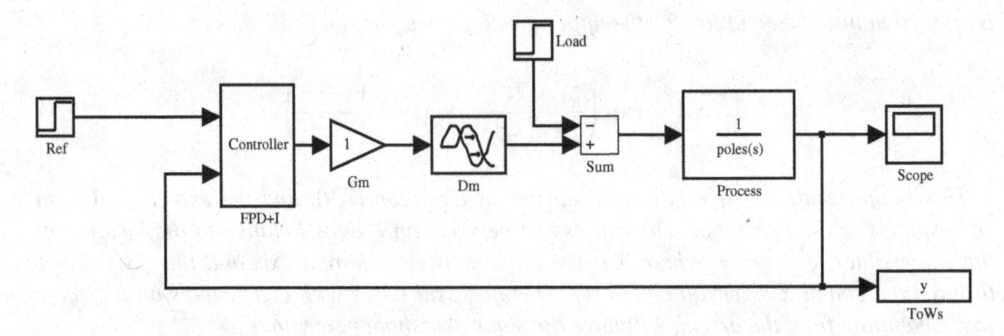

Figure 7.8 Simulink test bench with reference step and load step. Derivative action acts on the process output rather than the error. The controller is a table based FPD+I controller. (bench.mdl)

four surfaces. Each test consists of two runs: one test of the gain margin, and another of the delay margin.

7.5 Test of Four Difficult Processes

The test processes are mathematical simplifications. They perhaps lack realistic complications, but Laplace transfer function models are a convenient way to portray some typical behaviour.

Each of the four processes in the following, represented by its poles and zeros, is inserted one after the other in the Simulink block for the process (Figure 7.8). Simulink provides the differential equation solver (ode45), and the process output is observed on the scope. Each control surface is inserted in turn in the control loop without changing the tuning. Hand-tuning of a PID controller for each process provides the settings of the gains for all controllers. The hand-tuning is loose, because the objective is to compare the relative performance of each standard control surface rather than finding a controller that provides the best response. Table 7.2 contains all the test results.

7.5.1 Higher-Order Process

Integral action demands one state-variable, and a PID controller can therefore at most determine two eigenvalues of a second-order system. Thus, a second-order system is most likely completely controllable, but systems of higher order are more difficult to control owing to internal state-variables that are uncontrollable.

Take the process $P(s) = 1/(s+1)^3$ as a moderate example of a higher-order process. It is a stable process with three identical poles on the real axis, which indicates that its open-loop step response lacks oscillations. The PID gains in Table 7.2 are the ones found earlier using the Ziegler–Nichols tuning rules (Chapter 4).

Table 7.2 indicates that the deadzone surface has the best gain margin (3.2 times larger than the linear surface). Apparently, the deadzone characteristic – a large relative gain far from the settling point, and a lower relative gain around the settling point – provides a good gain margin. On the other hand, the delay margin is zero, indicating a high sensitivity to delays in

Table 7.2 Gain margin G_m and delay margin D_m for four processes in closed loop. Exclamation marks (!) indicate the best performance in each column. Each closed-loop system is tested with four control surfaces (*quantizer, saturation, deadzone,* and *linear*). The selected PID gains (K_p, T_i, T_d) determine the fuzzy gains (*GE, GCE, GIE, GU*), and the fuzzy controller is an FPD+I controller. All controllers include integral action in order to balance the load step.

Process $P(s)$	$\frac{1}{(s+1)^3}$	$\frac{1}{s^2}$	$\frac{e^{-4s}}{1+2s}$	$\frac{25}{s^2+s+25}$
K_p	4.8	100	0.2	0.5
T_i	15/8	0.4	2.5	2.0
T_d	15/32	0.1	0	0.5
GE	100	100	100	100
GCE	46.9	10	0	50
GIE	53.3	250	40	50
GU	0.048	1	0.0020	0.005
Quantizer	$D_m = 0$ $G_m = 19.0$	$D_m = 0.05$ $G_m = 1$	$D_m = 5.5$ $G_m = 2.91$	$D_m = 6.5$ $G_m = \infty$
Saturation	$D_m = 0$ $G_m = 6.42$	$D_m = 0.04$ $G_m = 0.142!$	$D_m = 5.5$ $G_m = 2.75$	$D_m = 6.5$ $G_m = \infty$
Deadzone	$D_m = 0$ $G_m = 49.9!$	$D_m = 0$ $G_m = 1$	$D_m = 5.6$ $G_m = 3.4!$	$D_m = 6.5$ $G_m = \infty$
Linear	$D_m = 0.32!$ $G_m = 15.5$	$D_m = 0.04$ $G_m = 0.235$	$D_m = 5.5$ $G_m = 2.92$	$D_m = 6.5$ $G_m = \infty$

the control signal path. The linear control surface has the best delay margin; in fact, it is the only surface with a non-zero delay margin.

If delay margin is important, then the linear surface was the best choice; if not, the deadzone surface had the largest gain margin.

7.5.2 *Double Integrator Process*

The double integrator represents processes where a physical force affects acceleration (Newton's second law). Such control objects are therefore rather common (robots, satellites, vehicles, ships).

The process $P(s) = 1/s^2$ is a simple model. It is an unstable process, in the sense that an impulse input will make the level of the second integrator, in a series connection of two integrators, grow to infinity. The PID gains are loosely selected by hand-tuning. It is a little difficult to apply common tuning rules, as the closed-loop system behaves oppositely to what might be expected. Thus, an increase of the proportional gain tends to make the system less oscillatory. The gain margin is less than one, and the lower the better.

Table 7.2 indicates that the saturation surface has the best gain margin (0.6 times the linear surface gain margin). Apparently, the higher relative gain near the settling point provides a good gain margin. None of the surfaces have a very good delay margin, if any at all.

In this case, the saturation surface was the best choice.

7.5.3 *Process with a Long Time Delay*

Time delays in the process are notoriously difficult to control. It occurs in processes with a transport component (conveyor belt, piping systems, chemical processes, stock control, measurement delay). By the time the result of a control action shows on the process output, the controller may be trying something different. Derivative action is of little use, and the stability can be upset relatively easy.

The process $P(s) = e^{-4s}/(1 + 2s)$ is an example of a first-order process with a time delay (FOTD). The time delay (4 s), with the transfer function e^{-4s}, is fairly long compared to the time constant (2 s). The PID gains are loosely selected by hand-tuning. The controller is a PI controller, because derivative action has little effect. In order to simulate the system, the time delay is approximated by a polynomial, which is the first-order Pade approximation (e.g. the function `pade` in MATLAB®), that is, $e^{-4s} \approx (-s + 0.5) / (s + 0.5)$.

Table 7.2 indicates that the deadzone surface has the best gain margin (1.16 times the linear surface gain margin). Apparently, it is a slight advantage to have a deadzone around the settling point in order not to disturb the process too much. All surfaces have more or less the same delay margin.

In this case, the deadzone surface was the best choice.

7.5.4 *Process with Oscillatory Modes*

An oscillatory mode is a fast oscillation often superimposed on a slower, more smooth behaviour. It occurs in processes with fast moving components or gases (flexible robot arms, flexible space structures, disk drives, combustion chambers). It is difficult to attenuate the fast oscillation and obtain a fast response at the same time.

The process $P(s) = 25/(s^2 + s + 25)$ is an example of an oscillatory process with relatively small damping ratio ($\zeta = 0.1$). A PI controller can achieve a fast response and also regulate load changes well. However, D-action is also necessary in order to attenuate the fast oscillation.

Table 7.2 indicates that all surfaces have the same gain margin (infinite) and the same delay margin. None of them performs differently from the linear surface.

In this case, the linear surface is the best choice; the performance is the same, but the linear surface is the simplest to work with and to analyse.

7.6 The Nyquist Criterion for Stability

As mentioned earlier (Section 4.10), the frequency response characterizes the dynamics by the way sine waves propagate through the loop. A Nyquist plot provides a complete description for the chosen frequencies.

Consider the system in Figure 7.8, and assume for a moment that everything is linear. The closed loop transfer function is

$$T = \frac{CP}{1 + CP}$$

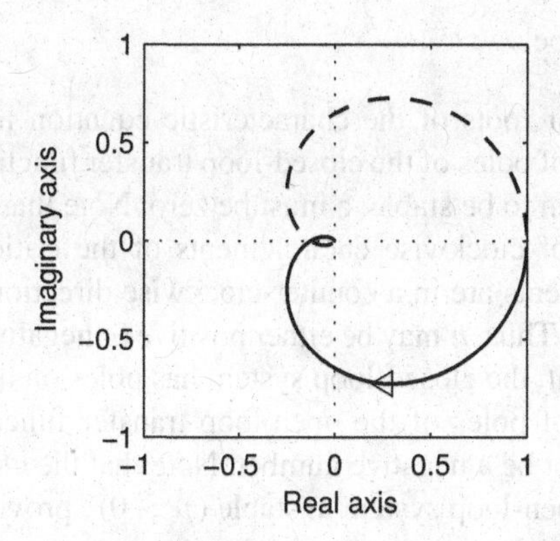

Figure 7.9 Nyquist diagram of process $P = 1/(s+1)^3$. (fignyquist.m)

where C is the controller transfer function and P is the process transfer function. The denominator is the characteristic polynomial, and the characteristic equation, which governs the stability, is

$$1 + CP = 0$$

The closed loop system is marginally stable when the magnitude frequency response $|CP| = 1$, while, at the same time, the phase frequency response $\angle CP$ is $\pm 180°$. This is equivalent to saying that the Nyquist diagram in the complex CP-plane passes through the *critical point* $(-1, 0)$.

Figure 7.9 shows the Nyquist diagram of the process $P = 1/(1+s)^3$. The curve does not pass through the critical point. The curve is drawn in the complex P-plane, which is a mapping of the complex s-plane. The plot is a polar plot, and each point on the curve represents a vector with magnitude $|P|$ and the angle $\angle P$ counted counter-clockwise from the positive real axis. The frequency is implicit, but the upper half of the diagram corresponds to negative frequencies $(\omega < 0)$ while the lower half corresponds to positive frequencies $(\omega > 0)$. The arrow shows the direction of increasing frequency. We focus on the lower half of the Nyquist diagram, because it is the plot of the frequency response $P(j\omega)$ with positive ω.

Inserting a proportional controller with the gain K_p in front of the process will change the diagram such that the new diagram of $K_p P$ will be a scaling of the P diagram by the factor K_p. If K_p is sufficiently large, corresponding to the ultimate gain K_u in the Ziegler–Nichols tuning method, the curve may pass through the critical point, and the system will be marginally stable. The frequency of oscillation will be the frequency that corresponds to the intersection point on $P(j\omega)$.

7.6.1 *Absolute Stability*

The Nyquist criterion relates the stability of the closed-loop system to the open-loop frequency response and open-loop pole locations. The Nyquist criterion is

$$z = n + p$$

The symbols are explained below:

- *Symbol* z. The number of roots of the characteristic equation in the right half-plane, or equivalently, the number of poles of the closed-loop transfer function in the right half-plane. For the closed-loop system to be stable, z must be zero. Note that z cannot be negative.
- *Symbol* n. The number of clockwise encirclements of the critical point by the Nyquist diagram. If the encirclements are in a counter-clockwise direction, n is the negative of the number of encirclements. Thus, n may be either positive or negative. If the Nyquist diagram intersects the critical point, the closed-loop system has poles on the $j\omega$-axis.
- *Symbol* p. The number of poles of the open-loop transfer function $L = CP$ in the right half-plane. Thus, p cannot be a negative number. Note that the closed-loop system may be stable ($z = 0$) with the open-loop system unstable $(p > 0)$, provided that $n = -p$.

In Figure 7.9, the Nyquist diagram does not encircle the critical point. Hence $n = 0$, and since there are no open-loop poles in the right half-plane $p = 0$. Thus $z = 0$ and the closed loop system is stable. On the other hand, if $C = K_p = 8$, the Nyquist diagram will pass through the critical point, and the closed loop system will be marginally stable.

7.6.2 Relative Stability

Since stability depends on the critical point, it is reasonable to assume that the stability margin depends on how closely the Nyquist curve passes the critical point. It follows that in order to be robust, the controller C should be designed such that the Nyquist plot of the open-loop transfer function $L = CP$ is sufficiently remote from the critical point. Conversely, given a number of alternative controller designs, the stability margin related to each controller is a measure of its performance. The so-called *maximum sensitivity* M_s is a stability margin, which is based on the distance to the critical point.

The sensitivity S of the closed-loop transfer function to changes in the controller, or the process, is the ratio of the closed-loop transfer function to the open-loop transfer function. The closed-loop transfer function is $L/(1 + L)$, and the *sensitivity function* is therefore defined as

$$S = \frac{1}{1 + L} \tag{7.12}$$

In the Nyquist diagram of L, the vector $1 + L$ points from the critical point to the moving point L, and its length is the moving point's distance from the critical point. When the distance is shortest, the sensitivity S is at its maximum M_s, or

$$\frac{1}{M_s} = \inf |1 + L(j\omega)|, \qquad 0 < \omega < \infty$$

A design criterion is therefore that the Nyquist curve of L touches a circle with its centre in the critical point and with a radius equal to a desired value of $1/M_s$. Typical values of M_s are in the range 1.2 to 2 (Åström & Hägglund 2006 p. 127), corresponding to a radius in the range 0.50 to 0.83. We can thus choose, say, radius 0.5 as our shortest acceptable distance to the critical point.

7.7 Relative Stability of the Standard Control Surfaces

This section is based on an approximate transfer function of the fuzzy controller. A later section explains how to achieve that, because it is rather mathematical in nature (Section 7.9). The results are useful for our purpose, however, and the current section therefore skips the mathematical detail.

A fuzzy controller does not have a transfer function, since it is nonlinear, but a so-called *describing function* approximates a linear transfer function. The approximation is a transfer function that depends on frequency as well as amplitude. Given a describing function of a fuzzy controller, especially the FPD+I controller, it is possible to draw a Nyquist plot when the process is linear. The result will be a picture of the relative stability in the 16 previous test cases (Section 7.5).

The performance is portrayed against the shortest acceptable distance to the critical point, which is chosen as $1/M_s = 0.5$. The farther the Nyquist curve is from the 0.5-circle, the more robust is the closed loop system. As previously, we are only changing the shape of the control surface, not the tuning, in order to make a fair comparison.

Figure 7.10 shows the results. In each figure, the frequency ω increases from low values to high values as the moving point moves towards the origin. Furthermore, Figure 7.11 is an extract of the actual stability margins $1/M_s$ in all 16 test cases, but they are normalized, such that the linear surface is associated with index 1. The larger the index, the larger is the distance

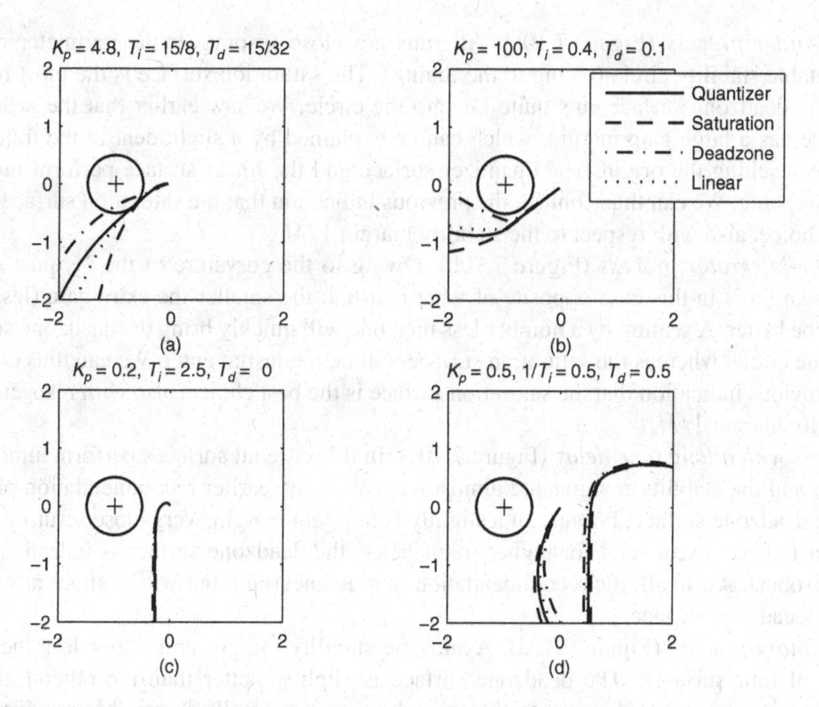

Figure 7.10 Nyquist plots related to the four test processes: (a) the third-order process, (b) the double-integrator process, (c) the FOTD, and (d) the oscillatory process. Each process is tested with the four standard controllers (quantizer, saturation, deadzone, and linear). The circle in each plot is a sensitivity circle with radius 0.5, which is chosen as the smallest acceptable distance from the critical point. (figdf.m)

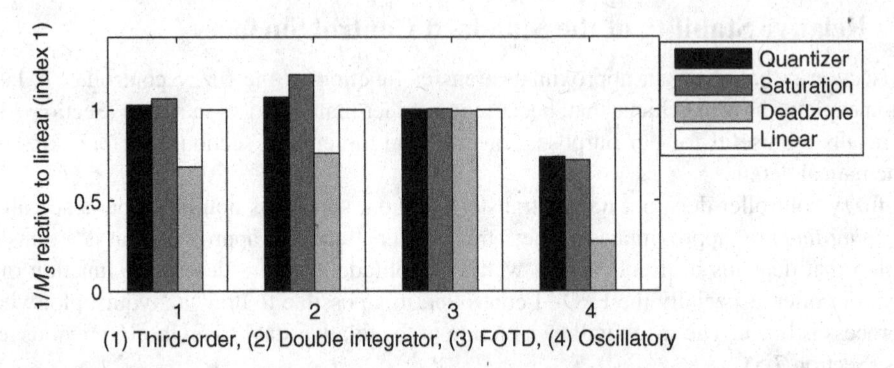

Figure 7.11 Stability margins $1/M_s$ relative to the linear surface. Each group of bars corresponds to one test process. Within each group are four test runs with different surfaces (quantizer, saturation, deadzone, linear). (figdf.m)

to the critical point, and the more robust is the system. As a first very quick assessment, the saturation surface wins over the others with the first two processes, and the deadzone surface has a slight advantage over the others with the fourth process. The following comments are based on the two figures.

- *Third-order process* (Figure 7.10a). All runs are close to or over the perimeter of our acceptable stability circle (owing to the tuning). The saturation surface is the most robust, and the deadzone surface cuts quite far into the circle. We saw earlier that the deadzone surface has a large gain margin, which can be explained by a slight dent in the trajectory before reaching the origin. The quantizer surface and the linear surface perform more or less the same. We can thus confirm the previous indication that the saturation surface is the best choice, also with respect to the stability margin $1/M_s$.
- *Double-integrator process* (Figure 7.10b). Owing to the curvature of the Nyquist curve, gain margin is in this case opposite of what is usual: the smaller the extra gain (less than one) the better. A scaling by a number less than one will quickly bring the deadzone surface into the circle, whereas the saturation surface will be the last to enter. We can thus confirm the previous indication that the saturation surface is the best choice, also with respect to the stability margin $1/M_s$.
- *Process with a long time delay* (Figure 7.10c). In this case all surfaces perform almost the same, and the stability margins are almost the same. The earlier recommendation pointed to the deadzone surface, because of a slightly better gain margin. Very close scrutiny of the Nyquist curve reveals that for higher frequencies, the deadzone surface is indeed slightly more robust. All in all, the recommendation here is uncertain, but with a slight advantage to the deadzone surface.
- *Oscillatory process* (Figure 7.10d). Again, the stability margin is more or less the same with all four surfaces. The deadzone surface is slightly better than the others, and for higher frequencies, its distance to the critical point is generally larger (but not for lower frequencies). All in all, the recommendation is uncertain, but the deadzone surface has perhaps a slight advantage. On the other hand, the deadzone causes a slow cycling around the settling point, which may be unacceptable.

It seems that the nonlinear surfaces provide only little improvement in the case of the FOTD process and the oscillatory process; other means must be used to cope with these processes. On the other hand, the nonlinear surfaces significantly affect the Nyquist diagram in the case of the third-order process and the double integrator, and a fuzzy controller likely performs better than a conventional PID controller, in these cases.

7.8 Summary

The analysis in this chapter provides a partial answer to the question of which control surface to apply, and when.

- *Higher-order process*. The saturation surface is the best choice.
- *Double-integrator process*. The saturation surface is the best choice. Note that gain margin is less than one for this process.
- *Process with a long time delay*. Undecided.
- *Process with oscillatory modes*. Undecided. At higher frequencies, deadzone has a higher damping and a lower gain compared to the linear controller, which should be an advantage. On the other hand, deadzone may cause limit cycles.

The analysis did not include step sizes of different amplitudes, and the results should therefore be taken with some caution.

7.9 Describing Functions*

A *describing function* is an approximate transfer function for a nonlinear element, and in general it depends on both frequency and amplitude. It can be viewed as an equivalent linear gain – amplitude and frequency dependent – which is optimal in the sense of a least-squares linearization of the nonlinearity. From the describing function, we achieve the *frequency response*.

A sinusoidal input to a linear system in steady state generates a sinusoidal response of the same frequency. Even though it is of the same frequency, it differs in amplitude and phase angle from the input. The difference is a function of the frequency, and the magnitude frequency response is the ratio of the output sinusoid magnitude to the input sinusoid magnitude. The phase response is the difference in phase angle between the output and the input sinusoids. The frequency response is powerful, since it describes the system performance completely.

The response of a *nonlinear* system is a composite function depending on frequency as well as amplitude. It is therefore necessary to examine the response by means of its expansion into a Fourier series, consisting of several sinusoidal terms.

The complex ratio of the *fundamental* frequency of the output to the sinusoidal input is the *describing function* for the nonlinear element.

The frequency response, particularly in the form of the Nyquist plot, provides (1) a clear criterion for a stability margin, (2) a prediction of limit cycles, and (3) a graphical visualization. These are useful analysis tools, but we are here mostly interested in the stability margin.

Certain assumptions have to be fulfilled in order to apply the describing function method (Slotine and Li 1991, Atherton 2011):

- *Single nonlinear element*. The method is developed for a single nonlinearity, and in our case the controller is the nonlinearity (Figure 7.8). The method requires that the model of the process be linear. If not, the process block must somehow be transformed into a nonlinear part and a linear part.
- *Time-invariant nonlinearity*. The Nyquist criterion applies only to linear, time-invariant systems. We shall restrict the analysis to the FPD+I controller. It is dynamic, since it depends on the derivative of the error and the integral error, but it does not change over time, and is thus time-invariant.
- *Filtering hypothesis*. The describing function is only valid for sinusoidal inputs. In the feedback loop, the output signal, which contains higher-order components, is fed back and mixed with the reference signal. It is therefore unclear whether the input to the controller is sinusoidal, but for a step on the reference the assumption is valid if the process acts like a filter. The output of the nonlinearity contains higher-order harmonics, besides the fundamental frequency, but, as an approximation, we consider only the fundamental frequency. The approximation is good when the process acts as a low-pass filter, suppressing higher-order frequencies. Many physical processes have this property (the degree of the numerator polynomial is less than the degree of the denominator polynomial). For example, if the process is a double integrator, the content of the nth (odd) harmonic fed back to the controller is $1/n^3$; the third harmonic content is thus only 3.7% of the fundamental component.

Let the input $x(t)$ to the nonlinearity be a sinusoid:

$$x(t) = A \sin(\omega t)$$

with amplitude A and angular frequency ω. The output $y(t)$ is periodic, but not sinusoidal. Its *Fourier series expansion*, however, is an infinite sum of sinusoidal signals:

$$\begin{aligned} y(t) &= \sum_{n=0}^{\infty} [a_n \cos(n\omega t) + b_n \sin(n\omega t)] \\ &= a_0 + \sum_{n=1}^{\infty} [a_n \cos(n\omega t) + b_n \sin(n\omega t)] \end{aligned}$$

The summation is over the n integer multiples of the fundamental frequency. The first term ($n = 0$), corresponding to the coefficient a_0, is the mean value of the signal $y(t)$, the next term ($n = 1$) is the *fundamental frequency*, and the remaining terms ($n > 1$) are higher *harmonics*.

By the filtering hypothesis stated earlier, we neglect all higher harmonics and consider only the fundamental,

$$y_1(t) = a_1 \cos(\omega t) + b_1 \sin(\omega t) = M \sin(\omega t + \varphi)$$

which is a sinusoid with the same frequency as the input signal and amplitude M. The ratio of the fundamental to the input is the describing function:

$$N(A, \omega) = \frac{y_1}{x}$$

Its magnitude is

$$|N(A, \omega)| = \frac{M}{A}$$

and its phase angle is

$$\angle N(A, \omega) = \angle y_1 - \angle x = \varphi$$

The Fourier series coefficients are the convolution integrals

$$a_0 = \frac{1}{2\pi} \int_0^{2\pi} x(t) \mathrm{d}(\omega t) \tag{7.13}$$

$$a_n = \frac{1}{\pi} \int_0^{2\pi} x(t) \cos(n\omega t) \mathrm{d}(\omega t) \tag{7.14}$$

$$b_n = \frac{1}{\pi} \int_0^{2\pi} x(t) \sin(n\omega t) \mathrm{d}(\omega t) \tag{7.15}$$

Intuitively, coefficient b_n is large if the signal $x(t)$ varies in phase with $\sin(n\omega t)$ and the area under the product curve is large; then the nth harmonic *in-phase component* (in phase with the input) is large. Similarly, coefficient a_n is large if $x(t)$ varies in phase with $\cos(n\omega t)$; then the nth harmonic *quadrature component* (meaning, 90° out of phase with the in-phase component) is large. Coefficient a_0 amounts to the mean value of the signal, which is often zero. Coefficient b_0 vanishes. Furthermore,

$$M = \sqrt{a_1^2 + b_1^2}$$

$$\varphi = tan^{-1} \left(\frac{a_1}{b_1} \right)$$

In general, the describing function depends on both amplitude A and frequency ω, but for static linearities it only depends on A.

7.9.1 Static Nonlinearity

Figure 7.12 shows an example of a static nonlinearity. It is a proportional controller with the gain k, but it saturates when the amplitude of the error signal is larger than a. It is called an *ideal saturation*, and is a *static nonlinearity*, because it depends on the error only, while a

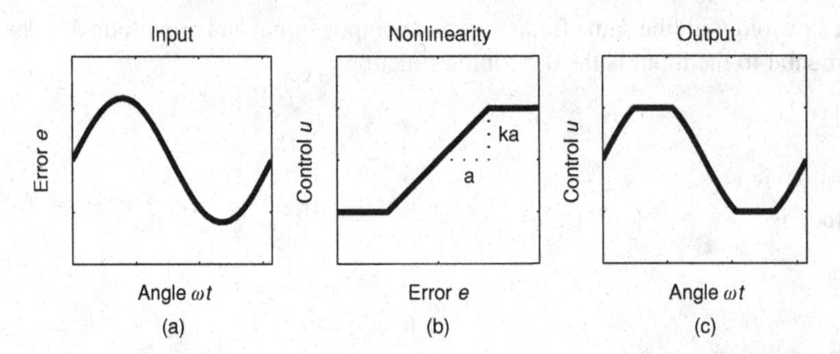

Figure 7.12 Ideal saturation. A sinusoidal input $e = Asin(\omega t)$ (a) is sent through the ideal saturation $u = \text{sat}(e)$ in (b). The saturation has the gain k in the linear region. The resulting nonlinear output is $u = \text{sat}(Asin(\omega t))$ in (c). The output (c) is the composition of the input (a) with the nonlinearity (b). (figsat.m)

dynamic nonlinearity would depend on the time derivative of its input signal. The input $e(t)$ to the nonlinearity is a sinusoid,

$$e(t) = A\sin(\omega t)$$

with amplitude A and angular frequency ω. The resulting output $u(t)$ is periodic, but not sinusoidal.

Figure 7.13 shows the *integrands* (the functions to be integrated under the integral sign) of Equations (7.14) and (7.15) of the two components of the fundamental. The in-phase integrand is positive everywhere, while the quadrature integrand is symmetric about the zero line. A general property of nonlinearities described by an odd function $f(-x) = -f(x)$ is that all the coefficients a_n will be zero. The average term a_0 is zero since the output signal u is symmetric about the zero line. What remains is to integrate the integrand related to the sin function.

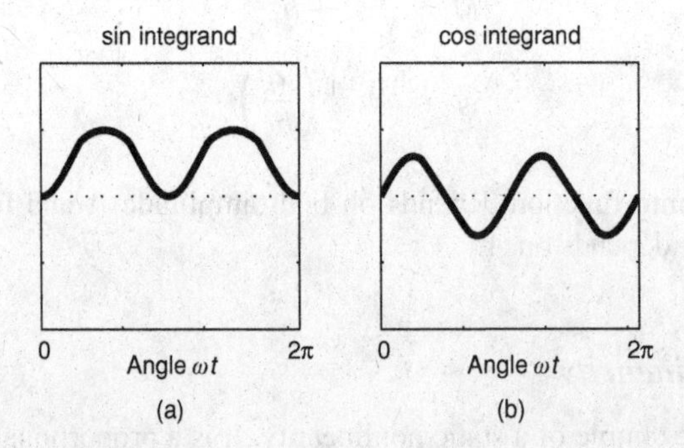

Figure 7.13 Fourier series integrands. The in-phase integrand of the fundamental frequency (a) and the quadrature integrand (b). (figsat.m)

An analytical solution exists for the ideal saturation; that is, it can be looked up in any textbook on nonlinear control (e.g. Slotine and Li 1991). The describing function for the ideal saturation is

$$N(A) = \frac{b_1}{A} = \frac{2k}{\pi}\left(\sin^{-1}\frac{a}{A} + \frac{a}{A}\sqrt{1-\frac{a^2}{A^2}}\right)$$

Notice that the describing function, in this case, does not depend on the frequency ω. A plot of $N(A)/k$ versus A/a will show (Slotine and Li 1991) that (1) $N(A) = k$ if the amplitude A is within the linear range of the saturation, (2) $N(A)$ decreases as the amplitude A increases, and (3) there is no phase shift. The first observation is expected, and can be used to check the correctness of the integration. The second is intuitively correct, since for large amplitudes, the output has constant amplitude, and therefore the gain decreases. The third is also intuitively correct, because saturation does not cause any delay.

Any signal can be expanded into a sum of sinusoidal components, and researchers have calculated analytically or charted the describing functions of many common linearities (see, for example, the tables in Atherton 1975, Šiljak 1968).

Numerical integration is another possibility. Although it is an approximation, rectangular integration can be a sufficiently good approximation if the sampling interval is small. Other, more accurate integration methods exist (trapezoidal rule, Simpson's rule, see also the function `quad` in MATLAB®).

7.9.2 Limit Cycle

A limit cycle is any isolated closed trajectory in the phase plane which is approached asymptotically both from within and without by other trajectories. A motion started on this curve will stay on it forever, circling periodically around the origin. The trajectory must be both closed, indicating periodicity, and isolated, indicating the limiting nature of the cycle. A limit cycle can be stable, unstable, or semi-stable.

Limit cycling is a unique feature of nonlinear systems, and the trajectory in the phase plane must enclose at least one equilibrium point. A limit cycle occurs if the closed loop denominator is zero:

$$1 + C(A, j\omega)P(j\omega) = 0 \tag{7.16}$$

Here $C(A, j\omega)$ is the frequency response of a nonlinear controller. When the frequency response locus $C(A, j\omega)P(j\omega)$ is plotted, any intersection with the critical point will be a solution of Equation (7.16). In general $C(A, j\omega)$ depends on both amplitude and frequency, so a family of curves can be drawn, each curve corresponding to a particular amplitude. The amplitude A_0 and the frequency ω_0 corresponding to an intersection is the amplitude and frequency of the oscillation. If n curves intersect, then the system has n possible limit cycles. The oscillation may be stable or unstable, where in this context semi-stable cases are classified as unstable. If Equation (7.16) does not have a solution, then the nonlinear system has no limit cycles.

7.10 Frequency Responses of the FPD and FPD+I Controllers*

We wish to find the describing function for the controller in Figure 7.8, but first the FPD version. It is a *dynamic nonlinearity*, since it depends on the derivative $\dot{e}$. The FPD+I also depends on the integrated error, but we will treat that case later. Whenever it is convenient, we shall use Laplace notation $s = \mathrm{d}/\mathrm{d}t$ for the differential operator.

A fuzzy PD controller is a nonlinearity $u = F(e, se)$ with two inputs, but the two inputs are dependent. We expect the describing function to be frequency dependent, owing to the differentiation, and therefore also the Fourier coefficient a_1 to be non-zero. Let the error signal be the sinusoid

$$e(t) = A\sin(\omega t)$$

then the derivative signal is

$$\dot{e}(t) = \omega A\cos(\omega t)$$

and the controller output is

$$u = F(e, se) = F(A\sin(\omega t), \omega A\cos(\omega t))$$

Now rewrite the fundamental to reflect differentiation:

$$\begin{aligned} u_1(t) &= a_1 \cos(\omega t) + b_1 \sin(\omega t) \\ &= a_1 \frac{1}{\omega}\frac{\mathrm{d}}{\mathrm{d}t}\sin(\omega t) + b_1 \sin(\omega t) \end{aligned}$$

Using the Laplace operator s, the describing function is

$$N(A, s) = \frac{u_1}{e} = \frac{b_1}{A} + \frac{a_1}{A}\frac{1}{\omega}s \tag{7.17}$$

This is a first-order transfer function that depends on ω. The first term is the in-phase component, and the second term is the quadrature component.

Its frequency response is obtained in the usual manner by the substitution $s = j\omega$, where j is the complex variable, $j^2 = -1$,

$$N(A, j\omega) = \frac{b_1}{A} + j\frac{a_1}{A} \tag{7.18}$$

A notational detail indicates the difference between the transfer function in Equation (7.17) and the frequency response in Equation (7.18): $N(A, s)$ refers to the transfer function, and $N(A, j\omega)$ refers to the frequency response. To recall, the frequency response is the steady state response of an element to a sinusoidal input; it is determined by a mapping of the transfer function by replacing s by $j\omega$.

Example 7.7 *Linear FPD*

Let us try to find the describing function for a linear fuzzy PD controller, for which we already know the result. Given a linear fuzzy controller that acts like a summation, the controller signal is

$$u = F(e, se) = (GE * e + GCE * se) * GU$$

Here GE and GCE are the input gains, and GU is the output gain. Since the controller is linear, its transfer function is

$$\frac{u}{e} = (GE + GCE * s) * GU \tag{7.19}$$

$$= GE * GU + GCE * GU * s \tag{7.20}$$

Given a sinusoidal input

$$e(t) = A\sin(\omega t)$$

the Fourier coefficient a_1 is from Equation (7.14),

$$\begin{aligned}
a_1 &= \frac{1}{\pi}\int_0^{2\pi} F(e, se)\cos(\omega t)\mathrm{d}(\omega t) \\
&= \frac{1}{\pi}\int_0^{2\pi} GE * GU * A\sin(\omega t)\cos(\omega t)\mathrm{d}(\omega t) \\
&\quad + \frac{1}{\pi}\int_0^{2\pi} GCE * GU * \omega A\cos(\omega t)\cos(\omega t)\mathrm{d}(\omega t) \\
&= 0 + GCE * GU * \omega A\frac{4}{\pi}\int_0^{\pi/2} \cos^2(\omega t)\mathrm{d}(\omega t) \\
&= GCE * GU * \omega A\frac{4}{\pi} * \left[\frac{1}{2}(\omega t) + \frac{1}{4}\sin(2\omega t)\right]_0^{\pi/2} \\
&= GCE * GU * A\omega
\end{aligned}$$

During integration ω is considered a constant scaling parameter, since we integrate one period with ωt going from 0 to 2π. The coefficient b_1 is from (7.15)

$$\begin{aligned}
b_1 &= \frac{1}{\pi}\int_0^{2\pi} F(e, se)\sin(\omega t)\mathrm{d}(\omega t) \\
&= \frac{1}{\pi}\int_0^{2\pi} GE * GU * A\sin(\omega t)\sin(\omega t)\mathrm{d}(\omega t) \\
&\quad + \frac{1}{\pi}\int_0^{2\pi} GCE * GU * \omega A\cos(\omega t)\sin(\omega t)\mathrm{d}(\omega t)
\end{aligned}$$

$$= GE * GU * A\frac{4}{\pi}\int_0^{\pi/2} \sin^2(\omega t)\mathrm{d}(\omega t) + 0$$
$$= GE * GU * A\frac{4}{\pi}\left[\frac{1}{2}(\omega t) - \frac{1}{4}\sin(2\omega t)\right]_0^{\pi/2}$$
$$= GE * GU * A$$

The describing function is thus

$$N(A, s) = \frac{b_1}{A} + \frac{a_1}{A}\frac{1}{\omega}s$$
$$= GE * GU + GCE * GUs$$

which is in agreement with Equation (7.20).

The example demonstrates how to evaluate the integrals analytically. It also shows that the describing function reduces to the linear transfer function, when the controller is linear.

7.10.1 FPD Frequency Response with a Linear Control Surface

The FPD input signals are confined to the input universes. We therefore expect a frequency response different from the crisp PD controller whenever saturation occurs. Let us examine the frequency response using numerical (rectangular) integration for convenience.

In the linear region, we derive from the previous example that the frequency response is

$$N(j\omega) = GE * GU + GCE * GU * (j\omega)$$

which is independent of amplitude. But when saturation is active, we expect the frequency response to be amplitude dependent. In terms of the Fourier coefficients, the frequency response is

$$N(A, j\omega) = \frac{b_1}{A} + j\frac{a_1}{A}$$
$$= \frac{1}{A}\frac{1}{\pi}\int_0^{2\pi} F(A\sin(\omega t), \omega A\cos(\omega t))\sin(\omega t)\mathrm{d}(\omega t)$$
$$+ j\frac{1}{A}\int_0^{2\pi} F(A\sin(\omega t), \omega A\cos(\omega t))\cos(\omega t)\mathrm{d}(\omega t)$$

With PD gains, say,

$$K_p = 1$$
$$T_d = 1$$

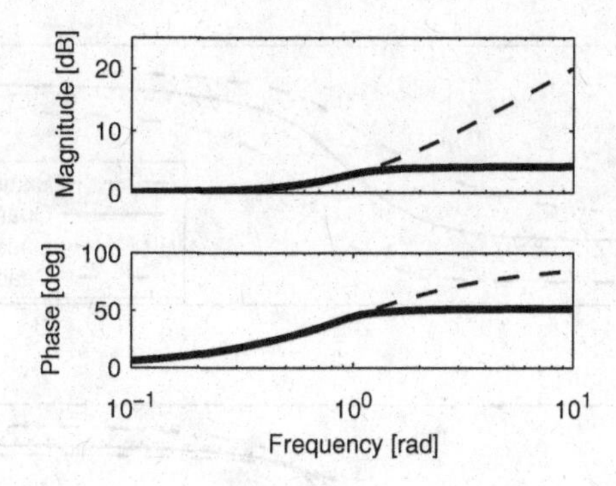

Figure 7.14 Bode diagram. Frequency response of a linear fuzzy PD controller (solid) in comparison with a crisp PD controller (dashed). The frequency axis is logarithmic and the magnitude response is in decibels (dB). (figfreqres.m)

a set of equivalent fuzzy gain settings are

$$GE = 100$$
$$GU = K_p/GE = 0.01$$
$$GCE = T_d/GU = 100$$

The plot in Figure 7.14 shows the magnitude frequency response $|N(A, j\omega)|$ and the phase frequency response $\angle N(A, j\omega)$ for the amplitude $A = 1$. In comparison with the corresponding PD controller, the FPD controller has a constant magnitude above the frequency $\omega = 1$, while the crisp PD controller magnitude is ever-increasing. The FPD controller has a maximum phase shift of about 52°, while the crisp PD has a maximum phase shift of 90° at high frequencies. This is due to the saturation of the change in error signal. At $\omega = 1$ the change in error is

$$CE = GCE * se = GCE * A\omega \cos(\omega t) = 100 * 1 * 1 * \sin(t)$$

The amplitude is 100, which is the limit of the universe, and the phase is 45° ($\omega = 1$). At higher frequencies, the saturation sets in.

The example clearly shows how the fuzzy PD cuts off at frequencies higher than a threshold determined by the input universe of change in error. The effect is useful for filtering out noise. It also implies that the fuzzy PD achieves less phase advance at higher frequencies. At low frequencies the performance is the same as the crisp PD controller.

7.10.2 FPD Frequency Response with Nonlinear Control Surfaces

We can now compare different nonlinear controllers with respect to their frequency response. Figure 7.15 shows the frequency responses of the four standard fuzzy controllers. The four

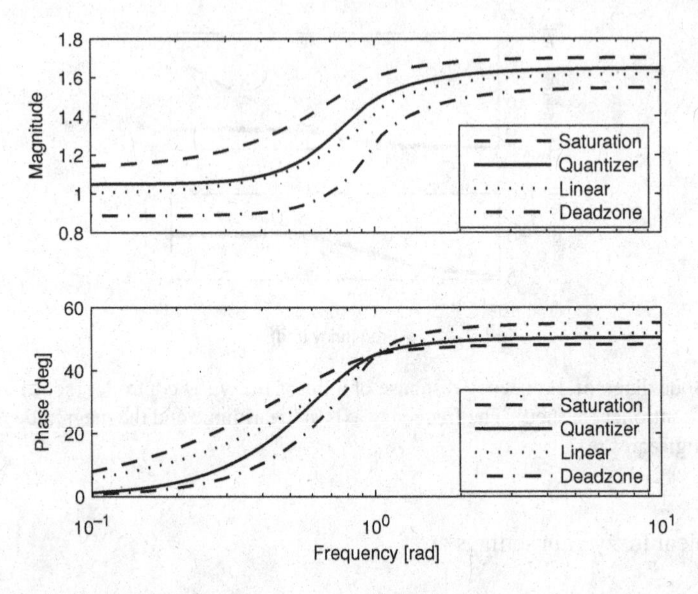

Figure 7.15 Bode diagram. Frequency response of the four standard control surfaces, one linear and three nonlinear. The frequency axis is logarithmic, but the magnitude and phase axes are linear. (figfreqnl.m)

controllers have the same fuzzy gains,

$$GE = 100$$
$$GCE = 100$$
$$GU = 0.01$$

The amplitude is $A = 1$. The four surfaces are the standard surfaces: linear, saturation, deadzone, and quantizer (Chapter 5).

Figure 7.15 shows that, overall, the saturation and the deadzone surfaces are the two extremes, with the linear and the quantizer in between. Relative to the linear surface, the saturation surface has a higher magnitude, while the deadzone surface has a lower magnitude. The phase advance for the saturation surface is high at low frequencies and low at higher frequencies, while for the deadzone surface the converse is true.

At $\omega = 1$ all surfaces have a phase advance of 45°, thus $a_1 = b_1$ ($\omega = 1$, $A = 1$ and $GE = GCE$).

Table 7.3 compares the frequency responses $N(A, j\omega)$ for the four surfaces numerically. The table reflects that the real part – corresponding to the proportional action – for each controller is independent of frequency. At high frequencies ($\omega = 10$) all surfaces have the same imaginary part, and at $\omega = 1$, all real parts are equal to the imaginary parts (45° phase).

Dividing the imaginary part by frequency in each case brings out each describing function.

The example shows that the four different surfaces have different frequency responses, and they all exhibit an upper cut-off frequency.

Table 7.3 Frequency responses at low amplitude $A = 1$, and at low, medium, and high frequency ω.

ω	PID	Linear	Saturation	Deadzone	Quantizer
0.1	$1 + j0.10$	$1 + j0.10$	$1.13 + j0.16$	$0.89 + j0.01$	$1.05 + j0.02$
1	$1 + j1$	$1 + j1$	$1.13 + j1.13$	$0.89 + j0.89$	$1.05 + j1.05$
10	$1 + j10$	$1 + j1.27$	$1.13 + j1.27$	$0.89 + j1.27$	$1.05 + j1.27$

The saturation surface provides large gain and large damping at low frequencies. It is therefore, generally speaking, suitable for a process which is unstable near the setpoint (compare the double-integrator test process). The deadzone surface provides less gain and damping at low frequencies. It is therefore, generally speaking, suitable for a process which is stable in a region around the settling point where limit cycling is acceptable. This is the case for a slow chemical process, such as a cement kiln.

7.10.3 The Fuzzy PD+I Controller

Integral action affects the frequency response in a known manner for the PID controller, and naturally we would like to include integral action in the describing function of a fuzzy controller. We shall restrict ourselves to the fuzzy PD+I controller, since it encompasses the crisp PID controller.

To begin with, let us examine the ideal crisp PID controller

$$u = K_p \left(e + \frac{1}{T_i} \int e(t)\,\mathrm{d}t + T_d \frac{\mathrm{d}e}{\mathrm{d}t} \right)$$

with the transfer function

$$C(s) = \frac{u}{e} = K_p \left(1 + \frac{1}{T_i} \frac{1}{s} + T_d s \right) \tag{7.21}$$

Its frequency response is obtained by the substitution $s = j\omega$:

$$\begin{aligned} C(j\omega) &= K_p \left(1 + \frac{1}{T_i} \frac{1}{j\omega} + T_d j\omega \right) \\ &= Kp + jK_p \left(T_d\omega - \frac{1}{T_i\omega} \right) \end{aligned}$$

In other words, the integral term contributes phase lag to the frequency response at lower frequencies.

The fuzzy PD+I controller delivers the control signal

$$U = \left(F(GE * e, GCE * se) + GIE * \frac{1}{s} e \right) * GU$$

It is a linear combination, therefore the Fourier integrals can be evaluated for each summand and the results added together. Its describing function is simply

$$N(A, s) = \frac{U_1}{e} = \frac{b_1}{A} + \frac{a_1}{A}\frac{1}{\omega}s + GIE * GU * \frac{1}{s} \tag{7.22}$$

where a_1 and b_1 are the Fourier coefficients for the fuzzy PD controller. Compared with crisp PID control (see Equation (7.21)), the gain $GIE * GU$ plays the same role as $K_p * 1/T_i$, which checks with the equivalent gains for the linear fuzzy controller. Furthermore, we can deduce that b_1/A corresponds to K_p, and $a_1/\omega A$ to $K_p T_d$.

Its frequency response is obtained by the substitution $s = j\omega$,

$$\begin{aligned} N(A, j\omega) &= \frac{b_1}{A} + \frac{a_1}{A}\frac{1}{\omega}(j\omega) + GIE * GU * \frac{1}{j\omega} \\ &= \frac{b_1}{A} + j\left(\frac{a_1}{A} - \frac{GIE * GU}{\omega}\right) \end{aligned}$$

To compare it with the frequency response of the fuzzy PD controller N_{FPD} the frequency response of the FPD+I controller can be written

$$N_{FPD+I} = N_{FPD} - j\frac{GIE * GU}{\omega} \tag{7.23}$$

The equation shows that the integral term contributes phase lag (a negative imaginary part) at lower frequencies, as expected. If we set $GIE = 0$, that is, we suppress integral action, the frequency response becomes identical to that of the FPD controller, as expected.

7.10.4 Limit Cycle

The describing function depends on amplitude, but the preceding diagrams were all based on the amplitude $A = 1$. The diagrams change with changing amplitude of the input sinusoid. Take for example the third-order process $P = 1/((s+1)^3)$ with the settings as previously (Table 7.2). However, now the amplitude is chosen to be $A = 0.45$. Figure 7.16 shows that the Nyquist diagram has moved closer to the critical point. In fact, the Nyquist curve corresponding to the deadzone surface passes through the critical point, and therefore the plot predicts a limit cycle with an amplitude of 0.45. The frequency corresponding to the frequency at the intersection of the critical point is $\omega = 1.04$ rad/s or 0.17 Hz (obtained from the program that produced the plot). The remaining three surfaces are more robust.

Figure 7.17 shows the step response with the deadzone controller, and indeed, a limit cycle is present. Closer inspection of the plot reveals that its amplitude is 0.46 (0.45 predicted) and its frequency is 0.17 Hz (0.17 Hz predicted).

The accuracy depends on whether the input to the nonlinearity is sinusoidal, or in other words, whether the filtering hypothesis is satisfied. The reference is constant, except for the initial step, therefore the error signal has the shape of the controlled output y (Figure 7.17, upper left), which in fact looks sinusoidal by inspection.

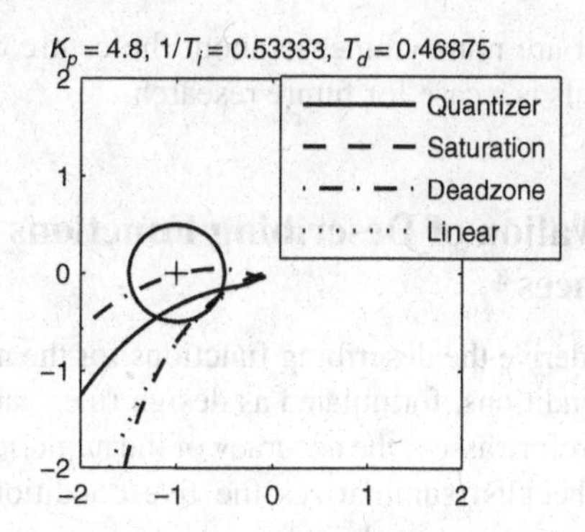

Figure 7.16 Nyquist diagram for the process $1/(s+1)^3$ with four FPD+I controllers and amplitude $A = 0.45$. (figdf.m)

The example shows that it was possible to predict a limit cycle, and that the deadzone surface in this case is less robust than the three other standard control surfaces. The predicted amplitude and frequency were quite accurate.

Since it is an approximation, the results may be inaccurate in certain cases. The applicability of the describing functions depends mainly on whether the input to the nonlinearity is really sinusoidal, during closed loop operation; this is *not* the case if, for instance, the reference varies in a non-sinusoidal manner. The results in the chapter are valid for symmetric, odd

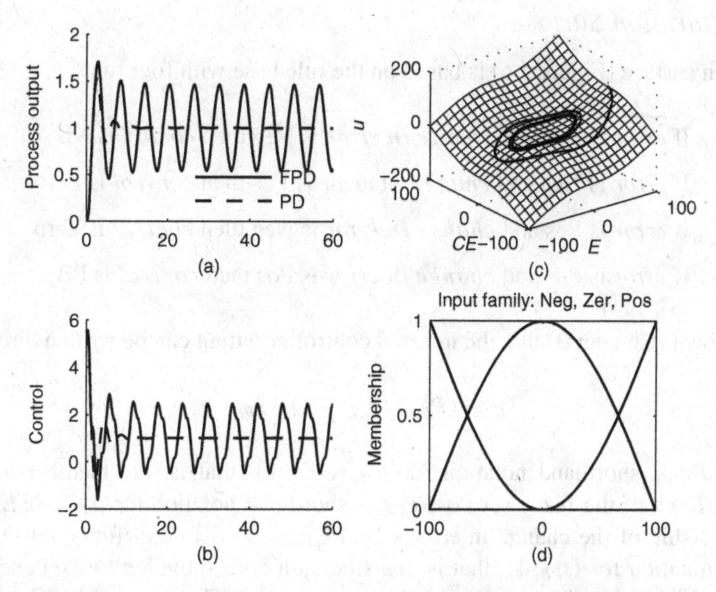

Figure 7.17 Step response (a) of the process $1/(s+1)^3$ and control signal (b). The equivalent PD controller is plotted with dashed lines. The controller is a nine-rules FPD controller with a deadzone surface (c). It is necessary to use three membership functions (d) to obtain the control surface. (figlimcyc.m)

functions only. Surfaces that are non-symmetric about the centre cause a non-zero average term in the Fourier series. This is a case for future research.

7.11 Analytical Derivation of Describing Functions for the Standard Surfaces*

It is possible to analytically derive the describing functions for the nonlinear standard control surfaces. Under the given conditions, formulated as design rules earlier, the Fourier integrals can be evaluated. We can therefore assess the accuracy of the numerically calculated frequency responses. The following checklist summarizes the five conditions, formulated as design choices, under which the four surfaces are built:

1. Use triangular premise sets that cross at membership $\mu = 0.5$.
2. Build a rule base containing all possible $\wedge-$ combinations of the premise terms.
3. Use multiplication $(*)$ for the $\wedge-$ connective.
4. Use conclusion singletons, positioned at the sum of the peak positions of the premise sets.
5. Use sum-accumulation and *COGS* defuzzification.

For the nonlinear rule bases, item 1 is relaxed such that the premise sets are smooth versions of triangular sets that cross at $\mu = 0.5$. Under these assumptions, we are able to analytically simplify the inference procedure, and this enables the analytical derivation of the describing functions.

7.11.1 Saturation Surface

The saturation surface (Figure 5.6) is based on the rule base with four rules,

If *error* is Neg and *change in error* is Neg then *control* is NB

If *error* is Neg and *change in error* is Pos then *control* is Zero

If *error* is Pos and *change in error* is Neg then *control* is Zero

If *error* is Pos and *change in error* is Pos then *control* is PB

and it was shown (Chapter 3) that the inferred controller output can be written directly as

$$u = (P_E + P_{CE} - 1) * s_{PB} \tag{7.24}$$

The symbol P_E is shorthand notation for $\mu_{Pos}(GE * e)$, that is, the membership of error signal $E = GE * e$ of the fuzzy set Pos; P_{CE} is shorthand notation for $\mu_{Pos}(GCE * se)$, that is, the membership of the change in error signal $CE = GCE * se$ of fuzzy set Pos; and s_{PB} is shorthand notation for $\langle s_{PB}, 1\rangle$, that is, the singleton corresponding to the conclusion PB. The value of s_{PB} is 200 when standard universes are used. The expression is valid even for nonlinear membership functions μ_{Pos} and μ_{Neg}, as long as $\mu_{Pos}(x) + \mu_{Neg}(x) = 1$. Therefore we can evaluate the Fourier integrals as a sum of integrals.

Take the in-phase coefficient first

$$\begin{aligned} b_1 &= \frac{1}{\pi}\int_0^{2\pi} F(e, se)\sin(\omega t)\mathrm{d}(\omega t) \\ &= \frac{1}{\pi}\int_0^{2\pi} (P_E + P_{CE} - 1) * s_{PB} * GU * \sin(\omega t)\mathrm{d}(\omega t) \\ &= \frac{1}{\pi}\int_0^{2\pi} \left(P_E - \frac{1}{2}\right) * s_{PB} * GU * \sin(\omega t)\mathrm{d}(\omega t) \\ &+ \frac{1}{\pi}\int_0^{2\pi} \left(P_{CE} - \frac{1}{2}\right) * s_{PB} * GU * \sin(\omega t)\mathrm{d}(\omega t) \end{aligned}$$

We notice in passing that the integral related to P_{CE} will vanish, owing to its cosine content. The membership function μ_{Pos} is defined as a smooth trapezoidal membership function (Chapter 2),

$$\mu_{STrapezoid}(x; a, b, c, d) = \left\{ \begin{array}{ll} 0, & x \le a \\ \frac{1}{2} + \frac{1}{2}\cos\left(\frac{x-b}{b-a}\pi\right), & a \le x \le b \\ 1, & b \le x \le c \\ \frac{1}{2} + \frac{1}{2}\cos\left(\frac{x-c}{d-c}\pi\right), & c \le x \le d \\ 0, & d \le x \end{array} \right\}, x \in \mathbb{R}$$

with the breakpoints $a = -100$, $b = 100$, $c = 100$, and $d = 100$. Using the standard universe $\mathcal{U} = [-100, 100]$, the fuzzy set Pos is thus defined by the membership function,

$$\begin{aligned} \mu_{Pos}(x) &= \frac{1}{2} + \frac{1}{2}\cos\left(\frac{x - 100}{100 - (-100)}\pi\right) \\ &= \frac{1}{2} + \frac{1}{2}\sin\left(\pi\frac{x}{200}\right) \end{aligned}$$

which is valid for the whole universe $-100 \le x \le 100$. Observe that the describing function for a sinusoidal (*harmonic*) nonlinearity can be looked up in a table.

The membership function μ_{Neg} is defined as a smooth trapezoidal membership function, with the breakpoints $a = -100$, $b = -100$, $c = -100$, and $d = 100$. Thus,

$$\begin{aligned} \mu_{Neg}(x) &= \frac{1}{2} + \frac{1}{2}\cos\left(\frac{x + 100}{100 - (-100)}\pi\right) \\ &= \frac{1}{2} - \frac{1}{2}\sin\left(\pi\frac{x}{200}\right) \end{aligned}$$

Indeed, the condition $\mu_{Pos}(x) + \mu_{Neg}(x) = 1$ is satisfied, since the sinusoidal terms cancel each other after summation.

We inject a sinusoidal input signal $e = A\sin(\omega t)$, $0 \leq \omega t \leq 2\pi$. Since $E = GE * e$, and $CE = GCE * se$, their firing strengths are $P_E = 1/2 + 1/2\sin(\pi * GE * e/200)$ and $P_{CE} = 1/2 + 1/2\sin(\pi * GCE * se/200)$. Thus,

$$\begin{aligned} b_1 &= \frac{1}{\pi}\int_0^{2\pi} \frac{1}{2}\sin\left(\frac{\pi * GE * e}{200}\right) * s_{PB} * GU * \sin(\omega t)\mathrm{d}(\omega t) \\ &+ \frac{1}{\pi}\int_0^{2\pi} \frac{1}{2}\sin\left(\frac{\pi * GCE * se}{200}\right) * s_{PB} * GU * \sin(\omega t)\mathrm{d}(\omega t) \\ &= s_{PB} * GU * \frac{1}{\pi}\int_0^{2\pi} \frac{1}{2}\sin\left(\frac{\pi * GE * e}{200}\right) * \sin(\omega t)\mathrm{d}(\omega t) \\ &= s_{PB} * GU * \frac{1}{\pi}\int_0^{2\pi} \frac{1}{2}\sin\left(\frac{\pi * GE}{200} A\sin(\omega t)\right) * \sin(\omega t)\mathrm{d}(\omega t) \end{aligned}$$

The integral occurs in the evaluation of the describing function for the harmonic nonlinearity (Atherton 2011), and we directly get

$$b_1 = s_{PB} * GU * \mathrm{J}_1\left(\frac{\pi}{2}\frac{GE}{100}A\right), \qquad -100 \leq GE * A \leq 100$$

Here J_1 is the so-called Bessel function of order 1, defined as the solution to the integral

$$\begin{aligned} \mathrm{J}_1(x) &= \frac{1}{2\pi}\int_0^{2\pi} \cos(\omega t - x\sin(\omega t))\mathrm{d}(\omega t) \\ &= \frac{1}{2\pi}\int_0^{2\pi} [\sin(\omega t)\sin(x\sin(\omega t)) + \cos(\omega t)\cos(x\sin(\omega t))]\,\mathrm{d}(\omega t) \end{aligned}$$

Its power series expansion is (see also the MATLAB® function `besselj`),

$$\mathrm{J}_1(x) = \frac{x}{2} - \frac{x^3}{2^3 * 1!2!} + \frac{x^5}{2^5 * 2!3!} - \frac{x^7}{2^7 * 3!4!} + \frac{x^9}{2^9 * 4!5!} - \cdots$$

We have only considered amplitudes within the operating region $-100 \leq GE * A \leq 100$, that is, we assume that there is no saturation in the input universe. It is possible to include larger amplitudes also (compare the introductory example of an ideal saturation), but it will complicate the expressions.

For the quadrature coefficient we get,

$$\begin{aligned} a_1 &= \frac{1}{\pi}\int_0^{2\pi} \frac{1}{2}\sin\left(\frac{\pi * GE * e}{200}\right) * s_{PB} * GU * \cos(\omega t)\mathrm{d}(\omega t) \\ &+ \frac{1}{\pi}\int_0^{2\pi} \frac{1}{2}\sin\left(\frac{\pi * GCE * se}{200}\right) * s_{PB} * GU * \cos(\omega t)\mathrm{d}(\omega t) \\ &= s_{PB} * GU * \frac{1}{\pi}\int_0^{2\pi} \frac{1}{2}\sin\left(\frac{\pi * GCE * se}{200}\right) * \cos(\omega t)\mathrm{d}(\omega t) \\ &= s_{PB} * GU * \frac{1}{\pi}\int_0^{2\pi} \frac{1}{2}\sin\left(\frac{\pi * GCE}{200}\omega A\cos(\omega t)\right) * \cos(\omega t)\mathrm{d}(\omega t) \end{aligned}$$

The integral corresponds to injecting a cosine with a frequency-dependent amplitude ωA into the harmonic nonlinearity. The solution is

$$a_1 = s_{PB} * GU * \mathrm{J}_1\left(\frac{\pi}{2}\frac{GCE}{100}\omega A\right), \qquad -100 \leq GCE * \omega A \leq 100$$

Again we have excluded, for simplicity, amplitudes where the signal saturates in the input universe.

We can now write the describing function in the form of Equation (7.17),

$$\begin{aligned} N(A, s) &= \frac{b_1}{A} + \frac{a_1}{A}\frac{1}{\omega}s \\ &= s_{PB} * GU * \frac{1}{A} * \mathrm{J}_1\left(\frac{\pi}{2}\frac{GE}{100}A\right) \\ &\quad + s_{PB} * GU * \frac{1}{\omega A} * \mathrm{J}_1\left(\frac{\pi}{2}\frac{GCE}{100}\omega A\right) * s \end{aligned}$$

valid for

$$-100 \leq GE * A \leq 100 \text{ and } -100 \leq GCE * \omega A \leq 100$$

with $\mathrm{J}_1(x)$ being the Bessel function of order 1 given above. The frequency response appears after the substitution $s = j\omega$,

$$\begin{aligned} N(A, j\omega) &= \frac{b_1}{A} + j\frac{a_1}{A} \\ &= \frac{1}{A} * s_{PB} * GU * \left[\mathrm{J}_1\left(\frac{\pi}{2}\frac{GE}{100}A\right) + j * \mathrm{J}_1\left(\frac{\pi}{2}\frac{GCE}{100}\omega A\right)\right] \end{aligned}$$

A comparison with the frequency response in Figure 7.15 – with $s_{PB} = 200$, $GU = 0.1$, $GE = 100$, $GCE = 100$, $A = 1$, and $0.1 \leq \omega \leq 1$ – showed negligible deviations, in the order of 10^{-15}.

7.11.2 *Deadzone Surface*

The deadzone surface (Figure 5.7) is based on the rule base with nine rules, and it was shown (Chapter 3) that under the above assumptions, the inferred controller output can be written directly as

$$u = \frac{1}{2}(P_E - N_E + P_{CE} - N_{CE})\, s_{PB}$$

Singleton $s_{PB} = 200$ when using standard universes. The expression is valid even for nonlinear fuzzy membership functions μ_{Pos}, μ_{Zero} and μ_{Neg}, as long as $\mu_{Pos}(x) + \mu_{Zero}(x) + \mu_{Neg}(x) = 1$.

The membership function μ_{Pos} is a smooth trapezoidal membership function with the breakpoints $a = 0$, $b = 200$, $c = 200$, and $d = 200$, and

$$\mu_{Pos}(x) = \begin{cases} 1 + \cos\left(\frac{x-200}{200-0}\pi\right) = 1 - \cos\left(\frac{x}{100}\frac{\pi}{2}\right) & \text{for } 0 \leq x \leq 100 \\ 0 & \text{for } -100 \leq x < 0 \end{cases}$$

The membership function μ_{Zero} is a smooth trapezoidal membership function with the breakpoints $a = -200$, $b = 0$, $c = 0$, and $d = 200$, and

$$\mu_{Zero}(x) = \cos\left(\frac{x}{100}\frac{\pi}{2}\right) \text{ for } -100 \leq x \leq 100$$

The membership function μ_{Neg} is a smooth trapezoidal membership function with the breakpoints $a = -200$, $b = -200$, $c = -200$, and $d = 0$.

$$\mu_{Neg}(x) = \begin{cases} 1 + \cos\left(\frac{x-(-200)}{0-(-200)}\frac{\pi}{2}\right) = 1 - \cos\left(\frac{x}{100}\frac{\pi}{2}\right) & \text{for } -100 \leq x \leq 0 \\ 0 & \text{for } 0 < x \leq 100 \end{cases}$$

It is clear that indeed $\mu_{Pos}(x) + \mu_{Zero}(x) + \mu_{Neg}(x) = 1$.

Now we can proceed with the Fourier coefficients. Take the in-phase coefficient first

$$\begin{aligned} b_1 &= \frac{1}{\pi}\int_0^{2\pi} F(e, se)\sin(\omega t)\mathrm{d}(\omega t) \\ &= \frac{1}{\pi}\int_0^{2\pi} \frac{1}{2}(P_E - N_E + P_{CE} - N_{CE})\, s_{PB} * GU * \sin(\omega t)\mathrm{d}(\omega t) \\ &= \frac{1}{\pi}\int_0^{2\pi} \frac{1}{2}(P_E - N_E) s_{PB} * GU * \sin(\omega t)\mathrm{d}(\omega t) \end{aligned}$$

The problem can be restated as that of finding the Fourier expansion for a nonlinearity of the form $\mathrm{sgn}(x)[1 - \cos(\frac{x}{100}\frac{\pi}{2})]$, where sgn is the signum function, which is -1 if x is negative and 1 otherwise. This is not straightforward, so we resort to a power series expansion. Owing to symmetry, we can shrink the integration limit to $\pi/2$, and

$$\begin{aligned} b_1 &= \frac{1}{\pi}4\int_0^{\pi/2} \frac{1}{2}P_E * s_{PB} * GU * \sin(\omega t)\mathrm{d}(\omega t) \\ &= s_{PB} * GU * \frac{2}{\pi}\int_0^{\pi/2}\left(1 - \cos\left(\frac{GE * A\sin(\omega t)}{100}\frac{\pi}{2}\right)\right) * \sin(\omega t)\mathrm{d}(\omega t) \end{aligned}$$

The cosine power series expansion is

$$\cos\varphi \approx 1 - \frac{\varphi^2}{2!} + \frac{\varphi^4}{4!} - \frac{\varphi^6}{6!} + \frac{\varphi^8}{8!} - \cdots \tag{7.25}$$

and thus the integral becomes an alternating series of powers of sinusoids of equal frequency, which makes the integration feasible, although it is an approximation. The error is arbitrarily small, however; it depends on the number of terms included in the series expansion, and the error increases as the magnitude of φ increases. For the eighth-degree polynomial shown, and the worst case where the amplitude of the sinusoidal input is maximum ($\varphi = \pm\pi/2$), the error is $2.5 * 10^{-5}$. With $\varphi = GE * A * \pi/200 * \sin(\omega t) = A_b \sin(\omega t)$ the coefficient is approximately

$$
\begin{aligned}
b_1 &\approx s_{PB} * GU * \frac{2}{\pi} \\
&\quad * \int_0^{\pi/2} \left(1 - \left(1 - \frac{\varphi^2}{2!} + \frac{\varphi^4}{4!} - \frac{\varphi^6}{6!} + \frac{\varphi^8}{8!}\right)\right) \sin(\omega t) \mathrm{d}(\omega t) \\
&= s_{PB} * GU * \frac{2}{\pi} \int_0^{\pi/2} \left(\frac{\varphi^2}{2!} + \frac{\varphi^4}{4!} - \frac{\varphi^6}{6!} + \frac{\varphi^8}{8!}\right) \sin(\omega t) \mathrm{d}(\omega t)
\end{aligned}
$$

Insertion of the value $\varphi = A_b \sin(\omega t)$ and evaluating is a complex task and prone to error, and we shall resort to a symbolic maths package (Maple, trademark of Waterloo Maple Inc). Thus,

$$
b_1 \approx s_{PB} * GU * \frac{2}{\pi} \left(\frac{1}{3} A_b^2 - \frac{1}{45} A_b^4 + \frac{1}{1575} A_b^6 - \frac{1}{99\,225} A_b^8\right),
$$

with

$$
A_b = \frac{GE * A}{100} \frac{\pi}{2}
$$

and

$$
-100 \leq GE * A \leq 100
$$

We have added the last constraint to avoid considering saturation in the input universe, for simplicity.

The quadrature coefficient is

$$
\begin{aligned}
a_1 &= \frac{2}{\pi} \int_0^{2\pi} F(e, se) \cos(\omega t) \mathrm{d}(\omega t) \\
&= \frac{1}{\pi} \int_0^{2\pi} \frac{1}{2} (P_E - N_E + P_{CE} - N_{CE}) * s_{PB} * GU * \cos(\omega t) \mathrm{d}(\omega t) \\
&= \frac{1}{\pi} \int_0^{2\pi} \frac{1}{2} (P_{CE} - N_{CE}) * s_{PB} * GU * \cos(\omega t) \mathrm{d}(\omega t) \\
&= s_{PB} * GU * \frac{2}{\pi} \int_0^{\pi/2} P_{CE} \cos(\omega t) \mathrm{d}(\omega t)
\end{aligned}
$$

The membership function P_{CE} is the same as P_E, but the input to the nonlinearity is now $\varphi = GCE * se = GCE * \pi/200 * \omega A \cos(\omega t) = A_a \cos(\omega t)$. Therefore we can approximate

$$a_1 \approx s_{PB} * GU * \frac{2}{\pi} \int_0^{\pi/2} \left(\frac{\varphi^2}{2!} + \frac{\varphi^4}{4!} - \frac{\varphi^6}{6!} + \frac{\varphi^8}{8!} \right) \cos(\omega t) \mathrm{d}(\omega t)$$

Using the symbolic maths package, with $\varphi = A_a \cos(\omega t)$, we have

$$a_1 \approx s_{PB} * GU * \frac{1}{\pi} \left(\frac{1}{3} A_a^2 - \frac{1}{45} A_a^4 + \frac{1}{1575} A_a^6 - \frac{1}{99\,225} A_a^8 \right),$$

with

$$A_a = \frac{GCE * \omega A}{100} \frac{\pi}{2}$$

and

$$-100 \leq GCE * \omega A \leq 100$$

We have added the last constraint to avoid considering saturation in the input universe, for simplicity.

Summarizing, we can write the describing function in the form of Equation (7.17),

$$N(A, s) = \frac{b_1}{A} + \frac{a_1}{A} \frac{1}{\omega} s \tag{7.26}$$

$$= \frac{1}{A} * s_{PB} * GU * \frac{2}{\pi} * S(A_b) \tag{7.27}$$

$$+ \frac{1}{\omega A} * s_{PB} * GU * \frac{2}{\pi} * S(A_a) * s \tag{7.28}$$

with

$$S(z) = \frac{z^2}{3} - \frac{z^4}{45} + \frac{z^6}{1575} - \frac{z^8}{99\,225}$$

$$A_b = \frac{GE * A}{100} \frac{\pi}{2}$$

$$A_a = \frac{GCE * \omega A}{100} \frac{\pi}{2}$$

The describing function is valid for

$$-100 \leq GE * A \leq 100 \text{ and } -100 \leq GCE * \omega A \leq 100$$

The frequency response appears after the substitution $s = j\omega$,

$$\begin{aligned} N(A, j\omega) &= \frac{b_1}{A} + j\frac{a_1}{A} \\ &= \frac{1}{A} * s_{PB} * GU * \frac{2}{\pi} * (S(A_b) + jS(A_a)) \end{aligned}$$

A comparison with the frequency response in Figure 7.15 – with $s_{PB} = 200$, $GU = 0.1$, $GE = 100$, $GCE = 100$, $A = 1$, and $0.1 \le \omega \le 1$ – showed a maximum deviation of 0.0013%, for both magnitude and phase.

7.11.3 Quantizer Surface

The quantizer surface (Figure 5.7) is based on the rule base with nine rules, and again we exploit the fact that the inferred controller output can be written directly as

$$u = \frac{1}{2}(P_E - N_E + P_{CE} - N_{CE})s_{PB}$$

Singleton $s_{PB} = 200$ when using standard universes. The expression is valid even for nonlinear fuzzy membership functions μ_{Pos}, μ_{Zero} and μ_{Neg}, as long as $\mu_{Pos}(x) + \mu_{Zero}(x) + \mu_{Neg}(x) = 1$.

The membership function μ_{Pos} is a smooth trapezoidal membership function with the breakpoints $a = 0$, $b = 100$, $c = 100$, and:

$$\mu_{Pos}(x) = \begin{cases} \frac{1}{2} + \frac{1}{2}\cos\left(\frac{x-100}{100-0}\pi\right) = \frac{1}{2} - \frac{1}{2}\cos\left(\frac{x}{100}\pi\right) = \sin^2\left(\frac{\pi}{2}\frac{x}{100}\right) & \text{for } 0 \le x \le 100 \\ 0 & \text{for } -100 \le x < 0 \end{cases}$$

The double angle relationship $\cos 2\alpha = 1 - \sin^2\alpha$ was applied in the above derivation. The membership function μ_{Zero} is a smooth trapezoidal membership function with the breakpoints $a = -100$, $b = 0$, $c = 0$, and $d = 100$,

$$\mu_{Zero}(x) = \frac{1}{2} + \frac{1}{2}\cos\left(\frac{x}{100}\pi\right) = \cos^2\left(\frac{\pi}{2}\frac{x}{100}\right) \quad \text{for } -100 \le x \le 100$$

Another double angle relationship $\cos 2\alpha = 2\cos^2\alpha - 1$ was applied above. The membership function μ_{Neg} is a smooth trapezoidal membership function with the breakpoints $a = -100$, $b = -100$, $c = -100$, and $d = 0$,

$$\mu_{Neg}(x) = \begin{cases} \frac{1}{2} + \frac{1}{2}\cos\left(\frac{x-(-100)}{0-(-100)}\pi\right) = \frac{1}{2} - \frac{1}{2}\cos\left(\frac{x}{100}\pi\right) = \sin^2\left(\frac{\pi}{2}\frac{x}{100}\right) & \text{for } -100 \le x \le 0 \\ 0 & \text{for } 0 < x \le 100 \end{cases}$$

By the Pythagorean relation $\sin^2\alpha + \cos^2\alpha = 1$ it is clear that indeed $\mu_{Pos}(x) + \mu_{Zero}(x) + \mu_{Neg}(x) = 1$.

Now we can proceed with the Fourier coefficients. Take the in-phase coefficient first:

$$\begin{aligned}
b_1 &= \frac{1}{\pi}\int_0^{2\pi} F(e, se)\sin(\omega t)\mathrm{d}(\omega t) \\
&= \frac{1}{\pi}\int_0^{2\pi} \frac{1}{2}(P_E - N_E + P_{CE} - N_{CE})\, s_{PB} * GU * \sin(\omega t)\mathrm{d}(\omega t) \\
&= \frac{1}{\pi}\int_0^{2\pi} \frac{1}{2}(P_E - N_E) s_{PB} * GU * \sin(\omega t)\mathrm{d}(\omega t)
\end{aligned}$$

The problem can be restated as that of finding the Fourier expansion for a nonlinearity of the form $\mathrm{sgn}(x)[\frac{1}{2} - \frac{1}{2}\cos(\frac{x}{100}\pi)]$, where sgn is the signum function. This is not straightforward, and therefore we resort to power series expansion. Owing to symmetry, we can shrink the integration limit to $\pi/2$, and

$$\begin{aligned}
b_1 &= \frac{1}{\pi}4\int_0^{\pi/2} \frac{1}{2} P_E * s_{PB} * GU * \sin(\omega t)\mathrm{d}(\omega t) \\
&= s_{PB} * GU * \frac{2}{\pi}\int_0^{\pi/2} \left(\frac{1}{2} - \frac{1}{2}\cos\left(\frac{GE * A\sin(\omega t)}{100}\pi\right)\right) * \sin(\omega t)\mathrm{d}(\omega t)
\end{aligned}$$

Using the cosine power series expansion again (Equation (7.25)) and with $\varphi = GE * A\sin(\omega t)\pi/100 = A_b\sin(\omega t)$, the coefficient is approximately

$$\begin{aligned}
b_1 &\approx s_{PB} * GU * \frac{2}{\pi} \\
&\quad * \int_0^{\pi/2} \left(\frac{1}{2} - \frac{1}{2}\left(1 - \frac{\varphi^2}{2!} + \frac{\varphi^4}{4!} - \frac{\varphi^6}{6!} + \frac{\varphi^8}{8!}\right)\right) \sin(\omega t)\mathrm{d}(\omega t) \\
&= s_{PB} * GU * \frac{1}{\pi}\int_0^{\pi/2} \left(\frac{\varphi^2}{2!} - \frac{\varphi^4}{4!} + \frac{\varphi^6}{6!} - \frac{\varphi^8}{8!}\right) \sin(\omega t)\mathrm{d}(\omega t)
\end{aligned}$$

We have evaluated the integral earlier, and get

$$b_1 \approx s_{PB} * GU * \frac{1}{\pi}\left(\frac{1}{3}A_b^2 - \frac{1}{45}A_b^4 + \frac{1}{1575}A_b^6 - \frac{1}{99\,225}A_b^8\right)$$

with

$$A_b = \frac{GE * A}{100}\pi$$

and

$$-100 \le GE * A \le 100$$

We have added the last constraint to avoid considering saturation in the input universe, for simplicity.

The quadrature coefficient is

$$\begin{aligned} a_1 &= \frac{1}{\pi}\int_0^{2\pi} F(e, se)\cos(\omega t)\mathrm{d}(\omega t) \\ &= \frac{1}{\pi}\int_0^{2\pi} \frac{1}{2}(P_E - N_E + P_{CE} - N_{CE}) * s_{PB} * GU * \cos(\omega t)\mathrm{d}(\omega t) \\ &= \frac{1}{\pi}\int_0^{2\pi} \frac{1}{2}(P_{CE} - N_{CE}) * s_{PB} * GU * \cos(\omega t)\mathrm{d}(\omega t) \\ &= s_{PB} * GU * \frac{2}{\pi}\int_0^{\pi/2} P_{CE}\cos(\omega t)\mathrm{d}(\omega t) \end{aligned}$$

The membership function P_{CE} is the same as P_E, but the input to the nonlinearity is now $\varphi = GCE * \omega A\cos(\omega t)\pi/100 = A_a\cos(\omega t)$. Therefore, we can approximate

$$a_1 \approx s_{PB} * GU * \frac{1}{\pi}\int_0^{\pi/2}\left(\frac{\varphi^2}{2!} - \frac{\varphi^4}{4!} + \frac{\varphi^6}{6!} - \frac{\varphi^8}{8!}\right)\cos(\omega t)\mathrm{d}(\omega t)$$

with $x = \omega A_a\cos(\omega t)$. Using the symbolic maths package,

$$a_1 \approx s_{PB} * GU * \frac{1}{\pi}\left(\frac{1}{3}A_a^2 - \frac{1}{45}A_a^4 + \frac{1}{1575}A_a^6 - \frac{1}{99\,225}A_a^8\right),$$

with

$$A_a = \frac{GCE * \omega A}{100}\pi$$

and

$$-100 \leq GCE * \omega A \leq 100$$

We have added the last constraint to avoid considering saturation in the input universe, for simplicity.

Summarizing, the describing function is,

$$N(A, s) = \frac{b_1}{A} + \frac{a_1}{A}\frac{1}{\omega}s \tag{7.29}$$

$$= \frac{1}{A} * s_{PB} * GU * \frac{1}{\pi} * S(A_b) \tag{7.30}$$

$$+ \frac{1}{\omega A} * s_{PB} * GU * \frac{1}{\pi} * S(A_a) * s \tag{7.31}$$

with

$$S(z) = \frac{z^2}{3} - \frac{z^4}{45} + \frac{z^6}{1575} - \frac{z^8}{99\,225}$$

$$A_b = \frac{GE * A}{100}\pi$$

$$A_a = \frac{GCE * \omega A}{100}\pi$$

The describing function is valid for

$$-100 \leq GE * A \leq 100 \text{ and } -100 \leq GCE * \omega A \leq 100$$

The frequency response appears after the substitution $s = j\omega$,

$$N(A, j\omega) = \frac{b_1}{A} + j\frac{a_1}{A}$$

$$= \frac{1}{A} * s_{PB} * GU * \frac{1}{\pi} * (S(A_b) + jS(A_a))$$

A comparison with the frequency response in Figure 7.15 – with $s_{PB} = 200$, $GU = 0.1$, $GE = 100$, $GCE = 100$, $A = 1$, and $0.1 \leq \omega \leq 1$ – showed a maximum deviation of 0.54% regarding both magnitude and phase.

Comparing the describing function of the quantizer with that of the deadzone (see Equations (7.26) and (7.29)), we find the expressions are identical. But the constants A_a, A_b for the quantizer are larger than the ones for the deadzone by a factor of two.

7.12 Notes and References*

Aracil and Gordillo (2004) show in an article how to apply describing functions to a saturation type of fuzzy PD controller, and Gordillo *et al.* (1997) to a quantizer type of controller. The article by Cuesta *et al.* (1999) is more rigorous.

For an easy introduction to describing functions, see the book by Slotine and Li (1991). For an impressive in-depth exposition, see Atherton (1975, 1982), and also his e-book (2011), which is a free downloadable version. Another option is the book by Gelb and Vander Velde (1968) available on-line. Treatments of dynamic nonlinearities are sparse, but Šiljak (1968, p. 126) mentions functions of the form $F(x, \dot{x})$, and Gelb and Vander Velde show the proportional plus derivative form of the describing function (1968, Ch. 2, Eq. 2.2-30).

Kickert and Mamdani (1978) attempted to apply describing functions on a fuzzy controller seen as a multi-level relay.

8

Fuzzy Gain Scheduling Control

A fuzzy gain scheduling controller changes the gains of a PID controller during operation. This is an advantage if the operating range is large. Several PID controllers are designed in strategic design points and joined together by smooth interpolation. A special case is to compensate a nonlinearity by its inverse, which can be built using fuzzy rules. The approach is demonstrated on an autopilot example with a large reference step to an unstable equilibrium. Fuzzy gain scheduling of a PID controller is an alternative to fuzzy PID controllers in the case of large operating ranges.

In order to cope with large variations, a PID controller can change gains depending on the operating point. For example, in level control of a surge tank, where an approximate level is sufficient, a small gain can be used when the level is close to the setpoint, and a large gain when the level is far from the setpoint. This may save energy, because the control effort is less compared to a linear controller with a high gain.

It is an example of *gain scheduling*, which is a control method that continuously changes the controller gains depending on the state of the process. It is a nonlinear control method of a special type, because it changes the parameters of a linear controller in a nonlinear manner. The changes occur according to the value of a *scheduling variable*. The scheduling variable is a measurement that changes as the process changes operating point. It could be, for example, the reference signal, the controlled output, or a performance index.

The four steps below describe the design procedure (after Rugh and Shamma 2000).

1. Develop a linear process model for a set of operating points.
2. Design a linear controller for each operating point.
3. Develop a schedule for the controller parameters, such that the transition from one controller to the other is smooth (bump-less).
4. Assess performance and relative stability, most likely by simulation.

The transition between the chosen operating points is particularly important. It is a major advantage of gain-scheduling that linear design procedures apply. Furthermore, the controller gains change quickly in contrast to the incremental updates that occur in adaptive systems.

Foundations of Fuzzy Control: A Practical Approach, Second Edition. Jan Jantzen.

8.1 Point Designs and Interpolation

In order to apply linear design methods, it is necessary to *linearize* the nonlinear process. The linearization is a Taylor series approximation of the process in a steady-state *operating point*, that is, a point where the process is at rest. Take, for example, the following first-order dynamical system (Bequette 2003):

$$\frac{\mathrm{d}x}{\mathrm{d}t} = f(x) \tag{8.1}$$

The first-order Taylor series expansion about the operating point x_0 is a simple approximation of the nonlinear function $f(x)$, that is,

$$f(x) \cong f(x_0) + \left.\frac{\partial f(x)}{\partial x}\right|_{x_0} (x - x_0) + \text{neglected higher order terms}$$

Since the operating point is a point of steady state, $f(x_0) = 0$ according to Equation (8.1). Furthermore, since x_0 is constant we can rewrite Equation (8.1) as follows:

$$\frac{\mathrm{d}(x - x_0)}{\mathrm{d}t} \cong \left.\frac{\partial f(x)}{\partial x}\right|_{x_0} (x - x_0)$$

or, by a variable substitution,

$$\frac{\mathrm{d}x'}{\mathrm{d}t} \cong a'x' \tag{8.2}$$

Here, $x' = x - x_0$ represents a *deviation variable*, and $a' = \partial f/\partial x|_{x_0}$ is the slope of the function at the operating point.

The new Equation (8.2) has the same form as Equation (8.1), but it is a relative description of the local dynamics in terms of the deviation from the operating point. If the deviation is sufficiently small, the approximation will be good. Figure 8.1 shows an operating point and its local first-order approximation. A change of operating point is associated with the *static gain*

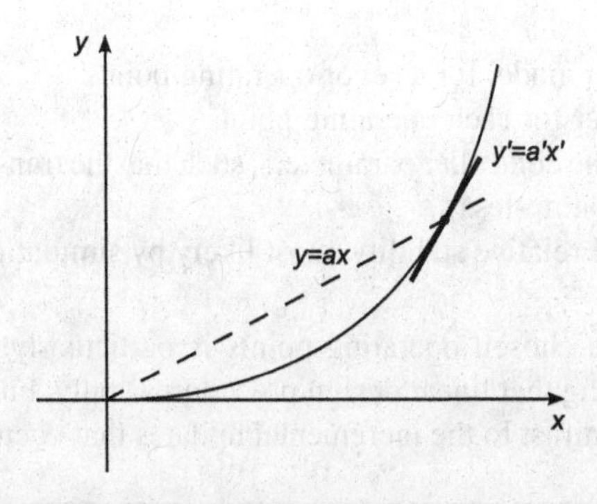

Figure 8.1 Operating point and a first-order deviation model. The gain a, related to large variations of the reference, is different from the gain a', related to small variations from the operating point.

a. Small deviations from the operating point, are associated with a *dynamic gain*, which is a'. The two gains are geometrically related by the shape of the curve. Controller settings that are optimal for a change of operating point (a reference step), may thus be less than optimal for regulating variations around the operating point (disturbance rejection), and vice versa.

With respect to gain scheduling, the linearization describes the local behaviour of the nonlinear process about the corresponding operating point, and this is defined by the fixed values of the scheduling variable σ. A set of so-called *point designs* of the controller are performed at the selected operating points. At each design point the controller and external signals are *trimmed* to supply a constant control signal to the process such that it remains steady. A family of linear controllers is computed, online, by interpolating the point designs.

Even if the point designs are stable, the process may be unstable in other points. Furthermore, the continuous change of gains introduces an additional feedback loop that affects the performance and relative stability. Thus, the scheduling variable should, preferably, be a slowly varying parameter in order to minimize the risk of upsetting the stability.

8.2 Fuzzy Gain Scheduling

A fuzzy controller can interpolate between crisp PID controllers. The resulting control signal is a weighted average depending on the overlap of the premise membership functions. The following is a sketch of such a controller:

$$\text{If } \sigma \text{ is Small then } u = K_p(\sigma_1)\left(e + \frac{1}{T_i(\sigma_1)}\int edt + T_d(\sigma_1)\dot{e}\right)$$

$$\text{If } \sigma \text{ is Large then } u = K_p(\sigma_2)\left(e + \frac{1}{T_i(\sigma_2)}\int edt + T_d(\sigma_2)\dot{e}\right)$$

The variable σ is the scheduling variable, which depends on the state of the closed-loop system. The values σ_1 and σ_2 are instances of σ that correspond to two operating points. A PID controller has been tuned at each of these operating points, resulting in six PID gain settings – two pairs of three – that depend on σ_1 and σ_2.

The rule base is of the Sugeno type (Section 3.2), and the premise side interpolates between the two PID controllers. The control signal is the activation weighted sum of the two controllers. When the membership functions of Small and Large overlap in such a way that the sum is always one, then the weights are normalized, and the control signal is always maximally defined with fuzzy membership 1.

Compared to the fuzzy PID controllers from the earlier chapters, this kind of controller avoids the universes around the controller input variables, and the fuzzy gains are obsolete. Contrary to the fuzzy PID controllers, the gain scheduling type uses error inputs that are unlimited over the whole region of operation, just like the crisp PID controller. The membership functions on the premise side must be defined on universes, nevertheless, but these can be defined at will. The standard universes are obsolete.

It is the gain factors which are interpolated, and all three PID gains can be interpolated in parallel. The interpolation can be linear as a special case; otherwise the fuzzy membership functions on the premise side control a nonlinear interpolation as shown previously (Chapter 3, Figure 3.5).

Example 8.1 *Gain scheduling of valve*

A nonlinear control valve is the actuator in a PI control loop that controls a linear process with the transfer function $P(s) = 10/(s+1)^3$ (Example 5.1). The model of the valve is

$$f(x) = x^4, 0 \leq x \leq 1$$

The previous PI controller had the gains $K_p = 0.10$, $T_i = 1$. The controller is tuned to give a good response in the lower region, but the system is unstable in the upper region (Figure 5.2 shows the response to three consecutive steps (2, 4, 6) of the reference). Can a fuzzy gain scheduler stabilize the system?

► *Solution*

The gain of the control valve increases rapidly with increasing x, and it would help to decrease the proportional gain in the upper region of operation. One strategy is to keep the static gain of the controller–valve combination the same. Thus the relative gain margin is the same in the design points.

The static gain of the process is 10, and thus the valve output must attain a value of one tenth of the reference when it is in trim. The reference is the scheduling variable. We choose two design points $f(x_1) = 0.2$ and $f(x_2) = 0.6$ corresponding to the lowest and the highest value of the steady-state flow from the valve.

Figure 8.2 shows the response. It is stable at all three levels of the reference, and the three step responses are somewhat oscillatory, but more or less similar. Notice that the control signal has a distinct downward dip when the reference changes, especially at the time t = 100. As soon as the reference steps up, the scheduler changes the proportional gain to a lower value in anticipation of the future steady state. It follows that the dynamics of the system are affected by the scheduler.

Figure 8.3 shows the schedule that changes the proportional gain. It is a soft z-curve as a result of the fuzzy membership functions. The second step at t = 50, where the process output

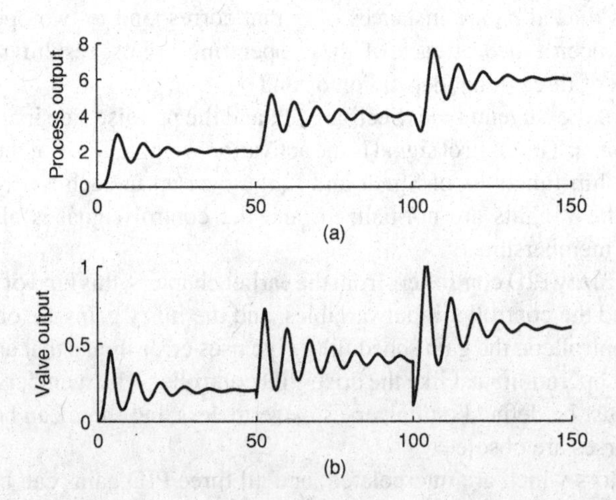

Figure 8.2 Fuzzy gain scheduling. The controller is a PI controller with $T_i = 1$, and the proportional gain K_p changes from 0.100 to 0.0439 during the three steps in the reference (a). The control signal (b) operates within the saturation limits of the valve. The process is $P(s) = 1/(s+1)^3$. (figvalvegain.m)

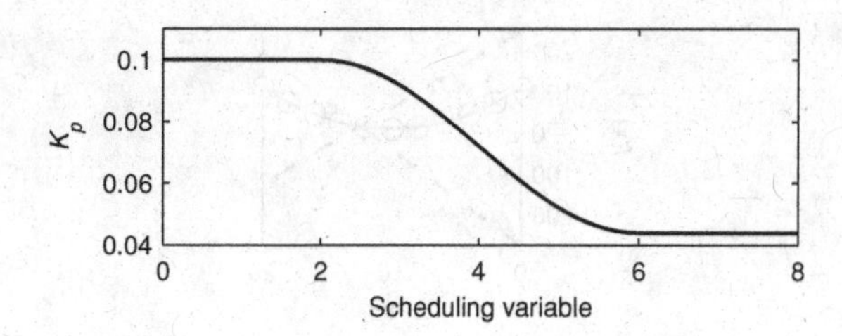

Figure 8.3 Fuzzy interpolation between the high and the low value of K_p. The reference is the scheduling variable. (figvalvegain.m)

settles at the value 4, corresponds to the middle of the sloping section of the z-curve. The scheduler thus covered three operating points based on just two design points.

The fuzzy gain scheduler consists of two rules, one for each design point, as follows:

$$\textit{If reference is Low then } K_p = 0.10$$

$$\textit{If reference is High then } K_p = 0.0439$$

The rule base interpolates between the gains according to the definition of the membership functions, namely: Low is a z-curve stretching from 2 to 6, and High $= 1 -$ *Low. The schedule in Figure 8.3 would have been piecewise linear, had the membership functions been linear.*

The value of K_p *on the conclusion side of the last rule was chosen to keep the open-loop static gain the same. The static gain of the process is 10. The valve contributes a factor* $g_1 = f(x_1)/x_1$ *to the static gain. In the lower operating point* $f(x_1) = 2/10$ *and therefore* $x_1 = \sqrt[4]{0.2}$. *In a similar manner we find* g_2 *related to the higher operating point. The guideline is to keep* $K_p(1)\,g_1 = K_p(2)\,g_2$, *and from this relation we find that* $K_p(2) = 0.0439$.

The dynamic gain changes, however, and that explains why the responses become slightly more oscillatory as the reference steps up.

8.3 Fuzzy Compensator Design

A previous example achieved a good result by compensating a nonlinear element by a function close to its inverse (Example 5.6). A fuzzy rule base can be made to approximate such a curve by *function approximation* techniques. An optimization routine adjusts the parameters of the rule base until the approximation is the closest possible according to a selected performance measure.

In the case of straight lines, the fuzzy rule base interpolates such that the *interpolant* (the fitted curve) is a smooth curve that falls between or on the lines that are to be approximated; the function approximation of the inverse of the control valve characteristic is an example (Figure 5.14).

Example 8.2 *Function approximation in MATLAB®*

Given a simple set of 21 data points that lie on a known parabola $f(x) = (x - 10)^2 - 10$, *what is the result of letting the MATLAB® Fuzzy Logic Toolbox find a fuzzy rule base that approximates the function?*

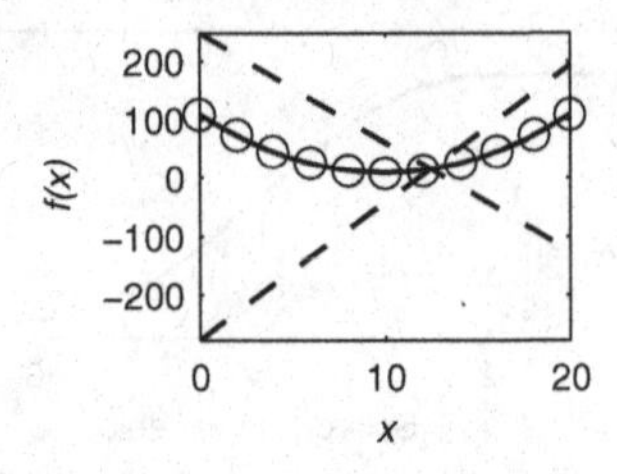

Figure 8.4 Function approximation by the Fuzzy Logic Toolbox. The toolbox finds the interpolant (solid line) given the data points (circles). The two straight lines (dashed) control the interpolant. (figparab.m)

► *Solution*

The toolbox finds the membership functions and the conclusion side of a first-order Sugeno rule base. That is, straight lines control the interpolant, and the toolbox iterates until a performance criterion is minimized. We specify to the toolbox that the rule base should consist of two rules, as sketched below:

If x is Low then Line 1

If x is High then Line 2

Figure 8.4 shows the result. The interpolant passes through all data points, and the approximation is perfect. The result is based on adjusted membership functions, and the toolbox did not 'know' that the data points were generated from a parabola. The toolbox adjusts the position and the slope of the lines in order to find the best fit to the data points. Notice that the lines cross at a point somewhat to the right of the middle; one would perhaps expect the crossing to be exactly in the middle owing to symmetry.

The toolbox also adjusts the membership functions on the premise side of the rules; see Figure 8.5. The book by Jang, Sun and Mizutani (1997) describes the details of the optimization routine.

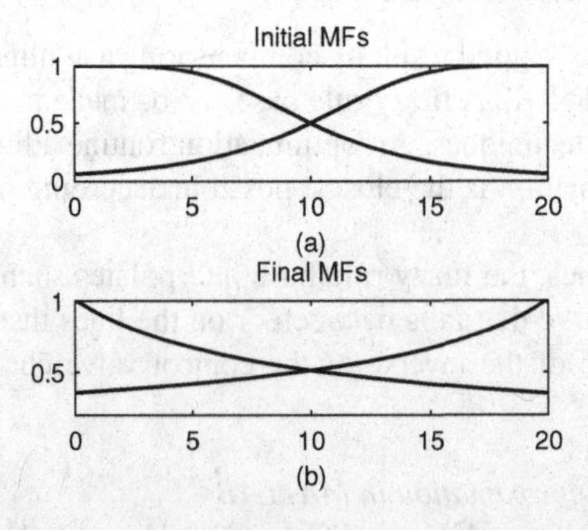

Figure 8.5 Adjusted membership functions. The toolbox starts with two default membership functions (a), and it adjusts the parameters until it finds the membership functions that provide the best performance (b). (figparab.m)

The example shows that an automatic function approximation may result in straight lines that are difficult to interpret. The optimizer is 'blind' in the sense that it adjusts the parameters to provide the best fit according to some performance criterion. The end result is difficult to interpret, unless some constraints are added.

We would rather prefer that each line represents the local behaviour as a *tangent* or *asymptote* (tangent at infinity) to the nonlinear function. According to the design of Sugeno rule bases, two rules interpolate between two lines in the following manner

$$y = \frac{\alpha_1 * l_1 + \alpha_2 * l_2}{\alpha_1 + \alpha_2}$$

where α_i is the firing strength of rule i, and l_i $(i = 1, 2)$ is just a convenient symbol for a line. If we include the constraint that membership functions sum to one over the whole universe, the result is $y = \alpha_1 * l_1 + (\alpha_1 - 1) * l_2$. Each line can be described by two adjustable parameters, and an optimization routine would then have to adjust five parameters.

The fewer parameters to adjust the better, and if we limit our view to the region of interest, it is possible to simplify the interpolation. Regard each line segment as a vector $\mathbf{v}_i$ $(i = 1, 2)$, starting from the intersection of the lines. The two vectors can be thought of as a *basis* for a coordinate system. Now relax the normalization constraint, such that the new interpolant is defined by $y = \alpha_1\mathbf{v}_1 + \alpha_2\mathbf{v}_2$. As α_1 and α_2 change in magnitude, the result will be a blend of the two vectors.

To simplify the interpolation even more, apply two linear (triangular) fuzzy membership functions, where each function spans the whole universe, but make these adjustable by means of an exponent c. When $c > 1$, this is equivalent to applying a linguistic hedge to the fuzzy set in the form of a *concentration* operator, which is described in an earlier section (Section 2.2.2). The rule base that performs the interpolation is now defined as:

$$\begin{aligned} &\text{If } x \text{ is Low}^c \text{ then } \mathbf{p} = \mathbf{v}_1 \\ &\text{If } x \text{ is High}^c \text{ then } \mathbf{p} = \mathbf{v}_2 \end{aligned} \tag{8.3}$$

The inference engine uses *-activation, sum-accumulation, and no defuzzification. Figure 8.6 shows the result with concentration $c = 2$ and the corresponding membership functions. The interpolant is attracted first to $\mathbf{v}_1$ and then $\mathbf{v}_2$ as x runs from left to right through the universe. On the left side of the universe, the membership function Low dominates. The location of the origin of the basis vectors is on the line $x = 0$. When moving towards the right, the membership function High gradually dominates, and the interpolant thus reflects the longer length of $\mathbf{v}_2$ compared to $\mathbf{v}_1$.

Higher concentration c makes the interpolant come closer to both vectors, and c thus controls the amount of attraction to the basis vectors. The membership functions are defined as $High = \mu_{Trapezoid}(x; -100, 100, 100, 100)$ and $Low = 1 - High$; see Equation (2.2).

Algorithm 8.1 summarizes the procedure. Once the two basis vectors have been chosen, then only one parameter, namely c, needs to be adjusted. The scope of the algorithm is a single turn of a curve, and it must be repeated in order to splice segments together to form a more complex curve. In that case all basis vectors should be joined into a path of control points.

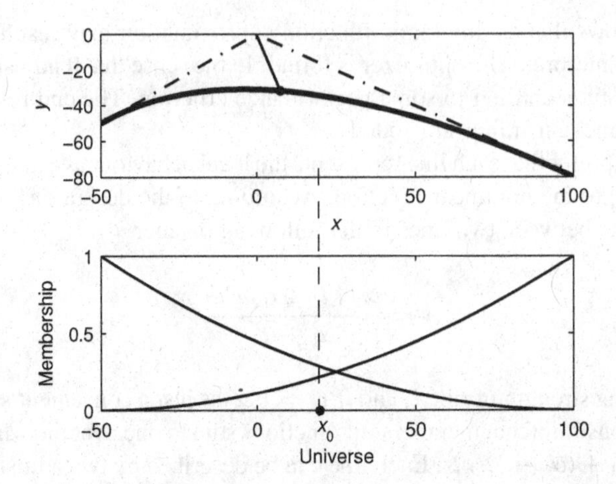

Figure 8.6 Interpolation of two vectors. A point x_0 on the universe corresponds to an image point on the interpolant defined by its locus vector $\mathbf{p}_0$. The membership functions are triangular functions raised to the power $c = 2$. (figvectorfit0.m)

Algorithm 8.1 Function approximation procedure.

1. Identify start and end tangents of the segment of the function to be approximated.
2. Define the basis vectors $\mathbf{v}_1$ and $\mathbf{v}_2$ as vectors originating at the intersection of the tangents, and extend them along the tangents to the end points of the segment to be approximated.
3. Interpolate, using Equation (8.3), with concentration $c = 2$.
4. Adjust c until the best approximation is achieved.

Example 8.3 *Approximate valve characteristic*

As previously, a nonlinear control valve is the actuator in a PI control loop that controls a linear process with the transfer function $P(s) = 10/(s+1)^3$ (Section 5.6). The following is a model of the valve:

$$f(x) = x^4, 0 \leq x \leq 1$$

Assume that Algorithm 8.1 only 'knows' the function $f(x)$ from a plot. How well does the algorithm approximate the inverse?

► *Solution*

The example in Section 5.6 tried to compensate by inverting two straight lines that were close to the characteristic $f(x)$. This time we choose two oriented vectors, lying on tangents to the inverse $f^{-1}(x)$.

We choose round numbers for the vectors, $\mathbf{v}_1 = (0, -0.7), \mathbf{v}_2 = (1, 0.3)$, and the origin of the vectors is in $O = (0, 0.7)$. Figure 8.7 shows the vectors and the resulting interpolant, together with the true inverse. The approximation is so good that it is difficult to visually

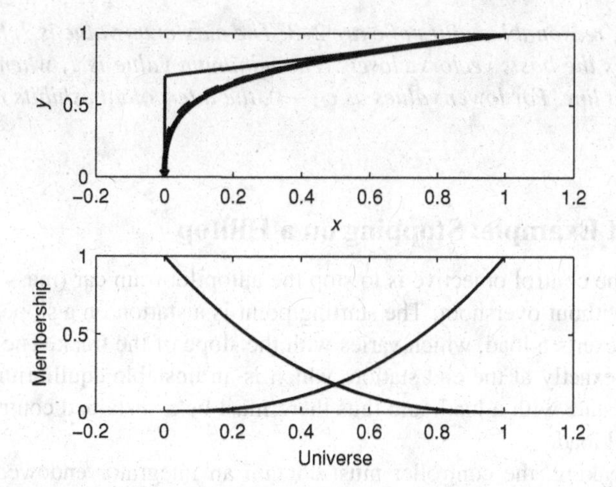

Figure 8.7 Approximation of the inverse valve characteristic (dashed line). The controlling vectors are tangential to the curve. The membership functions reflect that $c = 2.5$, and c was adjusted by hand-tuning. (figvectorfit.m)

distinguish the two from each other. The concentration is $c = 2.5$, which rules out the possibility that the approximation could be the exact inverse $\sqrt[4]{x}$.

Example 8.4 *Concentration parameter c*

What is the effect of the concentration parameter c, and what is its typical range of values?

► *Solution*

Concentration of a membership value μ results in a smaller membership value $\mu^c < \mu$ $(c > 1)$. It depresses the shape of a triangular membership function, and the ultimate membership function for $c \to \infty$ will be a singleton. Figure 8.8 shows the result of varying the parameter c. Higher values of c result in stronger attraction to the basis vectors.

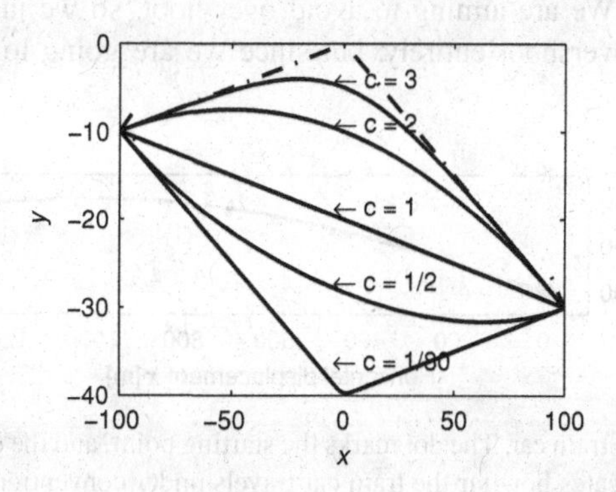

Figure 8.8 Five interpolations of one set of basis vectors. Higher concentration values of c result in more attraction to the basis vectors. Values of $c < 1$ result in repulsion. (figvectorfit0.m)

From the figure, reasonable values of c are 2–3. The maximum value is 7, because then the interpolant follows the basis vectors closely. The minimum value is 1, when the interpolant becomes a straight line. For lower values as $c \rightarrow 0$, the interpolant exhibits repulsion.

8.4 Autopilot Example: Stopping on a Hilltop

In this example, the control objective is to stop the autopilot train car (mass 1500 kg) on the top of a hill and without overshoot. The starting point is a station on a slope; see Figure 8.9. As usual, gravity exerts a load, which varies with the slope of the track. The major difficulty is to stop the car exactly at the end station, which is an unstable equilibrium. Furthermore, the car starts in a state with a load, and thus there must be a sustained control signal, which balances the initial load.

Technically speaking, the controller must contain an integrator endowed with an initial value. At the end station the load is zero, and in this state the integrator should also be zero. The two boundary conditions cause difficulties, because the train car will most likely overshoot owing to the integrator. Roughly speaking, the integrator will wind up to a large value, after the large reference step, and it will take time to empty the integrator again. It could possibly be done with a slow controller, but then the car must drive slowly.

Instead, we will try to achieve a faster response by scheduling the integral gain. The strategy is to have integral action at the starting point, but a zero integral gain at the top station, effectively suppressing any content of the integrator upon arrival. The design follows the design procedure for gain scheduling.

1. *Develop a linear process model for a set of operating points.* We skip this action, as we take the whole operating range as one step of the reference.
2. *Design a linear controller for each operating point.* The design is for the initial point. The reference steps from 0 to 1000 m, which is the x-coordinate of the end station. We use the desired settling time as a guiding parameter, and set arbitrarily $T_s = 400$ s. This corresponds to an average speed in the x-direction of 2.5 m/s, which is realistic. Using the previously developed tuning rules (Table 7.1), the derivative gain is thus set to $T_d = T_s/7.8 = 51.3$. We are aiming to avoid overshoot, so we just set $T_i = 4T_d = 205$. This will not avoid overshoot entirely, but since we are going to schedule this gain, an

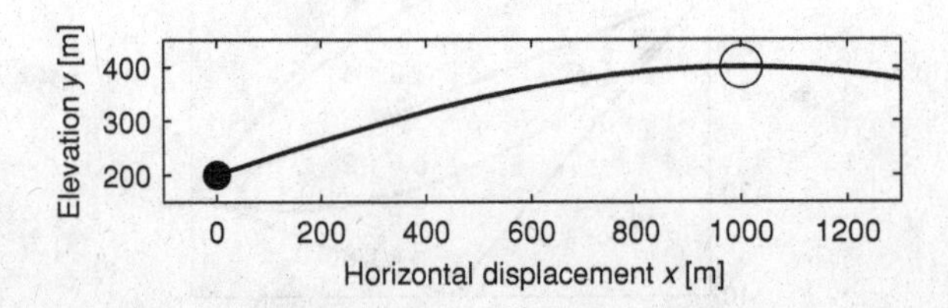

Figure 8.9 Position of the train car. The dot marks the starting point, and the circle is the stopping point (reference). The curve indicates how far the train car travels under conventional PID control, including overshoot. The track is sinusoidal. (figrunautopilot5.m)

approximate value will suffice. For the proportional gain we choose $K_p = 10 \times T_i T_d / K = 10 \times 205 \times 51.3/1500 = 70.1$, where the static gain of the process is related to the mass, that is, $K = 1/1500$. Finally, we have to choose the initial value of the integrator $I_0 = mg\cos(\pi/4) / (K_p/T_i) = 1500 \times 9.81 \times 0.707/(70.1/205) = 30\,400$. Here m is the mass, and g is the gravitational acceleration. The slope in the initial state is 45°, and the cosine is the projection of the tangential force acting on the car. The controller multiplies the integrated error by K_p/T_i, so we reduce the initial value by this amount to make the integrator balance the load. For the end point, we use the same controller settings, except that $1/T_i$ should be gain scheduled to zero here.

3. *Develop a schedule for the controller parameters, such that the transition from one controller to the other is smooth (bump-less).* We use a simple gain schedule based on the magnitude of the error, namely

$$\text{If abs(error) is large then } 1/T_i = 1/205$$
$$\text{If abs(error) is small then } 1/T_i = 0$$

   The schedule thus gradually suppresses integral action as the train car gets closer to the end station. Figure 8.10 shows the resulting functional dependency, which is a soft s-curve.

4. *Assess performance and relative stability, most likely by simulation.* Figure 8.11 shows the result of a simulation. The PID controlled response has a settling time, which more or less complies with the specification, but there is overshoot. The gain scheduled controller avoids the overshoot, and it has a much better settling time. The control signal is in both cases limited to 10 000 N to be realistic. The control signals show that the gain schedule reduces the control signal at a fairly early stage of the response, early enough to avoid overshoot.

The example shows that it was possible to avoid integrator windup by gradually suppressing the output from the integrator. It was rather simple to do in this case, compared with the conventional techniques for avoiding integrator windup. The gain schedule (Step 3) uses standard soft membership functions, resulting in a soft s-curve, but other and more complex cases could be included in the rule base.

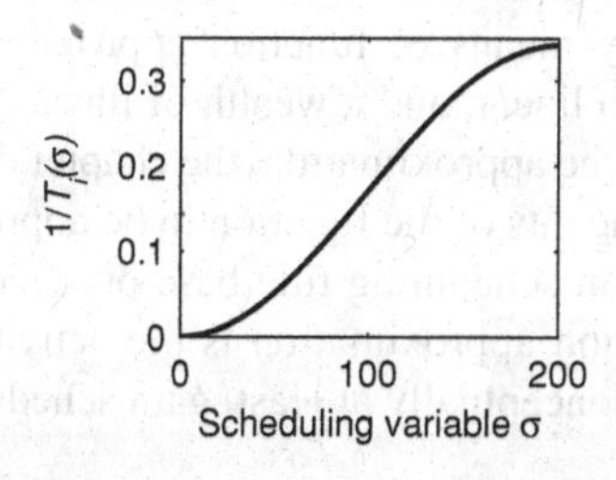

Figure 8.10 Schedule for the integral gain. The scheduling variable is the magnitude of the error $|e|$. Close to the end station, the integral action is more or less zero. (figrunautopilot5.m)

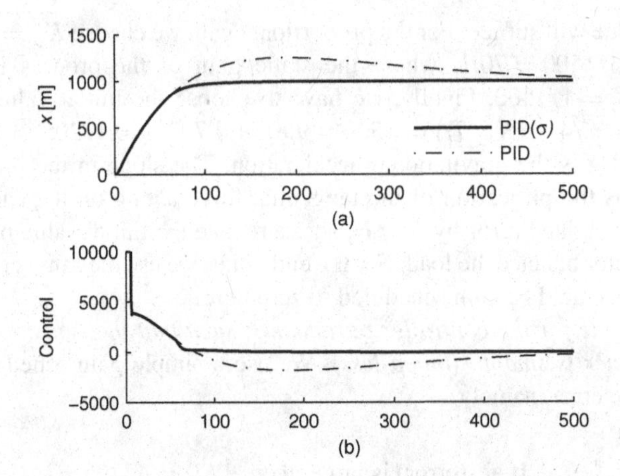

Figure 8.11 Responses. The train car position (a) exhibits overshoot with the conventional PID, but no overshoot with gain scheduling. The control signals (b) are limited to 10 000 N, and they are only slightly different from each other. (figrunautopilot5.m)

The example also shows that it was possible, even though this is a nonlinear case, to use the previously developed tuning rules (Table 7.1). All parameters depend on a single parameter, namely, the settling time, and it was easy to find satisfactory settings by hand-tuning.

8.5 Summary

The objective of gain scheduling is to enable linear controller design methods in a nonlinear system. At each design point, a linear controller is designed based on a linearization of the process. Several such controllers are then joined together by means of fuzzy interpolation. Although the individual designs are stable, there is no guarantee that the overall design is also stable, because the update mechanism itself may destabilize the system.

The advantage of using gain scheduling is that the standard universes and the fuzzy gains become obsolete. The PID controller at each design point has an unlimited range, in principle, just like conventional PID control. The actual range of operation is limited and controlled by the fuzzy membership functions on the premise side of the rules.

If it is possible to identify the characteristic of a nonlinearity, and if it is invertible, then a good inverse can be made by means of function approximation using fuzzy rules. The compensated function is close to linear, and a wealth of linear control design techniques can be applied. In order to construct the approximation, the chapter developed a simple piece-wise method, which preserves the tangents of the function to be approximated.

Geometrically speaking, a gain scheduling rule base produces a surface given a variation of the inputs. In contrast, function approximation is the activity of finding a rule base that approximates a given surface. Conceptually at least, gain scheduling is the inverse activity of function approximation.

The fuzzy PID controllers from previous chapters are suitable in the case of small deviations from the reference, while fuzzy gain scheduling is suitable when there are large deviations from the reference.

8.6 Case Study: the FLS Controller*

This section illustrates the problems in practical process control due to the complexity of multi-loop systems. The basic mechanism is gain scheduling of the incremental output of a possibly crisp PI controller. Although the section describes experiences of high practical value, it is not strictly necessary for the following chapters. Therefore the section can be skipped in a first reading (*-marked).

Human operators in the process industry are faced with nonlinear and time-varying behaviour, many inner loops, and much interaction between the control loops. Owing to sheer complexity it is impossible, or at least very expensive, to build a mathematical model of the plant, and furthermore the control is normally a combination of sequential, parallel, and feedback control actions.

Operators are nevertheless able to control complicated plants using rules based on their experience and training. The cement kiln controller by FLSmidth (FLS) was based on rules of thumb for manual control of a cement kiln. In a further development, several controllers were combined in a hierarchy by means of a priority system.

8.6.1 Cement Kiln Control

A cement kiln is a rotating chamber with a light slope. It is sometimes 160 m long, although modern kilns are shorter because of preheating. Ground limestone, clay, and sand react chemically at temperatures around 1430°C. The material advances slowly down the kiln in three to four hours. The process is difficult to control, because only a few measurements of the internal state are possible because of the heat and the complexity of the chemical reaction. Nevertheless, a skilled operator can be rather successful in maximizing clinker output (the product), while minimizing fuel and raw material consumption.

The operator monitors mainly four quantities: oxygen and carbon monoxide content in the exhaust gases, the temperature of the exhaust, the free-lime content (indicator of the temperature in the burning zone and the quality of the product), and the change in kiln drive torque. The operator then applies 40–50 rules of thumb to control the coal feed rate and the flow of air into the kiln. An example of a rule is the following:

> *If* the oxygen percentage is low, and the temperature is in the upper part of the range
>
> *then* decrease the air flow and reduce fuel slightly.

The FLS controller contains predefined fuzzy terms – *Low*, *High*, and *OK* – for the measured quantities. Similarly, the program contains terms for the control actions: *Medium*, *Negative*, *Large Negative*, *Small Negative*, *Medium Positive*, and so on. The operator can display the rules, and the rule above may appear in a program as

```
IF LOW(O2) AND (OK(TEMP) OR HIGH(TEMP))
THEN MNEG(DAMPER), MNEG(COAL)
```

The controller weights the control action from each rule depending on the degree of match. It determines the resulting control signal as a weighting of all actions dictated by the rules.

An operator screen and keyboard are normally placed in the control room as an integral part of the control desk. The operator can request colour displays of time series, selected measure points, alarm surveys, and diagrams. A printer produces 24-hour reports, alarm reports, and hard copies of graphical screens. In order to trace the data flow, one side of the screen displays the current value of each input or output variable, as well as the current firing strength of each rule. Changing a rule requires special authorization.

A skilled operator must tune the controller for a new kiln. Each kiln has its own operating behaviour, so the operator monitors its initial performance and adds or deletes rules as necessary. Since the rule language is somewhat close to natural language, the operator needs little or no computer expertise to instruct the controller.

Operators have reported savings in fuel and more stable product quality and operation. The controller also eliminates differences in operation among various operators. FLSmidth has implemented similar process controllers for a variety of other industrial processes.

Further details are available in the articles by Holmblad and Østergaard (1982, 1995).

Control tasks

To structure the process knowledge, a practical approach is to decompose the supervisory problem into hierarchical subsystems or *tasks*, each having a specific purpose. For instance, to reach a specific state of the plant, Yazdi (1997) specified a standard *control task* in terms of a set of necessary properties:

- *Name*. A name describing a task goal.
- *Goal*. For example: safe operation, high quality, low energy consumption.
- *Strategic conditions*. A condition set which only has to be valid for initiating the task. The strategic condition set is normally a process requirement before the task starts.
- *Execution conditions*. A set of conditions that have to be valid during task execution. The condition set is normally related to the operational constraints, for example, physical limitations and safety conditions. Information about these properties is derived by combining operator knowledge of the operational conditions with design knowledge.
- *Initial actions*. A set of actions that has to be carried out to prepare the task for control structuring. This property contains mostly binary actions (on/off, start/stop).
- *Control actions*. A set of sensors and actuators through which proper control can be designed in order to achieve the task objective. This property describes the resources of the control function.
- *Achievement indicator*. An indicator for the degree of goal achievement during the task operation.
- *Final action*. A task can end with a set of final actions, initiated when the task goal has been achieved.

A control task should be defined for each (fuzzy) controller. Each control task describes a detailed strategy for reaching a specific sub-goal. Execution of a control task is only allowed if *execution* conditions are valid at each sampling instant, while *strategic* conditions are checked only at the first sampling of the task.

8.6.2 *High-Level Fuzzy Control*

The overall control problem is to ensure that a particular goal is reached when all control tasks are combined. The FLSmidth design procedure, introduced in the company's second generation fuzzy controller, Fuzzy II (Østergaard 1990, 1996), includes, for this reason, a priority management system.

A *high-level* controller works on the level below that of the human operator. It coordinates control loop setpoints, which were previously coordinated by a human operator. The basic component of fuzzy high-level control is a set of rules for automatic operation based on practical experience and knowledge about manual control. Examples of fuzzy control rules are:

If Temperature is High and Pressure is OK *then* Medium Flow

If Temperature is OK and Pressure is OK *then* Small Flow

Since the condition of a rule is fulfilled to a degree, each rule will influence the result of the set of rules in accordance with its activation.

This heuristic design approach is useful when the process is partly unknown, difficult to describe by a mathematical model, if few measurements are available, or if the process is highly nonlinear.

High-level control performance

The meaning of improved performance is not always obvious, and it may depend on local conditions such as the present market situation, raw material costs, energy consumption, and overall strategic goals.

In most cases, performance will relate to profit in terms of reduced costs or increased productivity. The consumption of energy and raw materials depends on the supervisory control. In general, improved performance can be defined as

- having a more stable (i.e. steady) operation;
- running closer to the limits for acceptable product quality; and
- running closer to the environmental emission limits.

Steady operation is the most important key to improving performance. Oscillations require energy and raw materials, reduce the quality of the product, and increase emissions. If the process oscillates, average values must be kept within safe bounds. The standard deviation *STD* may be calculated on a daily basis, and an average standard deviation is then calculated for a period that is representative of the performance.

If the measurements and the quality parameters have target values, then a more feasible measure of steady operation is obtained by calculating the *target value deviation TVD* (Østergaard in Jantzen *et al.* 1999) around the target value *SP*, instead of the variations around the average value *AVR*, as in *STD*. The index is calculated as follows,

$$TVD = \sqrt{STD^2 + (SP - AVR)^2} \tag{8.4}$$

For a cement plant, it is realistic to expect reductions of 50%, or more in *STD* and/or *TVD*, as the result of a high-level control system.

High-level control configurations

Fuzzy controllers are combined with other controllers in various configurations as shown in Figure 8.12. The PID block consists of independent or coupled PID loops, and the Fuzzy block is a high-level control strategy. Normally, both the PID and Fuzzy blocks have more than one input and one output.

- *Fuzzy replaces PID*. In this configuration (Figure 8.12a), the operator may select between a high-level control strategy and conventional control loops. The operator has to decide which of the two most likely produces the best control performance.
- *Fuzzy replaces the operator*. This configuration (Figure 8.12b) represents the original high-level control idea, where manual control carried out by a human operator is replaced by automatic control. Normally, the existing control loops are still active, and the high-level control strategy makes adjustments of the controller setpoints in the same way as the operator does. Again it is up to the operator to decide whether manual or automatic control will result in the best possible operation of the process which, of course, may create conflicts.
- *Fuzzy adjusts the PID parameters*. In this configuration (Figure 8.12c), the high-level strategy adjusts the parameters of the conventional control loops. A common problem with

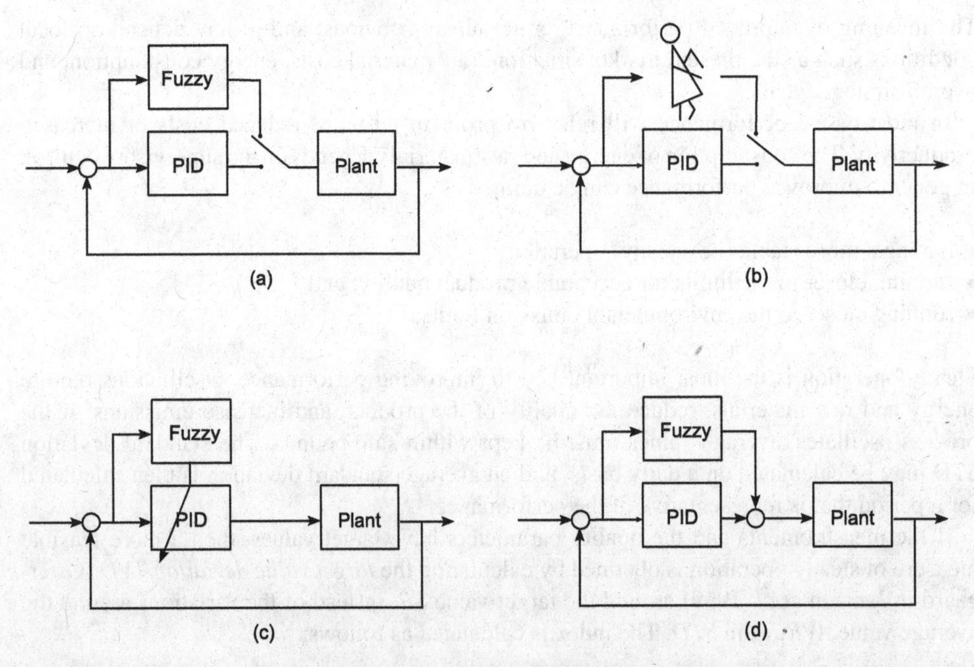

Figure 8.12 Fuzzy controller configurations. Fuzzy replaces PID (a), fuzzy replaces the operator (b), fuzzy adjusts the PID parameters (c), and fuzzy supplements PID control.

linear PID control of highly nonlinear processes is that the set of controller parameters is satisfactory only when the process is within a narrow operational window. Outside this, it is necessary to use other parameters or setpoints, and these adjustments may be done automatically by a high-level strategy.

- *Fuzzy supplements PID control.* Normally, control systems based on PID controllers are capable of controlling the process when the operation is steady and close to normal conditions. However, if sudden changes occur, or if the process enters abnormal states, then the configuration in Figure 8.12(d) may be applied to bring the process back to normal operation as fast as possible. For normal operation, the fuzzy contribution is zero, whereas the PID outputs are compensated in abnormal situations, often referred to as *abnormal situation management* (ASM).

Configurations (a) and (b) directly change the routines of the operator, which is a crucial point to take into account when the system is developed and installed.

8.6.3 The FLS Design Procedure

For an operator, control of the process consists of achieving various goals, more or less well-defined, such as, maximum output, minimum consumption of raw materials and energy, high product quality, and safe process operation. Different processes have different control objectives but, in general, good process control may be defined through a list of control objectives that should be fulfilled to the extent possible. The concept of control objectives is a key element in a high-level control strategy.

For a cement kiln, typical control objectives are: stable (steady) operation, good cement clinker quality, high production, complete combustion, low fuel consumption, and low energy consumption. As control objectives are frequently in conflict, high-level coordination is required.

In other words, priorities have to be assigned to the various control objectives, specifying which objectives are considered the most important to fulfil. The elements of the FLS design procedure for a process control strategy (Østergaard 1990, 1996) are as follows:

- *State index calculations.* Find the current process state. The calculation combines measurements into an index for the current process stability, product quality, production level, and so on. Normally, a state index combines various measurements into a single figure. The degree of process stability, the product quality, and the production level are typical examples of state indices for a kiln control strategy.
- *Control groups.* Arrange the overall control strategy into groups of control objectives. A control group, a subset of the control strategy, is a group of objectives that are related through priority numbers.
- *Priority management.* Determine the extent to which the control actions should be executed to fulfil the individual objectives. The priority management system manages the scheduling of control actions in order of importance.
- *Control objectives.* Specify the goals of the control strategy. A control objective consists of tasks.

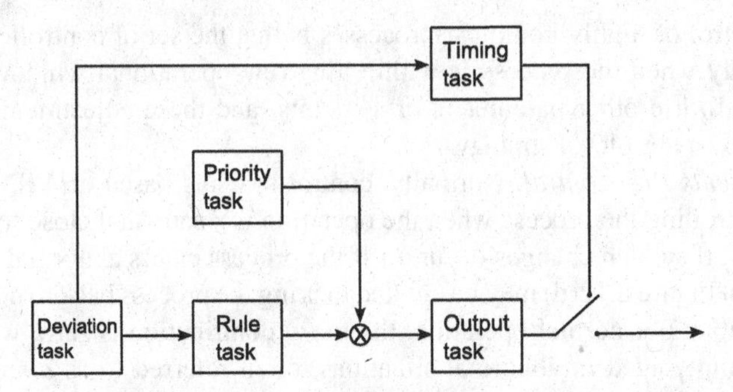

Figure 8.13 Control objective module in Fuzzy II.

The state indices are important to the structure of the FLS design scheme for a high-level control strategy, as they form the basis for dividing the overall strategy into control groups that can be treated independently. The state indices are used to coordinate control actions from the various control groups.

Every control objective is implemented in accordance with the so-called *objective module*, which consists of five tasks, as shown in Figure 8.13.

- *Deviation task.* This task involves calculating and evaluating the degree to which the objective is fulfilled. Normally, the calculation results in an error value $e_i \in [-1, 1]$, which expresses how far the current process state is from the objective; a value of 0 signifies that the objective is fulfilled.
- *Rule task.* Normally, this block holds a set of fuzzy control rules, and the output of the rule block is normally a *change* in action in the interval $[-1, 1]$. Other techniques may also be used, such as PID, neural nets, and mathematical models.
- *Priority management task.* The rule block for each control objective results in control actions, which are multiplied by a weight factor between 0 and 1. The weight factor w_i generated by objective i is calculated as

$$w_i = 1 - |e_i| \tag{8.5}$$

  The weight factor is thus a function of the deviation e_i from the objective. The smaller the weight factor, the more the lower control actions, below objective i, are suppressed. The total weighting of an objective's output is the product of all higher priority weight factors (Figure 8.14); it will be 1 if all objectives with a higher priority are fulfilled, and 0 if one or several objectives are not fulfilled. The priorities reflect built-in knowledge about optimal interaction of rule blocks.
- *Output task.* The output task involves evaluation of process constraints and selection among alternative control actions based upon the current index values; in this task, fuzzy output is converted to engineering units, that is, denormalized physical units. The logic for selecting alternative adjustments may be fuzzy or non-fuzzy depending on whether a gradual or a hard switch is the most appropriate. In most cases, the fuzzy-logic approach gives the best

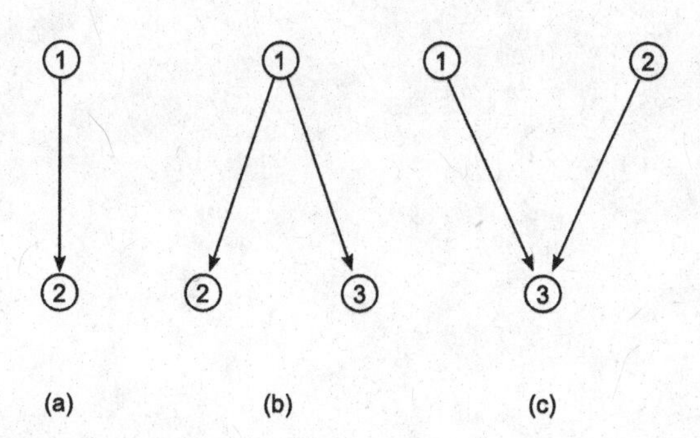

Figure 8.14 Priority management: (a) objective 1 higher than 2; (b) objective 1 affects two objectives on a lower level; (c) objectives 1 and 2 both affect an objective on a lower level.

control performance, simply because no physical process operates with sudden changes between alternative control actions.

- *Timing calculation task*. The timing calculation consists of determining when and how often control actions are to be executed. It is just as important as the rule block for determining the proper function of the control strategy. Even the timing calculation is normally fuzzy in the sense that the time interval between control actions changes gradually as a function of the deviation value. The larger the deviation, the more frequent the control actions.

Each objective has several tuning parameters: an output gain, input normalization, and tuning of the timing calculation. The control objective output must be a *change* in control action, that is, an incremental controller. A suppressive weight $w_i = 0$ thus results in no change, which is equivalent to maintaining the status quo.

8.7 Notes and References*

Rugh and Shamma note in a survey paper (2000) that it is only after the 1990s that research literature on gain scheduling starts to appear. They describe the general features of the approach and discuss recent research. Gain scheduling has a long application history though, notably within flight control and in the automotive industry.

A chapter on gain scheduling in Åström and Wittenmark (1995) describes the principle and design phases. The chapter presents briefly applications to ship steering, pH control in a stirred tank reactor, combustion control in a boiler, fuel–air control in a car engine, and flight control.

Bequette (2003) gives a brief application-oriented presentation of the approach, and the book has a teaching module that applies nonlinear PID control for level control in a surge vessel.

9

Fuzzy Models

The performance of a fuzzy rule base depends on a number of adjustable parameters, and they can be adjusted such that the rule base behaves in a more or less desired manner. In the simplest cases, a handmade model, in the shape of a rule base, suffices to approximate a curve or a surface. In more complex cases, based on measurement data, a machine-made model can approximate a function, even in many dimensions. Fuzzy rules can interpolate between local linear models in order to approximate a nonlinear function. Local linear models are conveniently built using a least-squares optimization. If the partitioning into local linear models cannot be done manually, the clustering algorithms *hard c-means* and *fuzzy c-means* can point to local clusters that are candidates for a linear model. A neuro-fuzzy model is more complex – perhaps also more capable in difficult cases – and it combines the training of a neural network model with the readability of a fuzzy rule base. The result is a rule base that 'learns' the underlying function in data. The inner product provides a common geometric foundation of the model architectures.

Mathematically speaking, this chapter concerns function approximation of both continuous functions and functions available only at discrete points.

In approximating continuous functions, the objective is usually to provide a simpler form than the original function. The approximation should be easier to handle analytically or easier to evaluate on a computer.

In the case of functions available only at discrete points, the chapter considers fuzzy membership functions. Actually, we have already introduced one variety of such an approximation: a model based on a weighting between two vectors (Chapter 8). The present chapter shows how simple approximations can smooth noisy experimental data. The function provides a curve or a surface instead of a collection of scattered points. The aim of such a *data-driven* machine-made model is to leave as much as possible of the routine work to the computer.

By far, the most widely used functions for data fitting are straight lines, and we will attempt to use these whenever possible. Multiple-input multiple-output (MIMO) models are assumed decomposed into a collection of multiple-input single-output (MISO) models that are local linear models.

Foundations of Fuzzy Control: A Practical Approach, Second Edition. Jan Jantzen.

9.1 Basis Function Architecture

The models in this chapter all spring from an inner product architecture. It is a mapping of an input vector **u** to the scalar output $\hat{y}$, which is an approximation to the true value y. The mapping is a weighted sum of *basis functions* f_j, as defined by the following expression:

$$\hat{y} = \sum_{j=0}^{M} w_j f_j(\mathbf{u}) = \mathbf{f}(\mathbf{u}) \cdot \mathbf{w} \tag{9.1}$$

The weighting is linear in the parameters w_j, while the basis functions must be nonlinear in order to realize a nonlinear mapping. The model includes an *offset* parameter (*bias*) that adjusts the zero-point: by letting $f_0 = 1$, its corresponding parameter w_0 implements the offset.

The basis functions can take many forms, but in this chapter they will be membership functions that return a membership value between zero and one. They are generally of the MISO type, since each function takes a whole vector as input, and it returns a single membership value. Given a collection of data points $(y, \mathbf{f}(\mathbf{u}))$, the objective of function approximation is to find a **w** that fits the data in the best way. Once the weights are determined, the right-hand side of Equation (9.1) interpolates between the given data points.

Example 9.1 *Geometry of the inner product*

The definition in Equation (9.1) contains an inner product. We can tell that the inner product is zero when **f** *and* **w** *are orthogonal, even in many dimensions. How can we interpret, geometrically, the whole equation in three dimensions?*

► *Solution*

Let us disregard the basis functions themselves, and restate Equation (9.1) in a generic form in three dimensions, that is,

$$z = w_1 x + w_2 y + w_0$$

or

$$w_1 x + w_2 y - z + w_0 = 0 \tag{9.2}$$

The following is the same equation in vector notation:

$$\begin{bmatrix} w_1 & w_2 & -1 \end{bmatrix} \begin{bmatrix} x \\ y \\ z - w_0 \end{bmatrix} = 0 \tag{9.3}$$

Since the product is zero, this equation shows that the vector $\mathbf{n} = \begin{bmatrix} w_1 & w_2 & -1 \end{bmatrix}^T$ *is a* normal *vector (perpendicular) to a plane in three dimensions, which passes through the point* $(x, y, z) = (0, 0, w_0)$.

Figure 9.1 shows the plane and the normal for the case $\mathbf{w} = \begin{bmatrix} -0.5 & -0.25 & -1 \end{bmatrix}^T$. *The weight* $w_0 = \mathbf{w}(3)$ *controls the offset in the z-direction. Thus, the weight vector* **w** *controls both the normal vector and the location, and it therefore defines the plane completely.*

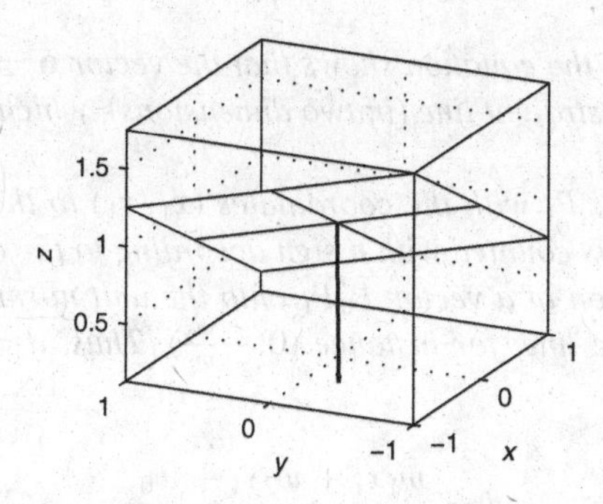

Figure 9.1 Plane determined by a normal vector, which is in turn determined by the weight vector **w**. (figinner.m)

The distance d of a point P_1 with the coordinates (x_1, y_1, z_1) to the plane is measured on the normal **n**, *and the distance is counted with a sign according to the orientation of the normal. The distance is the projection of a vector $\overrightarrow{P_0P_1}$ onto the unit normal vector, where P_0 is an arbitrary point lying in the plane, for instance $(0, 0, w_0)$. Thus, $d = \overrightarrow{P_0P_1} \cdot \mathbf{n} / \|\mathbf{n}\|$, and this gives:*

$$d = \frac{w_1x_1 + w_2y_1 - z_1 + w_0}{\sqrt{w_1^2 + w_2^2 + 1}}$$

In summary, we find the distance of a point to the plane by inserting the coordinates of the point in the left-hand side of Equation (9.2) and divide by the length of the normal vector **n**. *The plane is a switching plane, in the sense that the distance changes sign, when the trajectory of the point passes through the plane.*

Example 9.2 *Distance from a line*

We said previously, in connection with the PD controller (Section 5.11), that an inner product was related to the distance to a line. How can that be seen here?

► *Solution*

The previous example (Example 9.1) showed how to calculate the distance from a point to a plane in three dimensions. There is a similar mechanism in two dimensions. Given the equation

$$z = w_1x + w_2y + w_0 \tag{9.4}$$

we shall regard z as being the result of a given input point (x, y). Letting $z = 0$, we proceed as before (Example 9.1) to find the following:

$$\begin{aligned} 0 &= w_1x + w_2y + w_0 \\ &= w_1x + w_2\left(y + \frac{w_0}{w_2}\right) \\ &= \begin{bmatrix} w_1 & w_2 \end{bmatrix} \begin{bmatrix} x \\ y + \frac{w_0}{w_2} \end{bmatrix} \end{aligned}$$

Since the product is zero, the equation shows that the vector $\mathbf{n} = \begin{bmatrix} w_1 & w_2 \end{bmatrix}^T$ *is a normal vector (perpendicular) to a straight line (in two dimensions), which passes through the point* $(x, y) = (0, -\frac{w_0}{w_2})$.

The distance d of a point P_1 *with the coordinates* (x_1, y_1) *to the line is measured on the normal* **n**, *and the distance is counted with a sign according to the orientation of the normal. The distance is the projection of a vector* $\overrightarrow{P_0P_1}$ *onto the unit normal vector, where* P_0 *is an arbitrary point lying on the line, for instance* $(0, -\frac{w_0}{w_2})$. *Thus,* $d = \overrightarrow{P_0P_1} \cdot \mathbf{n} / \|\mathbf{n}\|$, *and this gives:*

$$d = \frac{w_1 x_1 + w_2 y_1 + w_0}{\sqrt{w_1^2 + w_2^2}}$$

Thus, we find the distance of a point to the line by inserting the coordinates of the point in the right-hand side of Equation (9.4) and divide by the length of the normal vector **n**. *The line is a switching line, in the sense that the distance changes sign, when the trajectory of the point crosses the line.*

In summary, given that $z = f(x, y)$ *by Equation (9.4), then* $z = \|\mathbf{n}\|\, d$. *In other words, the output of the process in Equation (9.2) is proportional to the perpendicular distance of the input point to a switching line determined by the weights. A three-dimensional mesh plot just maps the values of* z *along the vertical axis, for a range of input points* (x, y), *and the result is a surface.*

9.2 Handmade Models

By *handmade models* we shall mean models that are built by a human being as opposed to machine-made models. Technically speaking, such an approach is *supervised*, since it includes human intervention. A skilled person may build such a model based on knowledge about the function to be approximated. As an example, consider a graph on paper showing the decreasing efficiency of a solar hot water collector as the ambient temperature drops – our skilled person can provide a functional approximation of the graph for computer simulations.

9.2.1 *Approximating a Curve*

A singleton Sugeno type of rule base has a basis function architecture under certain conditions. Take, as an example, the rule base

$$\begin{aligned} &\text{If } x \text{ is Low}^c \text{ then } \hat{y} = k_1 \\ &\text{If } x \text{ is High}^c \text{ then } \hat{y} = k_2 \end{aligned} \tag{9.5}$$

The conclusion of a rule is just a constant (k_1, k_2). The adjustable parameter c is the *concentration exponent*, introduced previously (Chapter 8). The membership function on the premise side of, say, the first rule, is $\mu^c_{Low}(x)$. Thus x is the input, and the membership function μ^c

is the basis function. If the inference engine uses *-activation, sum-accumulation, and no defuzzification, the output is, simply,

$$\hat{y}(x) = \mu_{Low}^{c}(x)\,k_1 + \mu_{High}^{c}(x)\,k_2 \tag{9.6}$$

The membership function μ_{Low} ($\mu_{High} = 1 - \mu_{Low}$) could be anything, but even a simple triangular membership function can provide good results together with the adjustable concentration exponent c. The rule base can generate several kinds of nonlinearity as the following example illustrates.

Example 9.3 *Saturation and deadzone*

Referring to the previously mentioned one-dimensional standard nonlinear components (Figure 5.1), is it possible to generate soft versions of a deadzone, a saturation, and a quantizer?

► *Solution*

The nonlinearities are curves, and rule base (9.5) can generate various curves. We use a generic universe $[-1, 1]$, *and in order to normalize the output, we fix* $k_1 = -1, k_2 = 1$. *Figure 9.2 shows the result, and the comments below relate to the figure.*

- Linear (saturation) (a) and (b). *Linear (triangular) membership functions produce a linear component within the universe. Beyond the universe, the rule base saturates, and the component is effectively a linear saturation with a proportional band.*
- Saturation (c) and (d). *Changing membership functions from triangular to soft triangular generates a soft saturation. If the constants* (k_1, k_2) *were closer to zero, the curve would be closer to a relay characteristic.*
- Deadzone (e) and (f). *Linear, triangular membership functions, combined with a concentration exponent* $(c = 6)$, *generate a soft deadzone. The deadzone will tend towards a crisp deadzone for* $c \to \infty$.
- Quantizer (g) and (h). *Soft triangular membership functions, combined with a concentration exponent* $(c = 8)$, *generate a quantizer type of characteristic. The nearly horizontal middle level becomes wider with higher values of* c.

Note that all four are generated using the same rule base and the same underlying membership functions (triangular or soft triangular). The curves can be manipulated into non-symmetric curves, if that is desired, by adjusting the constants k_1, k_2.

Interpolating between vectors

Two curves can be joined together in order to form a more complex curve. In that case, it is convenient if the component curves can be spliced together along a common tangent.

A rule base can generate a piece of a curve from vectors that are tangential to a desired function. The following is an example (Chapter 8):

$$\begin{aligned}&\text{If } x \text{ is Low}^c \text{ then } \hat{\mathbf{p}} = \mathbf{v}_1\\&\text{If } x \text{ is High}^c \text{ then } \hat{\mathbf{p}} = \mathbf{v}_2\end{aligned} \tag{9.7}$$

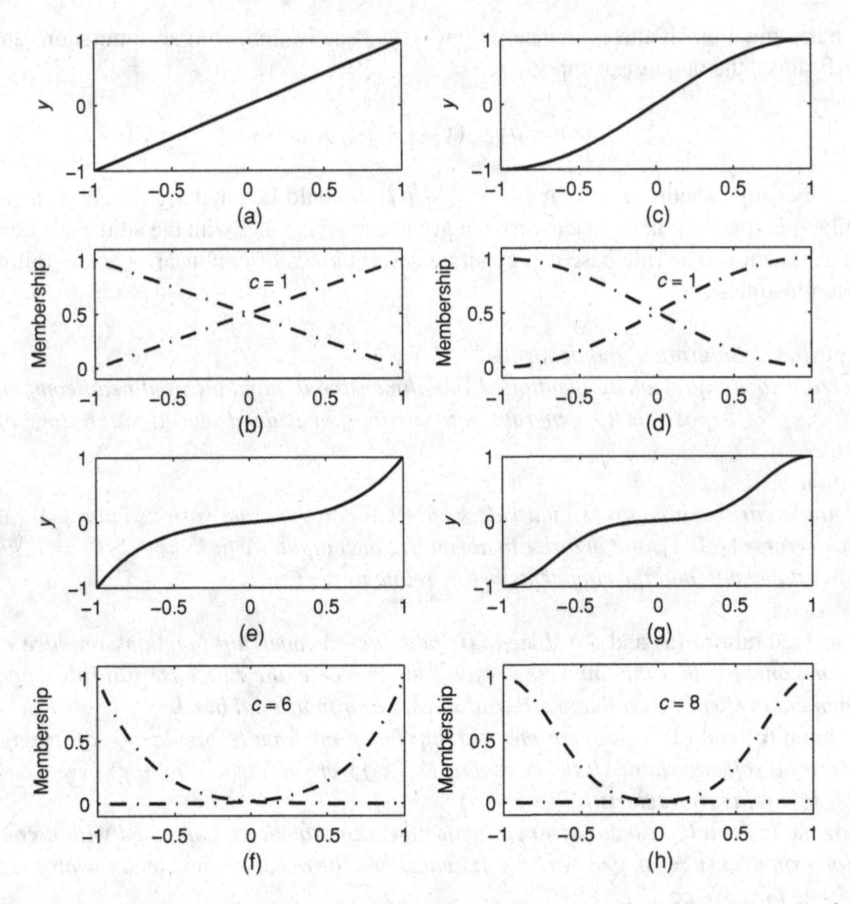

Figure 9.2 Function approximations: linear (a), soft saturation (c), soft deadzone (e), and soft quantizer (g). The membership functions that generated the curves are under each curve (dash-dot line). The parameter c is the concentration exponent. The rule base in Equation (9.5) generated all curves. (fignonlins.m)

As before, the inference engine uses *-activation, sum-accumulation, and no defuzzification, which means that the output is a combination of the conclusion vectors, that is,

$$\hat{\mathbf{p}}(x) = \mu^{c}_{Low}(x)\,\mathbf{v}_1 + \mu^{c}_{High}(x)\,\mathbf{v}_2$$

In this case, the weights, which are the vectors $\mathbf{v}_1$ and $\mathbf{v}_2$, are given, and the function interpolates between the vectors. The variable x is the independent variable defined on the universe of discourse. The vector $\hat{\mathbf{p}}$ is a point on the interpolant, defined by an x-coordinate and a y-coordinate, but its x-coordinate is generally different from the running input x-coordinate.

In a more general setting, the curve is a parametric curve, depending on an implicit parameter t $(0 \le t \le 1)$. To emphasize such a change of viewpoint, we substitute x by t, and thus the fuzzy set $High = t$ and $Low = 1 - t$. The interpolant is thus defined by the following expression:

$$\hat{\mathbf{p}}(t) = (1-t)^c\,\mathbf{v}_1 + t^c\mathbf{v}_2 \qquad (0 \le t \le 1) \tag{9.8}$$

The expression resembles the definition of a quadratic Bézier curve, but the concentration exponent $c > 0$ makes it different. If we regard $\mathbf{v}_1$ and $\mathbf{v}_2$ as a *vector basis* in a coordinate system, then $z_1 = (1-t)^c$ is the first coordinate and $z_2 = t^c$ is the second coordinate of the point $\hat{\mathbf{p}}$. As t runs from 0 to 1, the interpolant is attracted to the first basis vector, then the second. The vectors define a parallelogram inside which the interpolant will reside. The derivative of the interpolant is

$$\frac{\mathrm{d}\hat{\mathbf{p}}}{\mathrm{d}t} = c\left(-(1-t)^{c-1}\,\mathbf{v}_1 + t^{c-1}\mathbf{v}_2\right)$$

The derivative provides good insight into the workings of the interpolation, as the following observations show.

- For $t = 0$, the derivative is a vector in the opposite direction of $\mathbf{v}_1$, which means that $\mathbf{v}_1$ is tangential to the interpolant at the starting end.
- Similarly, for $t = 1$, the derivative is a vector proportional to $\mathbf{v}_2$, which indicates that $\mathbf{v}_2$ is tangential to the interpolant at the final end. That is reassuring, because the basis vectors were created along estimated tangents to the function to be approximated. Thus, the interpolant starts and ends in the desired directions.
- For $c = 1$, the derivative is the constant vector $\mathbf{v}_2 - \mathbf{v}_1$. Since the derivative is constant, the interpolant is a straight line, and the basis vectors are no longer tangents to the interpolant.

In summary, t is a running parameter, and c can be adjusted to make the fit closer to the basis vectors. The fit can be arbitrarily close to the basis vectors, but typically $c \in [2, 3]$ which provides a smooth transition.

Example 9.4 *Properties of concentration*

The interpolation mechanism in Equation (9.8) has a geometric foundation; are there any other geometric properties?

► *Solution*

The coordinates $z_1 = (1-t)^c$ and $z_2 = t^c$ satisfy the equation

$$z_1^{\frac{1}{c}} + z_2^{\frac{1}{c}} = 1$$

The equation provides the explicit dependency of, say, z_2 on z_1, that is,

$$z_2 = \left(1 - z_1^{\frac{1}{c}}\right)^p$$

When $\mathbf{v}_1$ and $\mathbf{v}_2$ are orthogonal, Equation (9.8) defines a so-called Lamé curve. *When $c = 1/2$, the equation defines an ellipse, particularly a circle when the basis vectors have the* *same length. When* $c = 2/5$, *the curve defines Piet Hein's* super-ellipse. *A Lamé curve can approximate a rectangle arbitrarily close. In our case, the basis vectors provide the option to transform the curve by choosing non-orthogonal basis vectors with arbitrary lengths. In the following equation,*

$$z_1^{\frac{1}{c}} + z_2^{\frac{1}{c}} = d^{\frac{1}{c}}$$

the number d is the Minkowski distance *of a given point* (z_1, z_2) *from the origin. For* $c = 1/2$, *the distance measure is equivalent to the* Euclidean distance *defined by the Pythagorean relationship between the sides of a right-angled triangle.*

Example 9.5 *Vector fit*

Given two vectors $\mathbf{v}_1 = \begin{bmatrix} -50 & -50 \end{bmatrix}^T$, $\mathbf{v}_2 = \begin{bmatrix} 100 & -80 \end{bmatrix}$ *located at the origin of the coordinate system, what does the interpolant look like?*

► *Solution*

The interpolant starts at the tip of $\mathbf{v}_1$ *and ends at the tip of* $\mathbf{v}_2$. *The result was shown earlier (Figure 8.6). Notice that a particular coordinate* $x = x_0$ *does not necessarily correspond to the x-coordinate of the interpolant* $\hat{\mathbf{p}}(x_0)$, *since x is a parameter for the vector function* $\hat{\mathbf{p}}(x)$.

Figure (8.8) showed the effect of concentration. As c increases $(c > 1)$ *the interpolant tends to the basis vectors. When c decreases* $(0 < c < 1)$ *the interpolant tends to a rectangle. The case* $c = 1$ *performs a straight interpolation between the basis vectors.*

Two curves can be joined into one in order to create more complex curves. The resulting curve will be smooth (differentiable) at the joining point, if the two pieces share a common tangent at the point.

Example 9.6 *Joining two curves*

Is it possible to approximate a curve with an inflection point, such as a second-order step response, given the end points and the inflection point?

► *Solution*

Such a curve contains two bends. Each of these can be approximated using two vectors, and Figure 9.3 shows the result. The two components meet at the inflection point where they share the same tangent. The concentration exponent c was adjusted manually to give a reasonable curve.

The shape of the interpolant can be modified by changing the end points and the slope in the inflection point.

9.2.2 *Approximating a Surface*

A lookup table is a common way to implement a nonlinear relationship in computer controlled systems. For example, the engine control unit for a spark ignition engine uses a table to control the timing of the ignition, depending on engine speed and load. In the case of a large table, it can be a taxing task to manually fill the table with numbers. Linear interpolation is a first solution to alleviate the task, but it will result in sharp edges and corners. A rule base can interpolate smoothly between fewer, but larger, granules of the lookup space.

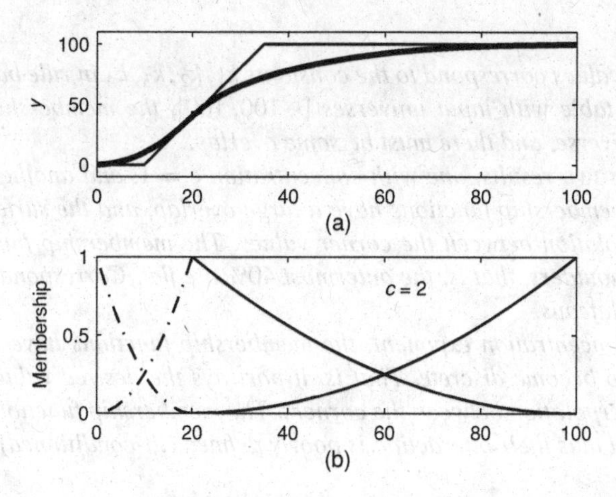

Figure 9.3 Approximation of a second-order response. The approximation is pieced together by two curves controlled by two sets of tangents that meet in the inflection point (a). Two sets of membership functions (b) generate the approximation. All are triangular, but modified by a common concentration exponent ($c = 2$). (figmodel.m)

In the example rule base

$$\begin{aligned}
&\text{If } x \text{ is Pos}^c \text{ and } y \text{ is Pos}^c \text{ then } \hat{z} = k_1\\
&\text{If } x \text{ is Pos}^c \text{ and } y \text{ is Neg}^c \text{ then } \hat{z} = k_2\\
&\text{If } x \text{ is Neg}^c \text{ and } y \text{ is Pos}^c \text{ then } \hat{z} = k_3\\
&\text{If } x \text{ is Neg}^c \text{ and } y \text{ is Neg}^c \text{ then } \hat{z} = k_4
\end{aligned} \tag{9.9}$$

each premise constitutes a basis function, which is two-dimensional in the inputs, since it depends on both x and y. As before, the membership functions are endowed with a concentration operator in the shape of the adjustable parameter c.

In order to achieve a smooth interpolant, the inference engine uses multiplication for 'and', *-activation, sum-accumulation, and no defuzzification. The output is then defined by:

$$\begin{aligned}
\hat{z} = {} & \mu^c_{Pos}(x)\,\mu^c_{Pos}(y)\,k_1 + \mu^c_{Pos}(x)\,\mu^c_{Neg}(y)\,k_2\\
& + \mu^c_{Neg}(x)\,\mu^c_{Pos}(y)\,k_3 + \mu^c_{Neg}(x)\,\mu^c_{Neg}(y)\,k_4
\end{aligned}$$

The product terms of the kind $\mu^c_{Pos}(x)\,\mu^c_{Pos}(y)$ are the basis functions. Each membership function μ is defined on an interval, and the product term is therefore defined on a rectangular area of the lookup table. The four rules can cover the whole table.

Example 9.7 *Surface interpolation*

Given the following four corner-values of a table $z(x, y)$*:* $z(100, 100) = 200$, $z(100, -100) = -25$, $z(-100, 100) = 100$, $z(-100, -100) = -200$, *is it possible to make an interpolated surface that spans the whole table?*

► *Solution*

The four corner values correspond to the constants k_1, k_2, k_3, k_4 in rule base (9.9). In order to span the whole table with input universes $[-100, 100]$, the membership functions must cover the whole universe, and there must be some overlap.

Figure 9.4 shows two results, one with concentration $c = 1$, and another with $c = 10$. In the first case, the membership functions have a large overlap, and the surface is a soft, two-dimensional interpolation between the corner values. The membership functions have been designed to have shoulders, that is, the outermost 40% are flat. Correspondingly, the surface contains four flat plateaus.

With the large concentration exponent, the membership functions have less overlap, and the surface tends to become discrete. That is, it portrays the desired values in the corners, but there is less interpolation between the corners. The membership functions are more local, almost disjoint, and thus their interaction is poorly defined (ill-conditioned).

The example shows that four rules can cover a whole table. The resulting surface, and its variation, depends on the overlap of the membership functions. Each product membership function constitutes a granule, and the variation in the surface depends on the number of granules.

The granules have a rectangular cross-section, but it is possible to construct other shapes. *Radial basis functions*, for instance, are defined by local polar coordinates, and the

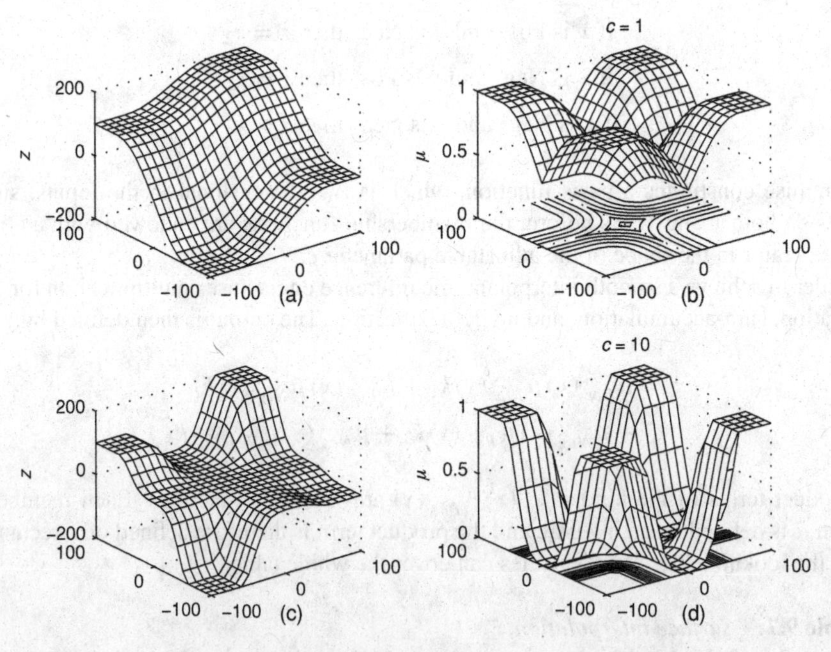

Figure 9.4 Control surface (a) and the basis functions that created it (b). The alternative surface in (c) is the result of applying a large concentration exponent ($c = 10$). The membership functions (d) have less overlap as a result of concentration. (figsurffit.m)

cross-section is thus circular or elliptic (Haykin 2009). Irregular shapes can be made by nesting of ellipses (Jakubek, Hametner, and Keuth 2008).

Interpolating between planes

Two planes can be interpolated in order to form a smooth transition from one to the other. Consider the following sketch of a rule base:

$$\begin{aligned} &\text{If } \sigma \text{ is } A_1 \text{ then } \hat{z} = w_{11}x + w_{12}y + w_{10} \\ &\text{If } \sigma \text{ is } A_2 \text{ then } \hat{z} = w_{21}x + w_{22}y + w_{20} \end{aligned} \tag{9.10}$$

The rule base performs an interpolation between two linear models. The premise side of the rules uses a scheduling variable σ, which can be a function of one of the other variables or a combination. The labels A_1 and A_2 refer to two fuzzy membership functions, such as Pos and Neg. The conclusion sides of the rules contain two linear models $\hat{z} = f(x, y)$ in the form of an inner product. The weights have indices such that the first index corresponds to the rule number, and the second index is our usual index for weights.

Such an inner product form is used to model sampled data systems. Take, for example, the model $y_{k+1} = ay_k + bu_k + c$, where k is the sample number (related to time), y is the process output, u is the process input, and c is a constant base input signal. This is a dynamic, linear, discrete-time model for a first-order process, and the task of the rule base is to interpolate between two such models in a smooth manner.

The geometric interpretation of the model is familiar (Example 9.1). In the first rule, for example, the vector $\mathbf{n}_1 = \begin{bmatrix} w_{11} & w_{12} & -1 \end{bmatrix}$ is the normal vector of a plane that passes through the point $(0, 0, -w_{10})$. Each rule thus defines a tangent plane to the interpolant surface. The fuzzy sets on the premise side control the location of the tangent plane. The simplest design, with triangular membership functions A_1 and $A_2 = 1 - A_1$, performs a linear gain scheduling on the weight vectors.

Example 9.8 *Interpolating two planes*

Is it possible to achieve a smooth surface by interpolating between two cutting planes?

► *Solution*

Using the sum of the inputs as the scheduling variable, an example rule base is:

$$\begin{aligned} &\textit{If } x + y \textit{ is Neg then } \hat{z} = \frac{1}{2}x + \frac{1}{2}y + \frac{1}{8} \\ &\textit{If } x + y \textit{ is Pos then } \hat{z} = \frac{1}{2}x - \frac{1}{4}y + \frac{1}{4} \end{aligned}$$

Figure 9.5(a) shows the result. The scheduling variable $x + y$ is maximally positive when $x = y = 100$. *In that corner of the surface, the second rule is fully activated, while the first rule is inactive. Likewise in the diagonally opposite corner. The normal vectors control the tangent planes in those two corners. The surface is smooth everywhere.*

The membership functions are triangular with universes $[-200, 200]$. *The implementation performs a gain scheduling on the weight vectors* $\mathbf{w}_1 = \begin{bmatrix} \frac{1}{2} & \frac{1}{2} & \frac{1}{8} \end{bmatrix}$, $w_2 = \begin{bmatrix} \frac{1}{2} & -\frac{1}{4} & \frac{1}{4} \end{bmatrix}$.

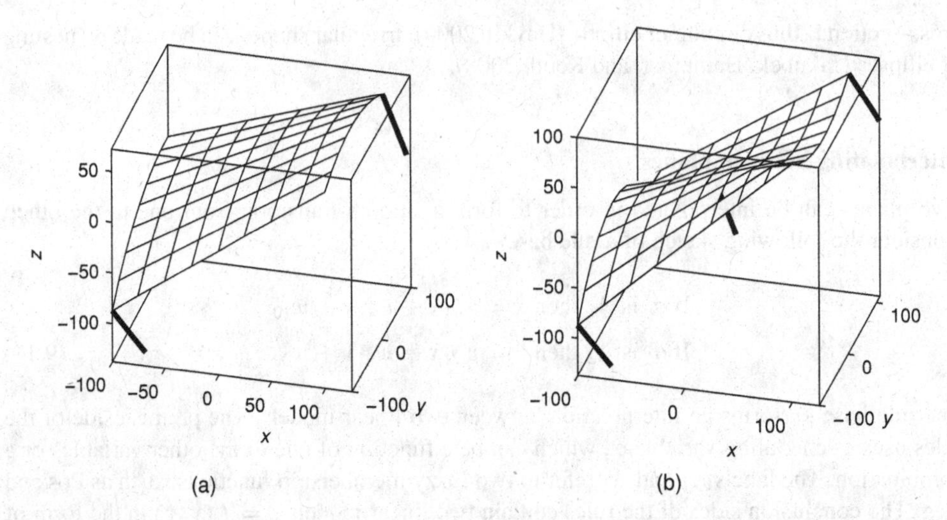

Figure 9.5 Interpolation between two planes (a), and join of two surfaces (b). Each local plane is determined by a normal vector (bold line). (figjoinsurf.m)

Introducing concentration of the membership functions ($c > 1$) will make the surface tend to a deadzone characteristic, with a nearly horizontal tangent plane in the origin of the coordinate system.

Example 9.9 *Joining two surfaces*

Is it possible to join two surfaces of the kind in the previous example (Example 9.8) over a common third surface?

► *Solution*

Using the sum of the inputs as the scheduling variable as previously, an example rule base is the following:

$$\textit{If } x + y \textit{ is Neg then } \hat{z} = \frac{1}{2}x + \frac{1}{2}y + \frac{1}{8}$$

$$\textit{If } x + y \textit{ is Zer then } \hat{z} = \frac{1}{2}x - \frac{1}{4}y + \frac{1}{4}$$

$$\textit{If } x + y \textit{ is Pos then } \hat{z} = \frac{1}{2}x + \frac{1}{4}y + \frac{1}{8}$$

Figure 9.5(b) shows the result. The conclusion sides of the first two rules are the same as previously (Example 9.8). The third rule contributes a third plane. The planes from rules 1 and 3 are joined together via the plane in rule 2.

The family of the three membership functions Neg, Zer, Pos cover the whole range $[-200, 200]$ of the scheduling variable $x + y$. The membership functions are triangular with peaks at -200, 0, and 200 respectively, they cross at $\mu = 0.5$, and pairwise they sum to one.

9.3 Machine-Made Models

If a program builds a model entirely from (measurement) data, the model is machine-made. Technically speaking, algorithms that build models without human intervention are *unsupervised*. For example, fitting a straight line to a collection of data is mostly unsupervised – 'mostly', because the program must be 'told' by a human being that it should look for a straight line, not a parabola, or something else.

The primary purpose of fitting a curve to discrete points is to find a function which cuts through the collection of data in an optimal way. Since measurement data usually include noise, the function must pass between points, in order to map the underlying functional relationship in the simplest possible way. The fitted function should not try to model the noise.

9.3.1 Least-Squares Line Fit

If the function to be approximated is y and the approximating function is $\hat{y}$, then the *least-squares* fit minimizes the difference between the two. The difference is minimized if the error measure $E = \frac{1}{2}\sum(y_k - \hat{y}_k)^2$ is minimized. The factor $\frac{1}{2}$ is unimportant; it is only there for cosmetic reasons in the likely event that E must be differentiated.

Relatively low-order polynomials are often the most useful for data fitting; higher-order polynomials tend to reproduce the noise in the data. The most widely used functions are straight lines, and data are often transformed and replotted – on logarithmic scales, for instance – until the data set assumes a form such that a straight line is a sufficient approximation.

For instance, suppose that the function $y = w_1 x + w_0$ relates an input x to a process output y in an experiment that generates noisy data. The objective is then to find the weights w_j that best fit the data. In fact, the experiment generates many samples of data, each giving rise to one such equation. There is only one set of weights for all equations, however. Generally, in vector-matrix notation, the relationship we are searching, is:

$$\hat{\mathbf{y}} = X\mathbf{w} \tag{9.11}$$

In order to determine $\mathbf{w}$ we insert the actual observations $\mathbf{y}$ in place of $\hat{\mathbf{y}}$. The length of $\mathbf{y}$ corresponds to the number of data samples, and the length of $\mathbf{w}$ corresponds to the number of measurement channels less the y-channel. The matrix X, which contains the actual input measurements, has a shape that conforms with the two vectors, and thus X has typically more rows than columns. When X is not square it cannot be inverted, unfortunately, but we can pre-multiply by its transpose to find a solution,

$$\begin{aligned} X^T\mathbf{y} &= X^T X\mathbf{w} \Leftrightarrow \\ \left(X^T X\right)^{-1} X^T\mathbf{y} &= \mathbf{w} \end{aligned} \tag{9.12}$$

provided $X^T X$ can be inverted. It can be shown that this solution does indeed minimize E, and the solution is unique (see, for example, Åström and Wittenmark 1995). The condition that $X^T X$ be invertible is called an *excitation* condition. It means that the experiment should be designed such that the measured data samples contain sufficient variation to avoid the

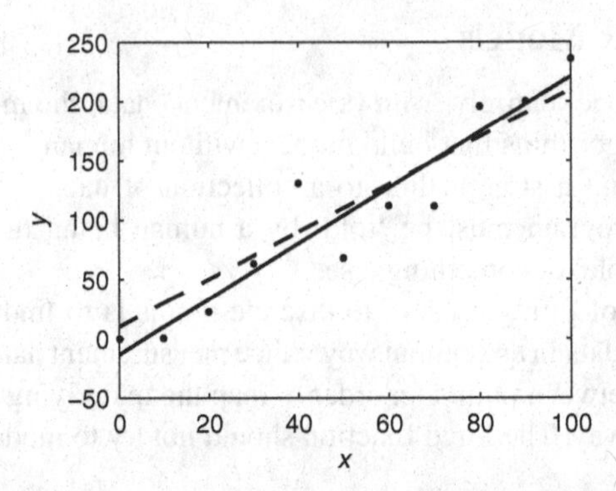

Figure 9.6 Least-squares fit of a straight line (solid line) to noisy data (dots). The true function is shown also (dashed line). (figls.m)

determinant of $X^T X$ becoming close to zero. As an example of how this might happen, take data that are sampled at a high rate from a slowly varying process; the data samples will be very similar, the rows of X will be almost linearly dependent, and $X^T X$ will be close to singular (*ill-conditioned*). The data should thus be selected intelligently, in order to portray the underlying function.

Example 9.10 *Least-squares line fit*

Given a set of data points, how can a machine make a straight line model?

► *Solution*

The starting point is Equation (9.11). MATLAB® has a function (`mldivide`) *that implements the least-squares fit given a matrix X and a vector* **y**. *A shorthand syntax is* $\mathbf{w} = X \setminus \mathbf{y}$ *which is to be understood as X divided into* **y** *('matrix left divide'). Assuming a model with the structure* $y = w_1 x + w_0$, *the matrix X consists of the input vector* **x** *in the first column, and a second column of ones to allow for the offset* w_0. *The vector* **y** *is the vector of noisy response measurements.*

Figure 9.6 shows the line fit. The data set consists of 11 points, and the straight line minimizes the sum of the squared vertical *distances from the points to the line. The data were constructed by the function* $y = 2x + 10 + n$, *where n is a normally distributed noise term with a standard deviation of 20. However, the least-squares optimizer does not 'know' this. In fact, it finds* $w_1 = 2.35$, $w_0 = -13.6$, *which is somewhat different from the true values (owing to the small size of the data set).*

9.3.2 *Least-Squares Basis Function Fit*

The previous section showed that the least-squares algorithm finds a model, almost on its own, given data and the structure of the model, which must be decided beforehand. It is a fairly general method, as there can be as many input channels as needed. Furthermore, the input channels can be functions of variables, such as, for instance, basis functions. The following example demonstrates how to fit a basis function model.

Example 9.11 *Basis function fit (figlsbasis.m)*

Given the same data points as previously (Example 9.10), is it possible to fit a Sugeno rule base?

► *Solution*

The Sugeno singleton model (Equation (9.5)) uses two adjustable constants on the conclusion side of the rules. They correspond to the weights in a basis function model, and they can be adjusted by the least-squares method.

The model is equivalent to the basis function form,

$$\hat{y}(x) = \mu^c_{Low}(x)\,k_1 + \mu^c_{High}(x)\,k_2 \qquad (c = 1)$$

We set the concentration exponent $c = 1$, because we are looking for a linear model. The universe is $x \in [0, 100]$, in accordance with the previous example, and we choose triangular membership functions for μ_{Low} and $\mu_{High} = 1 - \mu_{Low}$ that span the whole universe. For each data point, its x-coordinate generates a membership value. Referring to the least-squares model (Equation (9.11)), the X matrix consists of two columns:

$$X = \begin{bmatrix} \mu_{Low}(x_k) & \mu_{High}(x_k) \end{bmatrix}$$

The index k refers to the data sample ($k = 1, 2, \ldots, 11$). By means of MATLAB®'s backslash-operator (\), the solution to the least-squares problem (Equation (9.12)) is found to be

$$\begin{bmatrix} k_1 \\ k_2 \end{bmatrix} = \begin{bmatrix} -13.6 \\ 221 \end{bmatrix}$$

The solution is the same as in Figure (9.6). This is confirmed by inserting $x = 0$ and $x = 100$ in the equation for the previously found model $\hat{y} = 2.35x - 13.6$. The results check with the values in **w**. *Translating back to the rule base, the resulting model is*

$$\text{If } x \text{ is Low then } \hat{y} = -13.6$$

$$\text{If } x \text{ is High then } \hat{y} = 221$$

The example illustrates that the least-squares method is a useful 'workhorse', as it is well understood, and there are good computer implementations. It can adjust the singletons in a Sugeno rule base, in order to fit a straight line to nonlinear data samples. More complex models can be built by joining local, linear models piece by piece.

Generally speaking, the procedure for building a machine-made model consists of a forward pass and a backward pass. The forward pass determines the conclusion side of a rule base, by means of the least-squares method, while the backward pass adjusts the membership functions, by means of an optimization routine. Usually, the two passes must be repeated iteratively, in order to find the best fit.

Algorithm 9.1 is a relatively simple example of such a procedure. The algorithm joins two linear models based on a set of measured, noisy data points. The data set must be split at an appropriate place, and this is so far done manually (Step 1). The following example demonstrates the algorithm.

Algorithm 9.1 Joining two local linear models.

1. Split the data set in two, in an appropriate place.
2. Perform a linear line fit on each subset using the basis function form of the singleton rule base (Equation (9.6)).
3. The singletons define two tangents for the interpolant.
4. Define the origin of the basis vectors as the intersection of the two tangents.
5. Extend the vectors to the edges of the region of interest.
6. Apply the rule base with vector conclusions (Equation (9.7)) to derive the interpolant.
7. Adjust the concentration c to make the best fit. Use a routine for nonlinear optimization that minimizes an error function, for instance the sum of the squared errors.

Example 9.12 *Join of two line models*

Given the synthetic data set $y = \sin x + n$ $(0 \leq x \leq \pi)$, where n is normally distributed (Gaussian) noise with standard deviation 0.1. How does the modelling proceed, and how good is the model fit?

► *Solution*

Since we know the true function that generated the data, we can compare the true and the approximated function. It is a half sine wave with one peak, and we therefore try to join two linear models. Figure 9.7 shows the result. The procedure, in accordance with Algorithm 9.1, is the following:

1. *We split the data set at the x-coordinate 1.6. This is the x-coordinate of the data point which is closest to the middle of the x-range.*
2. *The approximation of the singleton rule base (Equation (9.6)) resulted in the conclusions $k_{11} = 0.096$, $k_{12} = 1.11$, for the left line and $k_{21} = 1.13$, $k_{22} = 0.095$ for the right line.*
3. *The conclusion singletons are associated with the end points of the two universes, and that determines the linear models.*

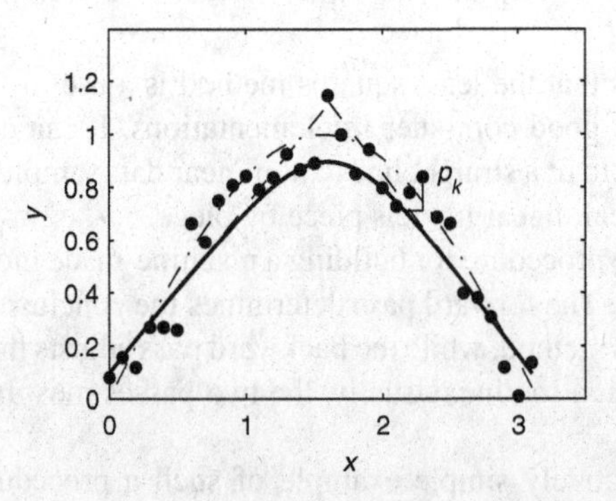

Figure 9.7 Model fit (solid line) of two local linear models (dashed lines). The true function is also shown (dash-dot line). The thin lines from the data point p_k indicate that the corresponding point of the model has a different x-coordinate. The model error is the vertical distance. (figls2basis.m)

4. *The intersection is found by equating the expressions for the tangents.*
5. *The end points of the basis vectors have x-coordinates corresponding to the outermost data points.*
6. *The basis vectors are inserted in the rule base in Equation (9.7) with concentration $c = 2$ as a starting value.*
7. *The MATLAB® optimizer* `fminsearch` *minimized the sum of the squared (vertical) errors, and found $c = 3.2$.*

The figure shows that the model does not quite follow the true function. The fit could be improved by including a third, horizontal linear model near the peak. This would turn the tangents more in the direction of the true tangents.

The example showed that it was possible to model a data set with very little human intervention, in this case splitting the data set. The accuracy can be improved by increasing the number of local models, or, in other words, performing more splits of the data. Although more local models will result in a better fit, they may start to model the noise by *overfitting* the data. Oppositely, *underfitting* may occur if there are too few local models. A compromise must be made in order to avoid either overfitting or underfitting.

9.4 Cluster Analysis

The objective of *cluster analysis* is to partition a data set into a number of natural and homogeneous subsets, where the elements of each subset are as similar to each other as possible, and at the same time as different from those of the other sets as possible. A *cluster* is such a grouping of objects. It consists of all objects close, in some sense, to the *cluster centre*. Cluster analysis can provide a partitioning for local linear models.

Figure 9.8 shows a cluster of 51 measurements of two variables x_1 and x_2, which are related by the following underlying process

$$x_2 = 0.5\,(x_1 + n) + 5 \qquad (5 \leq x \leq 10)$$

The input variable x_1 is contaminated by noise n, which is normally distributed with a standard deviation of 3. The noise propagates through the process, and since it is linear, the noise associated with x_2 is also normally distributed. Thus, the data points cluster around a centre in the scatter plot of x_2 against x_1. The first coordinate of the centre is the mean value m_1 of the measurements of x_1, and the second coordinate is the mean value m_2 of the measurements of x_2.

9.4.1 *Mahalanobis Distance*

A *bivariate* normal distribution is a two-dimensional version of the normal distribution with a remarkable property: its projections on the axes are normally distributed as well. For that reason, it is practical to assume a bivariate normal distribution for observations of two variables that are approximately normally distributed. Since x_2 depends on x_1, the cluster is not circular, but elliptical with axes (ξ_1, ξ_2) that are rotated and translated relative to the coordinate axes

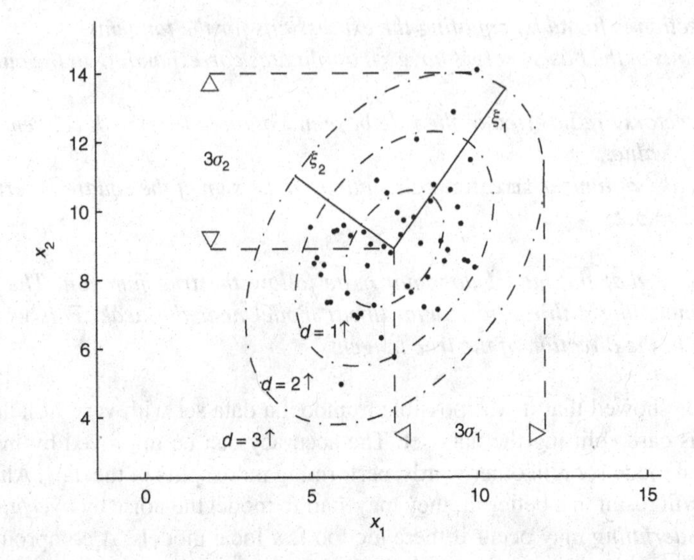

Figure 9.8 Scatter plot of data points around a cluster centre. The elliptical shape is defined by two principal axes (ξ_1, ξ_2). The ellipses (dash-dotted line) indicate Mahalanobis distances ($d = 1, 2, 3$) from the centre. (figmahalanobis.m)

(x_1, x_2). The level contours of a bivariate normal distribution are, as a consequence, ellipses (Duda, Hart, and Stork 2001).

Earlier we said that a cluster consists of all points close the cluster centre, but since the cluster is elliptical, the usual distance measure is impractical. For example, a change of measurement unit from, say, kilometres to metres increases one variable by three orders of magnitude which may dominate the distance measure. It is therefore more appropriate to measure the distance in terms of standard deviations, as in the *Mahalanobis distance* measure. Referring to Figure 9.8, three ellipses indicate the Mahalanobis distance $d = 1, 2, 3$ from the centre. In a large collection of normally distributed data points, 68% will have d less than one, 95% will have d less than two, and 99.7% will have d less than three. The figure also indicates the magnitude and location of three standard deviations ($3\sigma_1, 3\sigma_2$) of the measured variables (x_1, x_2); they are the projections of the ellipse corresponding to $d = 3$ on the x_1-axis and the x_2-axis, respectively. From the local viewpoint of the bivariate distribution, the intersections of the innermost ellipse ($d = 1$) with the axes (ξ_1, ξ_2) indicate the standard deviations of two local variables. It is possible to transform the x-coordinates to ξ-coordinates by a certain variable substitution, in order to work with variables that are statistically independent. However, the transformation obscures the physical meaning of the original coordinates.

In summary, given a set of measurements of normally distributed dependent variables, the cloud of data points can be approximated by an elongated ellipse (hyper-ellipse in many dimensions) with orthogonal *principal axes*. By disregarding the extent of the distribution in the direction of the shortest axis, the cluster centre and the longest axis determine a linear approximation to the data, which is our local linear model. The method is easily generalized to multiple dimensions, where *principal component analysis* (Duda, Hart, and Stork 2001) seeks a projection that best represents the data in a least-squares sense.

Example 9.13 *Major axis model*

Given the set of data points from Figure 9.8, how do we find the linear model?

► *Solution*

The principal component approach proceeds in the following manner. First, compute the mean values (m_1, m_2) *and the covariance matrix* S *for the data set. Next, compute the eigenvectors* $\mathbf{e}_i$ *and eigenvalues* λ_i *of* S*, and sort them according to decreasing eigenvalue* λ_1, λ_2*. Choose the eigenvector corresponding to the largest eigenvalue* λ_1*.*

The semi-major axis of an ellipse starts in the cluster centre (m_1, m_2)*, and extends to* $d\sqrt{\lambda_1}$*, where* d *is the Mahalanobis distance. The linear model is thus, on vector-matrix form,*

$$\hat{\mathbf{p}} = t\sqrt{\lambda_1}\mathbf{e}_1 + \mathbf{m} \qquad (-d \leq t \leq d)$$

The vector $\mathbf{m} = (m_1, m_2)$ *is the cluster centre, and* t *is a parameter. The limit* d *can be adjusted to fit the region of interest.*

The covariance matrix between the vector of input data $\mathbf{x}_1$ *and the vector of output data* $\mathbf{x}_2$ *can be computed by means of an outer product:*

$$S = \frac{1}{K-1}(\mathbf{x}_1 - m_1)(\mathbf{x}_2 - m_2)^T$$

where K *is the number of data samples (the length of* $\mathbf{x}_1$*).*

An ellipse is drawn by using the eigenvectors as an orthonormal basis for the local coordinate system. Thus, the unit circle will be mapped to an ellipse. Assuming that the eigenvalues are laid out in the diagonal of an otherwise zero matrix Λ*, and the corresponding eigenvectors are held in the columns of a matrix* V*, then the following mapping generates the ellipse for the Mahalanobis distance* d*:*

$$\hat{\mathbf{p}} = V\sqrt{\Lambda}\begin{bmatrix}\cos a \\ \sin a\end{bmatrix} d + \mathbf{m} \qquad (0 \leq a \leq 2\pi)$$

The square root just means that we take the square root of each diagonal element.

In general, the Mahalanobis distance from a data vector $\mathbf{x}$ to the cluster centre $\mathbf{m}$ is defined by

$$d^2 = (\mathbf{x} - \mathbf{m})^T S^{-1} (\mathbf{x} - \mathbf{m}) \tag{9.13}$$

In practice, the cluster centre $\mathbf{m}$ and the covariance matrix S may have been established already from previous data, which constitute a so-called *training set*. If $\mathbf{x} = \mathbf{x}'$ is a new data vector, then d is its distance to the already established cluster. If there are several clusters, the distance from $\mathbf{x}'$ to each cluster is calculated, where the distance depends on the shape of the cluster. The new point $\mathbf{x}'$ can thus be classified as belonging to the nearest cluster.

Example 9.14 *Geometry of the Mahalanobis distance*

Why is the Mahalanobis distance defined by means of an inverse matrix?

► *Solution*

The distance measure that Equation (9.13) defines is the square of a radial vector inside the cluster. This can be seen by a coordinate change.

The covariance matrix S describes the cluster. When it is constructed from noisy data, such that the two variables are not exactly dependent, the eigenvectors of S are orthogonal. The covariance matrix can be decomposed by its modal matrix$V$, *in which the columns are its eigenvectors. That is,*

$$SV = V\Lambda \Leftrightarrow S = V\Lambda V^{-1} \tag{9.14}$$

The matrix Λ contains the eigenvalues in the diagonal, with zeros elsewhere. The eigenvectors constitute an orthonormal basis for the cluster, owing to the symmetry of the covariance matrix; a point ξ, with local coordinates (ξ_1, ξ_2), corresponds to a point $\mathbf{x}$ in global coordinates. The following equations define the transformations back and forth between the two coordinate systems:

$$\begin{aligned} \mathbf{x} &= V\sqrt{\Lambda}\xi + \mathbf{m} \Leftrightarrow \\ \xi &= \sqrt{\Lambda^{-1}}V^{-1}(\mathbf{x} - \mathbf{m}) \end{aligned} \tag{9.15}$$

Here we included the translation by the vector $\mathbf{m}$ *to the cluster centre in the global coordinate system. The square roots scale the coordinates since Λ is diagonal. The point $(\xi_1, \xi_2) = (1, 0)$ is located on the first axis at the Mahalanobis distance 1, which is equivalent to one standard deviation locally. Similarly, the point $(\xi_1, \xi_2) = (0, 1)$ is located on the second axis, also at the Mahalanobis distance 1. The length in x-coordinates of the semi-axes of the $d = 1$ ellipse is the square root of the corresponding eigenvalues. Equation (9.15) shows that if a point (ξ_1, ξ_2) travels on the unit circle, the circle will be mapped to an ellipse by the matrix $V\sqrt{\Lambda}$, since a matrix always maps a circle to an ellipse, specifically a circle. The equation can be used to draw the ellipses on the computer screen.*

The squared length of a point in local coordinates is thus, in global coordinates,

$$\begin{aligned} \xi \cdot \xi &= (\mathbf{x} - \mathbf{m})^T \left(V^{-1}\right)^T \left(\sqrt{\Lambda^{-1}}\right)^T \sqrt{\Lambda^{-1}}V^{-1}(\mathbf{x} - \mathbf{m}) \\ &= (\mathbf{x} - \mathbf{m})^T \left[\left(V^{-1}\right)^T \Lambda^{-1} V^{-1}\right](\mathbf{x} - \mathbf{m}) \end{aligned}$$

Here we used $\left(\sqrt{\Lambda^{-1}}\right)^T \sqrt{\Lambda^{-1}} = \Lambda^{-1}$ as Λ is diagonal. The matrix in square brackets [ ] is itself a decomposition of a matrix, namely S^{-1}. Thus, the result is

$$d^2 = \xi \cdot \xi = (\mathbf{x} - \mathbf{m})^T S^{-1} (\mathbf{x} - \mathbf{m})$$

as in Equation (9.13). The final step used that $V^{-1} = V^T$, since the eigenvectors are orthonormal, and, from Equation (9.14), $S^{-1} = V\Lambda^{-1}V^{-1}$.

9.4.2 *Hard Clusters, HCM Algorithm*

The *hard c-means* (HCM) algorithm tries to locate clusters in a multi-dimensional space. The goal is to assign each point in the space to a particular cluster. The basic approach is as follows (Lewis 1990).

1. Manually seed the algorithm with c cluster centres, one for each cluster we are seeking. This requires prior information from the outside world of the number of clusters into which the points are to be divided; thus the algorithm is *supervised*, strictly speaking.
2. Assign each point to the cluster centre nearest to it.
3. A new cluster centre is computed for each class by taking the mean values of the coordinates of the points assigned to it.
4. If not finished, according to some stopping criterion (see later), go back to step 2.

Additional rules can be added that relax the requirement to specify precisely the desired number of clusters. Such rules allow nearby clusters to merge, or they allow clusters to split, if they have large standard deviations in the coordinates.

Example 9.15 *Two clusters*

Consider the data points plotted in Figure 9.7. Can the HCM algorithm find two clusters, that can be candidates for linear local models?

► *Solution*

The cluster algorithm might find one centre in the left half-plane and another in the right half-plane, if we specify that it is supposed to look for two clusters. The initial centres can be more or less in the middle of the plot, and during the iterations they should move towards their final positions. Afterwards, each point belongs to one or the other class, so the clusters are crisp. In general, the measurement data have to be normalized to roughly the same range, in order for the distance measure to work properly.

The hard c-means algorithm is based on a c*-partition* of the data space U into a family of clusters $\{C_i\}(i = 1, 2, \ldots, c)$, where the following set-theoretic equations apply,

$$\bigcup_{i=1}^{c} C_i = U \tag{9.16}$$

$$C_i \cap C_j = \emptyset \quad (i \neq j) \tag{9.17}$$

$$\emptyset \subset C_i \subset U \tag{9.18}$$

The set $U = \{\mathbf{u}_1, \mathbf{u}_2, \ldots, \mathbf{u}_K\}$ is a finite set of points in a multi-dimensional space, and c is the number of clusters. Equations (9.16)–(9.18) express the following: the clusters exhaust the whole universe, none of the clusters overlap, and a cluster can neither be empty nor contain all data samples. Note that

$$1 < c < K \tag{9.19}$$

since $c = K$ clusters just places each data sample into its own cluster, and $c = 1$ places all data samples into the same cluster.

Formally, the c-means algorithm finds a centre in each cluster that minimizes an objective function. The objective function depends on the distances between each vector $\mathbf{u}_k$ and the cluster centres $\mathbf{c}_i$, as follows:

$$J = \sum_{i=1}^{c} J_i = \sum_{i=1}^{c} \left(\sum_{k,\mathbf{u}_k \in C_i} \|\mathbf{u}_k - \mathbf{c}_i\|^2 \right) \tag{9.20}$$

Here, J_i is the objective function related to distances within cluster i.

The partitioned clusters are typically defined by a $c \times K$ binary characteristic matrix M, called the *membership matrix*, where each element m_{ik} is 1 if the kth data point $\mathbf{u}_k$ belongs to cluster i, and 0 otherwise. Since a data point can only belong to one cluster, the membership matrix M has the properties:

–the sum of each column is one; and
–the sum of all elements is K. (9.21)

If the cluster centres $\mathbf{c}_i$ are fixed, the m_{ik} that minimize J_i can be derived as follows:

$$m_{ik} = \begin{cases} 1 & \text{if } \|\mathbf{u}_k - \mathbf{c}_i\|^2 \leq \left\|\mathbf{u}_k - \mathbf{c}_j\right\|^2 \quad (j \neq i) \\ 0 & \text{otherwise} \end{cases} \tag{9.22}$$

That is, $\mathbf{u}_k$ belongs to cluster i, if $\mathbf{c}_i$ is the closest centre among all centres. If, on the other hand, m_{ik} is fixed, then the optimal centre $\mathbf{c}_i$ that minimizes (9.1) is the mean of all vectors in cluster i,

$$\mathbf{c}_i = \frac{1}{|C_i|} \sum_{k,\mathbf{u}_k \in C_i} \mathbf{u}_k \tag{9.23}$$

where $|C_i|$ is the number of objects in C_i, and the summation is an element-by-element summation of vectors.

The distance measure in Equation (9.22) is usually just the Euclidean distance between the data point and the centre. When the clusters differ in size or elongation, it is more appropriate to use the Mahalanobis distance.

Algorithm 9.2 Hard c-means (HCM).

---

1. Select c cluster centres $\mathbf{c}_i$ $(i = 1, 2, \ldots, c)$. For example, c random points from the set of data points.
2. Assign each object to a centre in M using Equation (9.22).
3. Compute new cluster centres $\mathbf{c}_i$ $(i = 1, 2, \ldots, c)$ using Equation (9.23).
4. Compute the objective function according to Equation (9.1). Stop, if either it is below a certain threshold level or its improvement over the previous iteration is below a certain tolerance. Otherwise go back to step 2.

Algorithm 9.2 summarizes the HCM clustering. Instead of Step 1, it is also possible to initialize a random membership matrix M in Step 2, and then follow the steps from there. The algorithm is iterative, and there is no guarantee that it will converge to an optimum solution. The performance depends on the initial positions of the cluster centres, and it is advisable to employ some method to find good initial cluster centres.

The performance greatly depends on the distance measure used in Step 2 when assigning each object to a centre. The demonstration example in Section 10.5 uses a single-coordinate distance, which separates the clusters in rectangular areas.

Example 9.16 *Two clusters, continued*

Consider again the data points plotted in Figure 9.7. Can the HCM algorithm partition the data set into two partitions, instead of having a human being do this?

► *Solution*

The algorithm moves the cluster centres around until the objective function is minimal. Figure 9.9 shows the result. The algorithm partitions the data into a cluster of 17 points and a cluster of 15 points. The algorithm was 'told' to find two centres, and it was initialized with two arbitrary centre locations, which are on the axes. They are unrelated to the final centres, and the algorithm converges towards a final solution.

The objective function, which is the sum of all squared distances, decreases rapidly and levels out, indicating convergence. When the objective function changes less than some threshold, the algorithm can be stopped.

In the example, the HCM algorithm found acceptable clusters that were rather similar to the split made by hand (Example 9.12). Each of these is a candidate for a linear model. It is a straightforward task to approximate a cluster by a line, either by taking the semi-major axis or by using a least-squares line fit (the two methods produce different results in general).

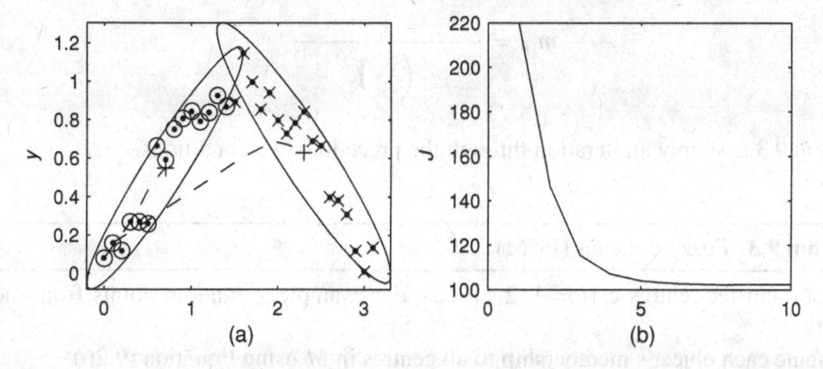

Figure 9.9 HCM clustering. The data points (a) have been partitioned into two clusters (symbols o, and x). The initial centres are two arbitrary data points, and they moved along two paths (dashed lines) to the final locations (+). The ellipses correspond to Mahalanobis distance $d = 2$. The plot of the objective function J in (b) shows that the algorithm could be stopped earlier, since there is very little change beyond the fifth iteration. (fighcm.m)

9.4.3 Fuzzy Clusters, FCM Algorithm

In the case of two overlapping clusters, it is reasonable to assume that points in the overlapping region have a fuzzy membership of *both* clusters. The fuzzy c-means algorithm (FCM) allows each data point to belong to a cluster to a degree specified by a membership grade, and thus each point may belong to several clusters (Bezdek in Ross 2010).

FCM partitions a collection of K data points specified by m-dimensional vectors $\mathbf{u}_k$ $(k = 1, 2, \dots, K)$ into c fuzzy clusters, and finds a cluster centre in each while minimizing an objective function. Fuzzy c-means is different from hard c-means, mainly because of a *fuzzy partitioning,* where a point can belong to several clusters with degrees of membership. This affects the calculation of cluster membership, cluster centres, and the objective function.

The membership matrix M is allowed to have elements in the range [0, 1]. A point's total membership of all clusters is equal to unity to maintain the properties (9.21) of the $\mathbf{M}$ matrix. The objective function is a generalization of (9.20):

$$J(M, \mathbf{c}_1, \mathbf{c}_2, \dots, \mathbf{c}_c) = \sum_{i=1}^{c} J_i = \sum_{i=1}^{c} \sum_{k=1}^{K} m_{ik}^q d_{ik}^2, \tag{9.24}$$

where m_{ik} is a membership between 0 and 1, $\mathbf{c}_i$ is the centre of cluster i, $d_{ik} = \|\mathbf{u}_k - \mathbf{c}_i\|$ is the distance of the ith cluster centre from the kth data point, and $q \in [1, \infty)$ is a *weighting parameter* (Bezdek in Ross 2010). There are two necessary conditions for J to reach a minimum,

$$\mathbf{c}_i = \frac{\sum_{k=1}^{K} m_{ik}^q \mathbf{u}_k}{\sum_{k=1}^{K} m_{ik}^q}, \tag{9.25}$$

and

$$m_{ik} = \frac{1}{\sum_{j=1}^{c} \left(\frac{d_{ik}}{d_{jk}} \right)^{2/(q-1)}} \tag{9.26}$$

Algorithm 9.3 is simply an iteration through the preceding two conditions.

Algorithm 9.3 Fuzzy c-means (FCM)

1. Select c cluster centres $\mathbf{c}_i$ $(i = 1, 2, \dots, c)$. For example, c random points from the data points.
2. Compute each object's membership to all centres in M using Equation (9.26).
3. Compute new cluster centres $\mathbf{c}_i$ $(i = 1, 2, \dots, c)$ using Equation (9.25).
4. Compute the objective function according to Equation (9.24). Stop if either it is below a certain threshold level or its improvement over the previous iteration is below a certain tolerance. Otherwise go back to step 2.

The membership matrix M can be initialized first, as an alternative, with random values between zero and one, observing the constraints in Equation (9.21). Then carry out the iterations (Steps 2–5). The algorithm may not converge to an optimum solution and the performance depends on the initial cluster centres, just as in the case of the hard c-means algorithm. There are a number of variants of FCM that concern the distance measure and the shape of clusters (Bezdek in Babuška 1998).

The HCM and FCM algorithms are similar, and they only differ in the calculation of cluster membership and distance. The HCM algorithm requires fewer iterations to converge, but the FCM algorithm allows a shared membership. Which one is best depends on the situation. The previous examples used HCM, and it was appropriate, because we were looking for hard partitions. In some classification problems, however, it is natural for the objects to have a shared membership of several classes.

Example 9.17 *Physical analogy to the membership matrix M in FCM*

The definition of the membership matrix M in Equation (9.26) is rather complex. Can it be explained in another way?

► *Solution*

It is difficult to find an explanation in the literature. There is, however, an analogy to the way bodies of mass attract each other by gravitational forces. Loosely explained, a point's membership of a cluster can be seen as its relative attraction with respect to the total attraction from all clusters – the closer, the larger. To demonstrate this, the first step is to expand the expression in Equation (9.26), as follows:

$$\begin{aligned} m_{ik} &= \frac{1}{\sum_{j=1}^{c}\left(\frac{d_{ik}}{d_{jk}}\right)^{2/(q-1)}} \\ &= \frac{1}{\left(\frac{d_{ik}}{d_{1k}}\right)^{2/(q-1)} + \left(\frac{d_{ik}}{d_{2k}}\right)^{2/(q-1)} + \cdots + \left(\frac{d_{ik}}{d_{ck}}\right)^{2/(q-1)}} \\ &= \frac{\frac{1}{d_{ik}^{2/(q-1)}}}{\frac{1}{d_{1k}^{2/(q-1)}} + \frac{1}{d_{2k}^{2/(q-1)}} + \cdots + \frac{1}{d_{ck}^{2/(q-1)}}} \end{aligned} \tag{9.27}$$

Notice that Equation (9.27) consists of inverse fractions. The numerator is the attraction to cluster i and the denominator is the sum of attractions to all clusters.

Figure 9.10 illustrates the interaction between a body with a mass m_k and two other bodies of mass m_1 and m_2. Only two others are included for simplicity, but generalization to c other bodies is straightforward. Distances are defined in the figure. According to Newton, any two bodies, say, number k and number 1, attract each other with a force according to the expression

$$F_{1k} = G\frac{m_k m_2}{d_{1k}^2} \tag{9.28}$$

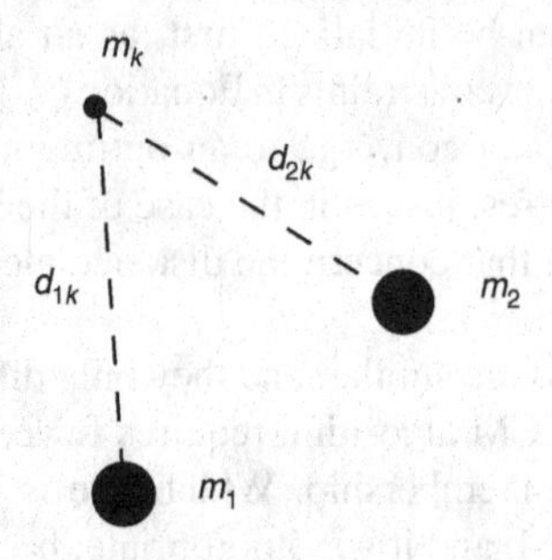

Figure 9.10 Newton's *Law of Universal Gravitation*. Physical bodies of mass interact with each other at a distance.

The symbol G stands for the gravitational constant. Accordingly, the sum of all forces is, in our case,

$$F = F_{1k} + F_{2k} = G\frac{m_k m_1}{d_{1k}^2} + G\frac{m_k m_2}{d_{2k}^2} \tag{9.29}$$

In order to see the analogy, we form the ratio of Equation (9.28) to Equation (9.29), whereby the gravitational constant cancels out:

$$\frac{F_{1k}}{F} = \frac{G\frac{m_k m_2}{d_{1k}^2}}{G\frac{m_k m_1}{d_{1k}^2} + G\frac{m_k m_2}{d_{2k}^2}} = \frac{\frac{m_2}{d_{1k}^2}}{\frac{m_1}{d_{1k}^2} + \frac{m_2}{d_{2k}^2}} \tag{9.30}$$

Setting masses $m_1 = m_2$, and extending to c masses, apart from m_k, then Equation (9.30) assumes a form similar to Equation (9.27) with $q = 2$. We thus interpret an object's membership of a cluster as the attraction to that cluster relative to the total attraction acting on the object, as determined by the distances.

Example 9.18 *Weighting parameter q*

What is the effect of q, and what is a good value?

► *Solution*

A typical value is $q = 2$. In that case, distances are squared, which is the usual practice.

Figure 9.11 is a plot of the membership of a data point to four cluster centres $\mathbf{c}_i$ $(i = 1, 2, 3, 4)$ at increasing distances. The figure confirms first of all that the membership decreases as distance increases. With a weighting q close to unity ($q = 1.1$), the membership of the nearest cluster is almost one, while the membership of each of the remaining clusters is zero. With more weighting ($q = 2$), memberships decay less with distance. As q increases further, the object's memberships tend to even out, such that they become more or less the same for all clusters. In all cases, the object's memberships of the four clusters sum to one.

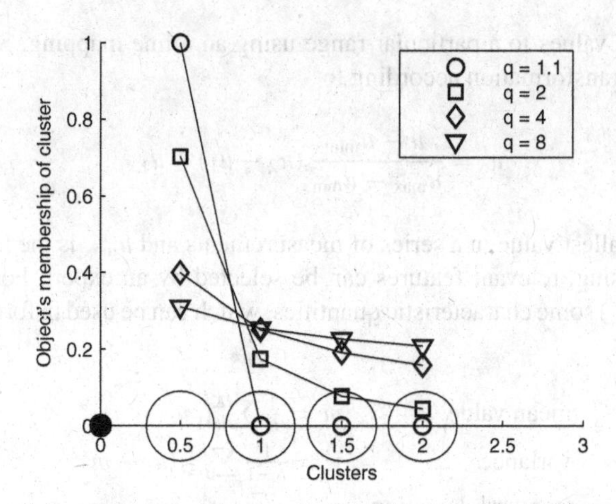

Figure 9.11 Membership of four clusters for various values of the weighting parameter q. The object (dot) is at the origin. The large circles indicate four clusters at increasing distances (0.5, 1, 1.5, 2) from the object. (figfcmq)

9.5 Training and Testing

It may seem relatively easy to obtain a model of an unknown process or component. In practice, however, several difficulties hinder the development. There is no rigid mathematics regarding such design choices, but this section provides a few rules of thumb.

In general, data analysis concerns objects which are described by *features*. The features form axes of an abstract *feature space* in which each object is represented by a point. Features can be selected directly, or they can be generated by a suitable combination of features. The latter option is necessary when a large number of features have to be reduced to a smaller number of features. Objects are fuzzy when one or more features are described in fuzzy terms. An example is a vehicle with a 'very fast' car engine, rather than top speed equal to some crisp number. Clusters are fuzzy when each object is associated with a degree of membership, rather than a crisp membership. An example is the cluster of 'sports cars'; a particular car would be a member to a degree depending on its top speed, air resistance, and weight.

Data conditioning

In practice, data conditioning (pre-processing) may be necessary.

Standardization transforms the mean of the set of feature values to zero, and the standard deviation to one. If the data are distributed according to the normal distribution with a mean value of m and standard deviation σ, the standardized values are found as:

$$u^* = \frac{u - m}{\sigma} \tag{9.31}$$

This is the basis for the Mahalanobis distance, since it measures the distance of a data value from the mean in terms of standard deviations.

Scaling takes the values to a particular range using an affine mapping. Scaling to a range $[u_1, u_2]$ is a linear transformation according to

$$u' = \frac{u - u_{\min}}{u_{\max} - u_{\min}}(u_2 - u_1) + u_1 \tag{9.32}$$

Here $u_{\min}$ is the smallest value in a series of measurements and $u_{\max}$ is the largest.

After pre-processing, relevant features can be selected by an expert. For a series of data $\mathbf{u} = (u_1, u_2, \ldots, u_K)$ some characteristic quantities, which can be used in forming the features, are

$$\begin{array}{ll} \text{mean value} & m = \frac{1}{K}\sum_{i=1}^{K} u_i \\ \text{variance} & v = \frac{1}{K-1}\sum_{i=1}^{K}(u_i - m)^2 \\ \text{standard deviation} & \sigma = \sqrt{v} \\ \text{range} & s = u_{\max} - u_{\min} \end{array} \tag{9.33}$$

Graphical representations, such as plots of frequency spectra and histograms, can support the selection.

Features may be correlated, that is, a change in one feature X is associated with a similar change in another feature Y. A *correlation coefficient* can be calculated as

$$r = \frac{\sum_{i=1}^{K}(x_i - m_1)(y_i - m_2)}{\sqrt{\sum_{i=1}^{K}(x_i - m_1)^2 \sum_{i=1}^{K}(y_i - m_2)^2}} \tag{9.34}$$

Here m_1 is the mean value of all the values x_i of feature X, and m_2 is the mean of all y_i values of the Y feature. The correlation coefficient assumes values in the interval $[-1, 1]$. When $r = -1$ there is a strong negative correlation between X and Y, when $r = 1$ there is a strong positive correlation, and when $r = 0$ there is no correlation at all. Strongly correlated features may indicate a linear dependency. If two features are linearly dependent, one of them is *redundant* (unnecessary); it is sufficient to select just one of them as a feature.

The *covariance matrix* S is the (square) matrix whose ijth element σ_{ij} is the covariance of observations x and y. Placing the observations of x in a vector $\mathbf{x}$ and the observations of y in $\mathbf{y}$, the covariance matrix is the outer product of a column-vector and a row-vector:

$$S = \frac{1}{N-1}(\mathbf{x} - m_1)(\mathbf{y} - m_2)^T$$

where N is the number of data samples (the length of $\mathbf{x}$). The matrix is symmetric, and the diagonal is a vector of variances. The off-diagonal elements are the covariances. If we first subtract the mean values $\mathbf{x}' = \mathbf{x} - m_1, \mathbf{y}' = \mathbf{y} - m_2$ and concatenate the two vectors into a matrix $X' = [\mathbf{x} \quad \mathbf{y}']$, then

$$S = \frac{1}{N-1}X'^T X'$$

This computes the outer product for each variable in a compact manner. MATLAB® has a function `cov` that finds the covariance matrix.

Stopping criteria

In principle, training continues until the weights stabilize and the average squared error over the entire training set converges to a minimum value. There are actually no well-defined criteria for stopping the algorithm's operation, but there are some reasonable criteria.

It is logical to look for a zero gradient of the error surface, since the gradient is zero in a minimum. A stopping criterion could therefore be to stop when the change of the objective function is less than a threshold. Training may take a long time, however. A typical threshold is 0.1 to 1% per training round (*epoch*); sometimes a value as small as 0.01% per epoch is used. A simple, empirical rule is to inspect the plot of the training curve to establish at which point the error starts to decrease more slowly, and stop the training manually. The simplest solution of all is to stop the training after a set number of epochs.

The number of training examples

If the architecture of the model is fixed, then an appropriate size for the training set must be determined. A practical rule of thumb is to require $K \geq 10 \times W$, where K is the number of training examples, and W is the number of parameters that must be adjusted. If the required K is infeasible in practice, then one option is to partition the training data and reuse them as in *multifold cross-validation* (Haykin 2009).

Training, test, and validation set

Training of a model starts with presenting a *training set* to the network. The objective is to train the model on the training set, such that it is able to make predictions based on 'unseen', future data. The network *generalizes* well when its output is nearly correct when presented with a test data set never used in training.

It is therefore necessary to split the available data into a training set and a *test set*. The model can be trained on the training set to reach a small error, but it is the error on the test set which is important. The goal is to minimize the objective function using test data, not training data. One strategy is therefore to train, test, adjust settings, and repeat. The training can stop when the test error is minimal, which can be seen on a plot of the objective function (for the test data).

The final test is to present the model with a third unseen set of data called the *validation set*. The model should not be allowed to update its parameters based on the validation set.

Overfitting and underfitting

Figure 9.12 shows examples of *overfitting* and *underfitting*. Overfitting occurs if the model contains too many adjustable parameters, such that the model order is higher than required or the model tries to fit to the noise in the training data. When this happens the error is low.

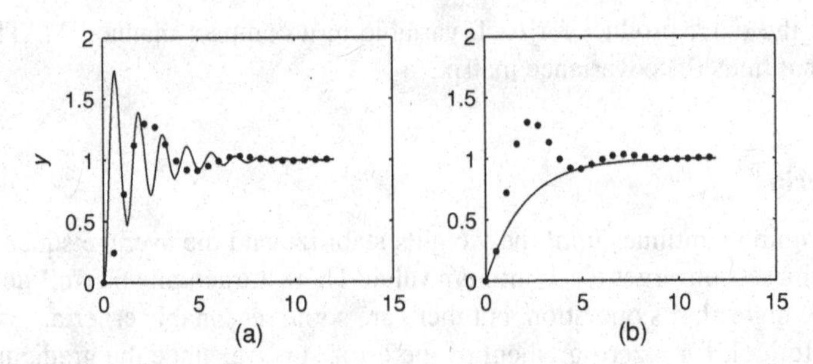

Figure 9.12 Overfitting (a) and underfitting (b) of data points (dots). The approximation (solid line) is inadequate in both cases. (figfitorder.m)

It looks like a desirable situation, but it is not; in the presence of noise it is undesirable to perfectly fit the data points. Instead a smoother approximation, that cuts through the data set, is more correct. A high error goal, as in *early stopping*, prevents a model from overfitting, because it gives the model some tolerance on the error.

The opposite effect, underfitting, is also possible. This occurs when there are too few parameters to adjust, then the model cannot produce outputs reasonably close to the desired outputs.

9.6 Summary

Fuzzy models can be applied to the compensation of nonlinear elements, gain scheduling, and system identification. Once a model is established, it can be used for computer simulations and even directly in the control loop, as in gain scheduling.

There are other ways to build models, even nonlinear models, that do not depend on fuzzy sets. These are highly developed and well tested, so why deploy the armoury of fuzzy logic? One justification for a fuzzy model is the rule based interface; it makes the model easier to understand for the layman. Furthermore, such a model can be merged with existing expert rules.

If a process is so well understood, that it can be modelled mathematically, the mathematical model will likely perform better, and it may even function globally; in that case, there is little justification for using a fuzzy model. However, until the process is mathematically understood – this can take decades – a fuzzy model can be an intermediate step.

It is intuitively appealing to have a model structure, such as the following

$$\text{If } \sigma \text{ is } A_1 \text{ then } \hat{z} = f_1(x, y)$$
$$\text{If } \sigma \text{ is } A_2 \text{ then } \hat{z} = f_2(x, y)$$
$$\vdots$$

Here σ is viewed as a scheduling variable, but the left-hand side could contain any inputs. When this type of structure is equivalent to a basis function architecture, then the methods in the chapter apply. It is the 'intuitive appeal' that makes a fuzzy model different from purely mathematical methods. The fuzzy model is closer to the reality seen by operators and engineers, who work with the problem at hand.

There are many examples of how the physical relevance of a model can be important with respect to development costs, time, and skills.

9.7 Neuro-Fuzzy Models*

A neural network can 'learn' to model a nonlinear process. The network is trained by means of feedback of the model error until the error is small. The result is a network of (many) adjusted parameters. The model is a so-called *black-box model* since the parameters and the network lack physical relevance. Oppositely, a fuzzy rule base consists of readable if–then rules – almost natural language – but it cannot learn the rules itself. The two are combined in *neuro-fuzzy systems* in order to achieve a readable model that can learn. The obtained rules may reveal insight into the data that generated the model, and they can certainly be merged with human expert rules established by operators.

9.7.1 Neural Networks

An artificial neural network, or just *neural network*, is a model structure paired with an algorithm for fitting the model to a set of given data. The network approach to modelling a process uses a standard nonlinear component, and it allows all the parameters to be adjusted. *Learning* or *training* (depending on the viewpoint) is the activity that optimizes a neural network to represent a nonlinear process. Neural network models are now widely accepted, and excellent books give them a comprehensive treatment (Haykin 2009; Duda, Hart, and Stork 2001). Here, our purpose is to illuminate a few fundamental concepts.

The *neuron* is the processing unit of a neural network. Figure 9.13 shows a simple example that consists of three basic elements:

1. A set of connecting *links*; each link carries a *weight*(w_0, w_1, w_2).
2. A *summation point* sums the input signals after they are multiplied by their respective weights.
3. An *activation function* $f(v)$ limits the output of the neuron. When the neuron is used for classification problems, the output is typically limited to the interval [0, 1] or alternatively [−1, 1].

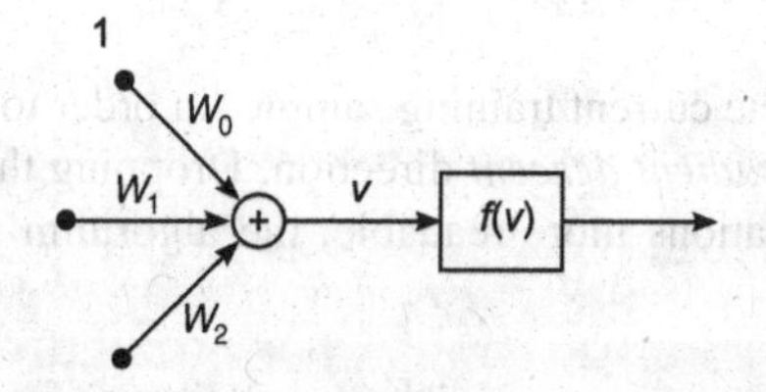

Figure 9.13 Block diagram of a neuron.

The summation in the neuron includes an offset w_0 by fixing the corresponding input at one. All inputs to the neuron are represented in the vector $\mathbf{u} = \begin{bmatrix} 1 & u_1 & u_2 \end{bmatrix}^T$, and the neuron output is a scalar $y = f(v)$. The weights of the connections are represented by the vector $\mathbf{w} = \begin{bmatrix} w_0 & w_1 & w_2 \end{bmatrix}^T$. The neuron thus performs the following function approximation:

$$\hat{y} = f(\mathbf{u} \cdot \mathbf{w}) \tag{9.35}$$

The activation function f is an arbitrary, typically nonlinear, function. A unit relay switch (hard limiter) results in an output equal to +1 or −1, which is useful for classification problems. A neuron with a relay activation function is called a *perceptron*.

Again, the inner product $\mathbf{u} \cdot \mathbf{w}$ is the foundation of the architecture. In the case of basis functions, the nonlinear functions act on the inputs, before the inner product is performed, while the neuron applies the nonlinearity after the inner product. Thus, both are nonlinear variants of the inner product (Example 9.1).

By means of a *training set* of data, an algorithm adjusts the weights such that the image f of the hyper-plane, that $\mathbf{w}$ defines, minimizes an objective function based on distance. A training set consists of, say, K samples of the input vector $\mathbf{u}$, together with each sample's corresponding output $\hat{y}$. A presentation of the complete training set to the model is called an *epoch*. The training is continued epoch after epoch, until the objective function is steady, or the algorithm reaches some other stop criterion.

Several neurons can be combined into a network. The simplest form of a layered network has just an input layer of source nodes that connect to an output layer of neurons, but not vice versa.

9.7.2 Gradient Descent Algorithm

Training is, strictly speaking, an optimization problem. As before, y is the desired output of the model and $\hat{y}$ is the actual model output. An objective function is the squared error,

$$E = \frac{1}{2}e^2 \tag{9.36}$$

where $e = y - \hat{y}$ is the error. The factor $1/2$ is insignificant, but it cancels the factor 2 that appears after differentiating E. The model is trained by minimizing E with respect to the weights. The weights are adjusted in the following incremental manner, according to the MIT rule (Section 6.7.1):

$$\mathbf{w}_{k+1} = \mathbf{w}_k + \Delta\mathbf{w}_k \tag{9.37}$$

The index k corresponds to the current training sample. In order to decrease E, the algorithm adjusts the weights $\mathbf{w}$ in a *gradient descent* direction. Dropping the index k for the moment, to make the following derivations more readable, the algorithm increments the weights as follows:

$$\Delta\mathbf{w} = -\eta\frac{\partial E}{\partial \mathbf{w}} = -\eta e\frac{\partial e}{\partial \mathbf{w}} \tag{9.38}$$

The vector $\partial E/\partial \mathbf{w}$ is the gradient of the objective function, and $\partial e/\partial \mathbf{w}$ is the *sensitivity derivative*, which is a vector of partial derivatives containing error sensitivities towards (small) changes in the weights.

The algorithm requires two passes, a *forward pass* and a *backward pass*. In the forward pass the weights remain unchanged. The forward pass begins at the input terminals of the network, and terminates at the output terminal by computing the error signal for the output. The backward pass starts at the output terminal by passing the error backwards through the model, in order to find the weight updates. This is called *error back-propagation*. We must calculate the derivative $f'(v) = \partial f/\partial v$ in order to find an analytical solution, and thus we prefer that the activation function is differentiable. Otherwise, the derivative must be approximated. The sensitivity derivative is the following:

$$\frac{\partial e}{\partial \mathbf{w}} = \frac{\partial (y - \hat{y})}{\partial \mathbf{w}} = -\frac{\partial \hat{y}}{\partial \mathbf{w}} = -\frac{\partial \hat{y}}{\partial v}\frac{\partial v}{\partial \mathbf{w}} = -f'\mathbf{u}$$

The last derivative $\partial v/\partial \mathbf{w}$ makes use of the inner product $v = \mathbf{u} \cdot \mathbf{w}$, where $\mathbf{u}$ is held constant after the forward pass. By inserting this result into Equation (9.38), inserting that into Equation (9.37), and switching the indices on again, then the weight adjustment mechanism is the following:

$$\mathbf{w}_{k+1} = \mathbf{w}_k + \eta e_k f' \mathbf{u}_k \tag{9.39}$$

Algorithm 9.4 summarizes the procedure. It concerns a single neuron only, and the generalized version is more complex, but it uses the same skeleton (Haykin 2009).

Algorithm 9.4 Gradient descent for one neuron.

1. *Initialize weights*. Set all weights to small random values, preferably within the interval $[-1, 1]$.
2. *Prepare inputs and desired outputs (training set)*. Collect a set of input vectors $\mathbf{u}$ and the corresponding outputs y.
3. *Forward pass*. Calculate the model outputs $\hat{y}$ by successive use of Equation (9.35).
4. *Backward pass*. Adjust weights by Equation (9.39). Choose a small training gain η.
5. *Return*. Calculate the value of objective function J. Stop if J is steady or if some other stop criterion is reached. Otherwise, go back to Step 2.

With respect to the objective function, which is a hyper-surface in general, the training starts from an arbitrary point on the surface, determined by the initial weights and the initial inputs. The algorithm moves towards a minimum in a step-by-step fashion. A solution is feasible in the linear case, whereas in the nonlinear case the algorithm may get trapped in a local minimum, never reaching a global minimum.

Example 9.19 *Error surface*

What is the nature of the error surface, given a simple neuron?

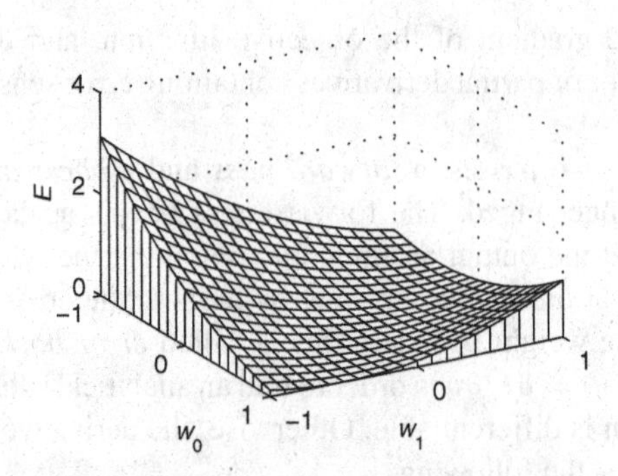

Figure 9.14 Error surface of a two-input, linear neuron. (fignnsurf.m)

► *Solution*

The objective function E (Equation (9.36)) is quadratic, so we expect to find a parabolic surface. Let the neuron have two inputs – of which one is an offset – and a linear activation function. The neuron output is

$$\hat{y} = f(\mathbf{u} \cdot \mathbf{w}) = \mathbf{u} \cdot \mathbf{w} = w_1 u_1 + w_0$$

Let $u_1 = 1$ *and the desired output* $y = 0.5$. *Then the error function becomes*

$$E = \frac{1}{2}e^2 = \frac{1}{2}(y - \hat{y})^2 = \frac{1}{2}(y - (w_1 u_1 + w_0))^2 = \frac{1}{2}(0.5 - (w_1 + w_0))^2$$

This is the square of a plane $z = w_1 + w_0$ *displaced from the origin by* 0.5, *and scaled by* 1/2.

Figure 9.14 shows the resulting surface. Its shape is like a trough with parabolic sides. Note, however, that the bottom is a line defined by $w_1 + w_0 = 0.5$. *This is unfortunate, because there are many optimal solutions. The surface depends on the input, and another training example will push the optimizer toward the global minimum.*

The example shows that the set of training data must be larger than the dimension of the search space, and the data must contain enough variation for the optimizer to find a globally best fit. In matrix terms, the data matrix must have full rank, and the covariance matrix should have a determinant far from zero (not ill-conditioned).

Example 9.20 *Single neuron line fit*

Given the same data as the least-squares line fit (Example 9.10), does the neuron find the same solution?

► *Solution*

It is a linear problem, and the neuron optimizes the squared error between model output and data y-values. The least-squares routine optimizes this; therefore we should expect the same result.

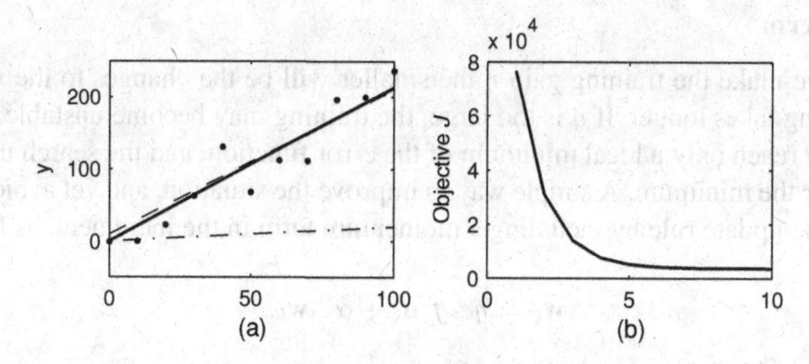

Figure 9.15 Single neuron line fit (a). The starting line is low (dash-dotted line) and the final result (solid line) is close to the true line (dashed). The objective function (b) shows convergence after about six iterations. (fignnline.m)

Figure 9.15 shows the result. It was produced in accordance with Algorithm 9.4:

1. Initialize weights. *We set arbitrarily* $w_0 = 0.3$, $w_1 = 0.2$.
2. Prepare inputs and desired outputs. *We take the data from Example 9.10; there are 11 points.*
3. Forward pass. *For all 11 samples of* u_1 *we calculate* $\hat{y} = w_1 u_1 + w_0$. *The result is a vector* $\hat{\mathbf{y}}$ *of 11 elements. We can now calculate 11 errors* $\mathbf{e} = \mathbf{y} - \hat{\mathbf{y}}$.
4. Backward pass. *The weights are adjusted by Equation (9.39). We set* $\eta = 0.0001$ *after some trial and error. The update was made at the end of the epoch by summing the 11 updates related to each sampling (batch update). If* η *is increased by a factor of ten, the algorithm becomes unstable, and the objective function tends to infinity.*
5. Return. *The objective function is* $J = 0.5 \times \mathbf{e} \cdot \mathbf{e}$. *The plot in Figure 9.15(b) shows that the algorithm can be stopped after 5–6 epochs.*

The data set consists of 11 points, and the neuron starts with a line which is more or less horizontal (Figure 9.15(a)). The neuron finds the weights $w_1 = 2.13$, $w_0 = 0.32$.

This is somewhat different from the values found by MATLAB®'s \-operator: $w_1 = 2.35$, $w_0 = -13.6$. *Extremely slow convergence causes the discrepancy; if the algorithm is allowed to continue for, say,* 10^5 *epochs, then the neuron will agree with MATLAB®'s \-operator solution. The training gain* η *is large enough during the first five epochs, but for the remaining epochs it is too small.*

In the example, the neuron quickly found an approximate solution, but a good solution required many training epochs. The root of the problem is a steep slope of the objective function at the beginning of the search for a minimum, while the bottom is shallow. The neuron works better on numbers in the interval $[-1, 1]$, and the typical solution is therefore to standardize all input data such that their mean value is zero and their standard deviation is one. Consequently, the result must be transformed back to the original domain. An activation function that is s-shaped, such as the hyperbolic tangent function tanh, will effectively provide a variable update gain on the weight increment; its derivative depends on the operating point of the neuron.

Momentum term

The smaller we make the training gain η the smaller will be the changes to the weights, but then the training takes longer. If η is too large, the training may become unstable. The search algorithm may reach only a local minimum of the error function, and the search may be slow, especially near the minimum. A simple way to improve the situation, and yet avoid instability, is to modify the update rule by including a momentum term in the increment, as follows:

$$\Delta \mathbf{w}_k = \eta e_k f' \mathbf{u}_k + \alpha \Delta \mathbf{w}_{k-1}$$

The gain factor α is usually a positive constant called the *momentum constant*. The effect is a low-pass filtering of the increments of the weight updates, thus reducing the risk of getting stuck in a local minimum. The following rules of thumb apply.

- For the training to be stable, the momentum constant must be in the range $0 \leq |\alpha| < 1$. When α is zero the back-propagation algorithm operates without momentum. Note that α can be negative, although it is unlikely to be used in practice.
- When the partial derivative $\partial E / \partial w_i$ has the same sign on consecutive iterations, the sum Δw_i grows in magnitude, and so the weight is adjusted by a large amount. Hence the momentum term tends to accelerate the descent.
- When the partial derivative $\partial E / \partial w_i$ has opposite signs on consecutive iterations, the sum Δw_i shrinks in magnitude, and the weight is adjusted by a small amount. Hence the momentum term has a stabilizing effect in directions that alternate in sign.

The momentum term may also prevent the training from terminating in a shallow local minimum on the error surface. The training parameter η has been assumed constant, but in reality it should be connection dependent. We may in fact constrain any number of weights to remain fixed by simply making the training gain η_i for weight w_i equal to zero.

Pattern and batch mode updates

The weight updates may proceed in one of two basic ways:

- *Pattern mode*. Weights are updated after each training example; that is how it was presented here. To be precise, consider an epoch consisting of K training examples (patterns) arranged in the order $[\mathbf{u}_1, \mathbf{y}_1], [\mathbf{u}_2, \mathbf{y}_2], \ldots, [\mathbf{u}_K, \mathbf{y}_K]$. The first example is presented to the model, the forward and backward computations are performed, and the weights are updated. Then the second example is presented, the forward and backward computations are performed, and the weights updated. This continues until the last example.
- *Batch mode*. Weights are updated after presenting all training examples that constitute an epoch. The weights are adjusted by the average of the updates associated with every training examples. The weight adjustment is thus made only after the entire training set has been presented to the network. The convergence is faster if the update after the epoch is the plain sum of the individual updates per training example, but the update mechanism may become unstable.

From an on-line point of view, where the data are presented as time series, the pattern mode uses less storage. The batch mode, however, is appropriate off-line, providing a more accurate estimate of the network's overall gradient vector.

9.7.3 Adaptive Neuro-Fuzzy Inference System (ANFIS)

ANFIS (Adaptive Neuro-Fuzzy Inference System) is an architecture which is functionally equivalent to a Sugeno type of rule base (Jang, Sun, and Mizutani 1997, Jang and Sun 1995). It appears in MATLAB®'s Fuzzy Logic Toolbox (MathWorks 2012). Loosely speaking, ANFIS is a method for tuning an existing rule base with a training algorithm based on a collection of training data. This allows the rule base to adapt.

Without loss of generality we assume two inputs, u_1 and u_2, and one output, y. Assume further a first-order Sugeno type of rule base with the following two rules,

$$\text{If } u_1 \text{ is } A_1 \text{ and } u_2 \text{ is } B_1 \text{ then } y_1 = c_{11}u_1 + c_{12}u_2 + c_{10} \tag{9.40}$$

$$\text{If } u_1 \text{ is } A_2 \text{ and } u_2 \text{ is } B_2 \text{ then } y_2 = c_{21}u_1 + c_{22}u_2 + c_{20} \tag{9.41}$$

If the firing strengths of the rules are α_1 and α_2 respectively, for two particular values of the inputs u_1 and u_2, then the output is computed as a weighted average,

$$y = \frac{\alpha_1 y_1 + \alpha_2 y_2}{\alpha_1 + \alpha_2} = \bar{\alpha}_1 y_1 + \bar{\alpha}_2 y_2 \tag{9.42}$$

The corresponding ANFIS network is shown in Figure 9.16. A description of the layers in the network follows.

1. Each neuron i in layer 1 is adaptive with a parametric activation function. Its output is the grade of membership to which the given input satisfies the membership function, i.e., $\mu_{A_1}(u_1)$, $\mu_{B_1}(u_2)$, $\mu_{A_2}(u_1)$, or $\mu_{B_2}(u_2)$. An example of a membership function is the generalized *bell function*,

$$\mu(x) = \frac{1}{1 + \left|\frac{x-c}{a}\right|^{2b}} \tag{9.43}$$

   where $\{a, b, c\}$ is the parameter set. As the values of the parameters change, the shape of the bell-shaped function varies. Parameters in that layer are called *premise parameters*.
2. Every node in layer 2 is a fixed node, whose output is the product of all incoming signals. In general, any other fuzzy AND operation can be used. Each node output represents the firing strength α_i of the ith rule.
3. Every node in layer 3 is a fixed node which calculates the ratio of the ith rule's firing strength relative to the sum of all rule's firing strengths,

$$\bar{\alpha}_i = \frac{\alpha_i}{\alpha_1 + \alpha_2}, \qquad i = 1, 2 \tag{9.44}$$

   The result is a *normalized firing strength*.

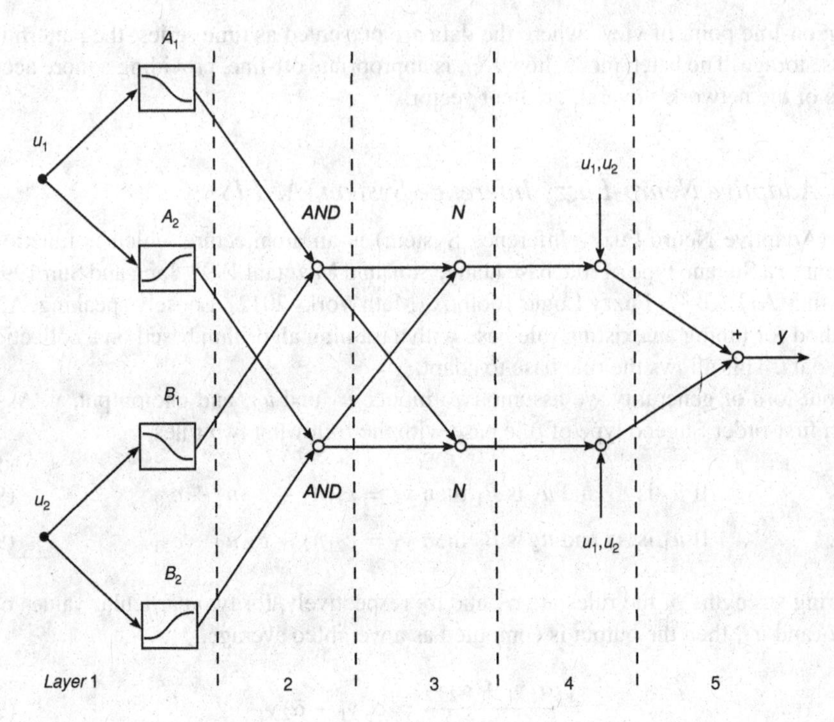

Figure 9.16 ANFIS architecture of a two-input Sugeno model with two rules (after Jang, Sun, and Mizutani 1997)

4. Every node in layer 4 is an adaptive node with a node output

$$\bar{\alpha}_i y_i = \bar{\alpha}_i (c_{i1}u_1 + c_{i2}u_2 + c_{i0}), \qquad i = 1, 2 \tag{9.45}$$

   where $\bar{\alpha}_i$ is the normalized firing strength from layer 3 and $\{c_{i1}, c_{i2}, c_{i0}\}$ is the parameter set of this node. Parameters in this layer are called *consequent parameters*.
5. Every node in layer 5 is a fixed node which sums all incoming signals.

It is straightforward to generalize the ANFIS architecture to a rule base with more than two rules.

The ANFIS training algorithm

When the premise parameters are fixed, the overall output is a linear combination of the consequent parameters. In symbols, the output y can be written as

$$y = \frac{\alpha_1}{\alpha_1 + \alpha_2} y_1 + \frac{\alpha_2}{\alpha_1 + \alpha_2} y_2 \tag{9.46}$$

$$= \bar{\alpha}_1 (c_{11}u_1 + c_{12}u_2 + c_{10}) + \bar{\alpha}_2 (c_{21}u_1 + c_{22}u_2 + c_{20}) \tag{9.47}$$

$$= (\bar{\alpha}_1 u_1) c_{11} + (\bar{\alpha}_1 u_2) c_{12} + \bar{\alpha}_1 c_{10} + (\bar{\alpha}_2 u_2) c_{21} + (\bar{\alpha}_2 u_2) c_{22} + \bar{\alpha}_2 c_{20} \tag{9.48}$$

which is linear in the consequent parameters c_{ij} $(i = 1, 2; j = 0, 1, 2)$. A hybrid algorithm adjusts the consequent parameters c_{ij} in a forward pass and the premise parameters $\{a_i, b_i, c_i\}$ in a backward pass (Jang *et al.* 1997). In the forward pass the network inputs propagate forward until layer 4, where the consequent parameters are identified by the least-squares method. In the backward pass, the error signals propagate backwards and the premise parameters are updated by gradient descent.

9.8 Notes and References*

Babuška has studied fuzzy models carefully (1998), and his *product-space clustering* method is based on the philosophy that the local models should be local linearizations. He also makes the point, that the fuzzy c-means algorithm, as it is presented here, prefers clusters that are spherical; elongated clusters can be difficult to detect, especially if they are not perpendicular to the feature axes. He proposes to use a cluster algorithm which adapts to the shape of the cluster (Gustafson and Kessel in Babuška 1998), and this is in turn based on the Mahalanobis distance. His book is a thorough treatment of the aspects that are relevant for models based on product-space clustering, which implies that the fuzzy membership functions are defined on rectangular surfaces of the input space. He developed a companion toolbox for MATLAB®, which is available from his university website.

Nonlinear system identification is treated thoroughly by Nelles (2001). He examines fuzzy basis function models, and, as a result, proposes a model algorithm called LOLIMOT: the Local Linear Model Tree algorithm. The LOLIMOT afterwards gained a reputation in the neuro-fuzzy community. For an introduction to neuro-fuzzy systems, see the book by Nauck, Klawonn, and Kruse (1997). For an overview of neuro-fuzzy control and the related optimization techniques, see the book by Michels, Klawonn, Kruse, and Nürnberger (2006).

Local model networks is a promising approach for the automotive industry (Hametner and Jakubek 2013; Jakubek, Hametner, and Keuth 2008).

Extracting rules from data is a form of modelling activity within pattern recognition, data analysis, or data mining, also referred to as the search for structure in data (Bezdek and Pal 1992). The goal is to reduce the complexity in a problem, or to reduce the amount of data associated with a problem. The book by Duda, Hart, and Stork (2001) can be used as an excellent reference book within the field of pattern recognition. The book discusses, among many other topics, Mahalanobis distance, decision surfaces, linear classifiers, and minimum distance classifiers. If the focus is on neural networks, including the field of *machine learning*, the book by Haykin (2009) is a major reference.

Neural network research started in the 1940s, fuzzy logic research in the 1960s, and the neuro-fuzzy research area is relatively new on that background. In 1995 came the Fuzzy Logic Toolbox for MATLAB®, which includes the ANFIS neuro-fuzzy architecture (Jang, Sun, and Mizutani 1997), which became widely known.

For a stringent treatment of the geometry of vector spaces and optimization, the book by Luenberger (1969) is still relevant reading. Much of the geometric arguments are founded on the so-called *projection theorem*, which is the basis for the least-squares method.

10

Demonstration Examples

This chapter provides examples that demonstrate how to design fuzzy controlled systems in practice. They are worked through examples, with solutions, and thus they provide a selection of teaching modules. The examples illustrate the use of a deadzone surface to save energy in a domestic hot water heater, gain scheduling control of an unstable chemical reactor with a cascade controller, integrator anti-windup for controlling the engine speed in a car, a saturation surface for controlling a multi-variable unstable ball-balancer, and finally fuzzy clustering for building a model of a first-order system with a nonlinear actuator. All the control examples start with a linear design, which is then improved by fuzzy nonlinearities. As a general rule of thumb, gain scheduling is useful for large signal reference tracking problems, while the fuzzy control surfaces are useful for small signal regulation problems.

10.1 Hot Water Heater

Within the European Union, energy efficiency in households has improved by 12% since the year 2000 (up until 2009[1]), and the political goal is to save 20% by the year 2020. Household energy comprises space heating, water heating, and electric appliances, while this example just focuses on saving energy for the heating of hot water.

Consider an electric tank heater for domestic hot water. Its volume is 110 litres, and it is equipped with a relay controller that keeps the temperature within a band during idle periods.Table 10.1 contains further details about the tank and definitions of the symbols.

Assume that the relay has a hysteresis of ± 4 K, such that it switches off when the tank temperature T is 4 above the setpoint, and it switches on again when T is 4 below the setpoint (degrees kelvin, K, refer to temperature differences). The tank is insulated with polyurethane foam, but the insulation is not perfect. There are continuous losses (UA), which depend on the current temperature difference between the inside and the outside of the tank. There is also a load on the system, which is a daily demand for hot water by the occupants. At seven o'clock

[1] www.odyssee-indicators.org/reports/household/household9.pdf

Foundations of Fuzzy Control: A Practical Approach, Second Edition. Jan Jantzen.

Table 10.1 Tank specifications.

Parameter	Symbol	Value
Volume	V	0.110 m^3
Power supply	P_i	3000 W
Thermal conductivity	UA	1.14 W/K
Density	ρ	1000 kg/m^3
Heat capacity	c_p	1 kcal/kg · K
Ambient temperature	T_a	21°C
Feed temperature	T_f	10°C
Feed flow	F_f	8 l/min
Reference temperature	T_r	51°C
Hysteresis	h	±4 K

in the morning there is a demand for hot water from the tank, and it is simultaneously replaced by cold water flowing into the tank. The duration is 7 minutes corresponding to maybe two showers. The following two questions are relevant for a home owner in order to save energy:

1. Does it pay to install a timer, that switches the heater off during the night, when the tank is idle?
2. Will a more sophisticated controller save energy?

The questions have a broader scope, since the system could be scaled to a larger size, in order to model the heating of a whole house. For example, the case of a ground source heat pump or an oil furnace that heats water in a buffer tank which feeds the radiators.

Figure 10.1 shows the diagram of a Simulink model that simulates the tank temperature over an adjustable period of time, say, 24 hours corresponding to one day, or 8760 hours corresponding to a whole year. Apart from the tank itself, the Simulink model includes three controllers: a relay, a crisp P-controller, and a fuzzy FP-controller. The three controllers can be switched in and out, one after the other, in order to compare their performance.

10.1.1 Installing a Timer Switch

Assume for the moment that the hot water tank is idle ($F_f = 0$). The only load on the system is the heat transfer through the walls. Figure 10.2 shows how the temperature evolves. The relay controller kicks in when $T = T_r - |h| = 51 - 4 = 47$, and thereafter the water heats up rather quickly. The losses are relatively small, and apparently the period for a heating cycle, when idling, is approximately $30\frac{1}{2}$ hours. It will make no difference if a timer turns the heater off for eight hours during the night; the controller already shuts it off for a longer period. The short answer to the first question is therefore: 'no'. For longer periods, say, one week, it does pay to turn the heater off. In that case, the heater will demand 3.5 kilowatt-hours to reheat the water at the end of the period, while leaving it on would require 5.2 kilowatt-hours, and 33% energy is then saved.

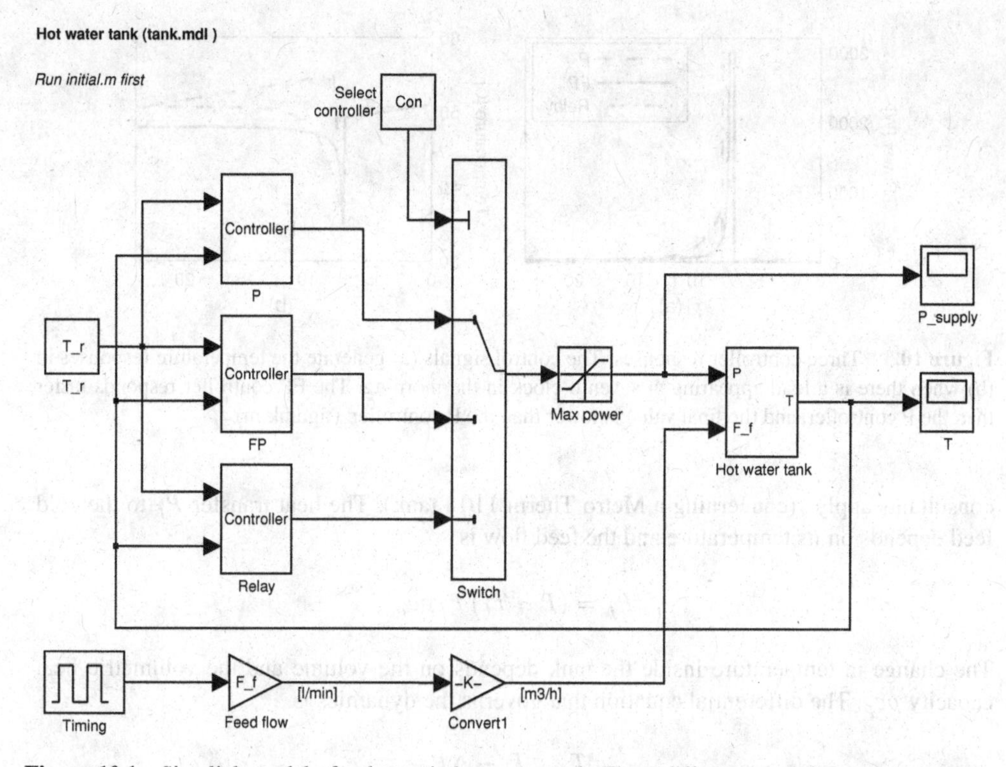

Figure 10.1 Simulink model of a domestic hot water tank. Three different controllers can be tested. The load corresponds to maybe two showers in the morning.

The input to the model is the power P_i supplied by the electric heater minus the heat loss ΔP through the wall, which depends on the temperature difference, as follows:

$$\Delta P = (T - T_a) U A$$

Here, U is the heat transfer coefficient for the wall, and A is the wall area. Numbers are a little difficult to find, and we use instead the estimate $UA = 1.14$ W/K that Danish energy

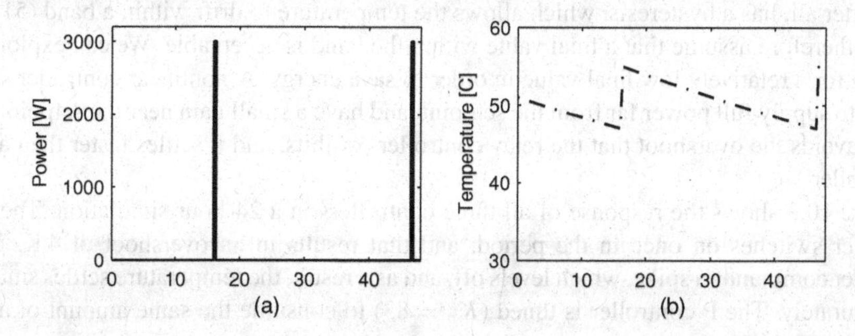

Figure 10.2 Power supply (a) and temperature (b) when the hot water tank is idle. The simulation lasts two days (48 h). (figtank.m)

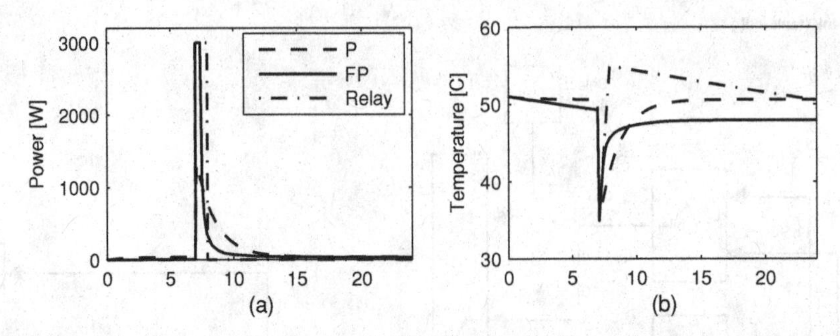

Figure 10.3 Three controller responses. The control signals (a) generate the temperature responses in (b) when there is a load appearing at seven o'clock in the morning. The FP controller responds faster than the P controller, and the final value is lower than the P controller. (figtank.m)

consultants apply[2] (concerning a Metro Therm 110 l tank). The heat transfer P_f to the cold feed depends on its temperature and the feed flow is

$$P_f = \left(T - T_f\right) F_f \rho c_p$$

The change in temperature inside the tank depends on the volume and the volumetric heat capacity ρc_p. The differential equation that governs the dynamics is

$$\frac{\mathrm{d}T}{\mathrm{d}t} = \frac{P_i - \Delta P}{V \rho c_p}$$

Simulink integrates this equation step by step, and the result is the tank temperature over time, as a function of the power supply, the heat loss, and the feed water.

10.1.2 Fuzzy P Controller

When a P controller replaces the relay, the control will be better in the sense that the result will be closer to the setpoint. However, in this case the final temperature is not critical. The relay, after all, has a hysteresis, which allows the temperature to drift within a band (51 ± 4), and we therefore assume that a final value within the band is acceptable. We can exploit this, and aim for a relatively low final value in order to save energy. A nonlinear controller should be able to supply full power far from the setpoint, and have a small gain near the setpoint. This way, it avoids the overshoot that the relay controller exhibits, and it settles faster than a crisp P controller.

Figure 10.3 shows the response of all three controllers in a 24-hour simulation. The relay controller switches on once in the period, and that results in an overshoot of 4 K. The P controller commands a spike, which levels off, and as a result, the temperature settles smoothly and accurately. The P controller is tuned ($K_p = 83$) to consume the same amount of annual

[2] www.maerkdinbygning.dk

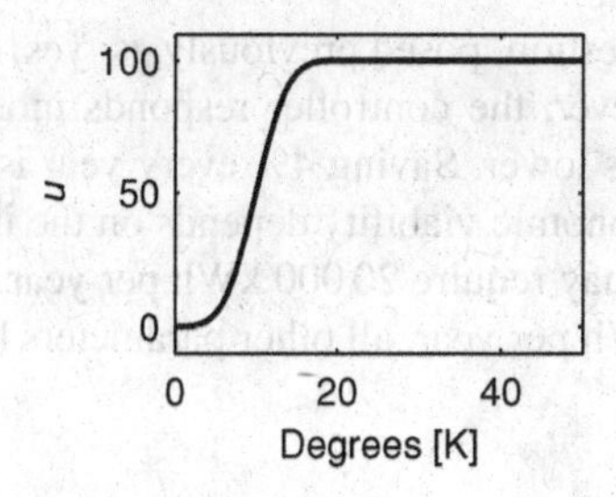

Figure 10.4 Characteristic of the FP controller. (figtank.m)

energy as the relay controller. With the equivalent setting, the figure shows that the FP controller settles faster than the crisp P controller, and at a lower final temperature. The final temperature is within the operating range of the relay.

The fuzzy controller consists of two rules

$$\text{If } e \text{ is Low}^c \text{ then } u = 0$$
$$\text{If } e \text{ is High}^c \text{ then } u = 100$$

Here, e is the setpoint error ($e = T_r - T$). It is a Sugeno rule base with weighted average defuzzification, and Figure 10.4 shows its characteristic. It is an s-curve specifically crafted for the deviations around 4 K, and $GE = 1$ ($GU = K_p = 83$) such that the characteristic concerns temperature differences directly. Excursions from the setpoint larger than 9 K, will command full power. However, a few degrees before the setpoint, the controller will command almost zero power, and the final temperature value will be relatively low, which saves energy. Still, the controller responds relatively fast.

The performance numbers in Table 10.2 show that the FP controller takes one and a half hours to reach $T = 47°C$ after the load comes on, which is average of the two other controllers. The concentration parameter ($c = 1.7$) was adjusted to achieve a compromise between a fast response and large annual energy savings. The FP controller saves 43 kWh annually, which corresponds to 4.1% saved.

Table 10.2 Test results. Time to reach $T = 47°C$ (t_{47}) after the load comes on, annual energy consumption (E_{365}), and relative consumption (%). The concentration exponent ($c = 1.7$) of the FP controller was adjusted in order to place FP between the two others with regard to t_{47}.

Controller	t_{47} [h]	E_{365} [kWh]	%
Relay	0.6	1048	100
P	2.4	1048	100
FP	1.5	1005	95.9

The answer to the second question, posed previously, is: yes, a sophisticated controller can save around 4% energy. However, the controller responds more slowly and the final value of the controlled temperature is lower. Saving 4% every year is quite good, considering that savings are tax free, but the economic viability depends on the investment cost.

Heating a house of 140 m^2 may require 20 000 kWh per year (in Denmark), and the scaled savings then amount to 800 kWh per year, all other parameters being equal.

10.2 Temperature Control of a Tank Reactor

Hydrolysis of propylene oxide, with sulphuric acid as catalyst, produces propylene glycol. The liquid raw mix is fed into a *continuous-flow stirred tank reactor* (CSTR) where a reaction takes place. The tank operates in a temperature range above room temperature and below the boiling point of water. In order to start the reaction, the tank must be heated. However, when the reaction starts, it develops reaction heat itself (*exothermic* reaction), and the tank must be cooled rather than heated. Figure 10.5 shows a diagram of the system, and it defines the symbols used.

The control objective is to minimize the start-up time, and thereafter to keep the reaction temperature steady, despite disturbances in feed concentration and feed temperature. The problem is nonlinear and it has multiple equilibrium states due to the exothermic reaction.

Propylene glycol is used as an antifreeze fluid in car engine coolers, solar panels as protection against night frost, and for aircraft de-icing. It is also used as a solvent for drugs or moisturizers, and it can be used as artificial smoke or fog for fire-fighter training or theatrical productions. The chemical reaction is the following: $(CH_2 - O - CH - CH_3) + H_2O \rightarrow (CH_2OH - \mathrm{CH}_2O - CH_3)$.

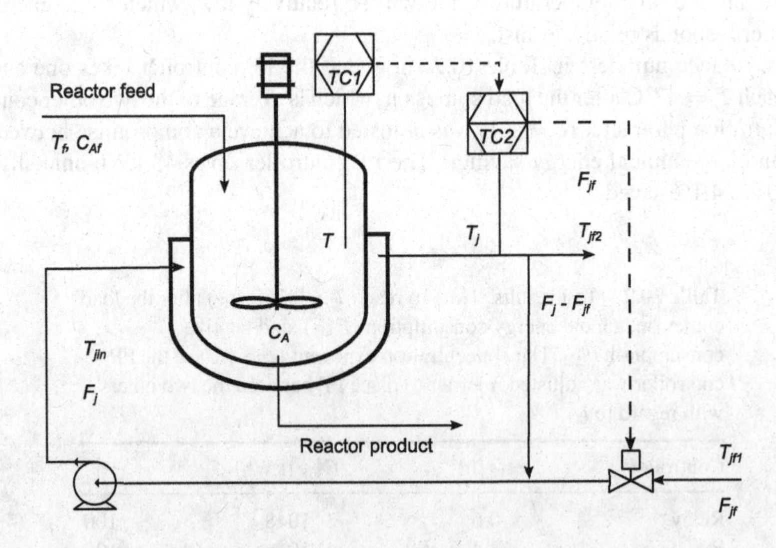

Figure 10.5 Piping and instrumentation diagram of the CSTR. The temperature controller TC2 (inner loop) controls the jacket temperature T_j, while TC1 (outer loop) controls the reaction temperature T.

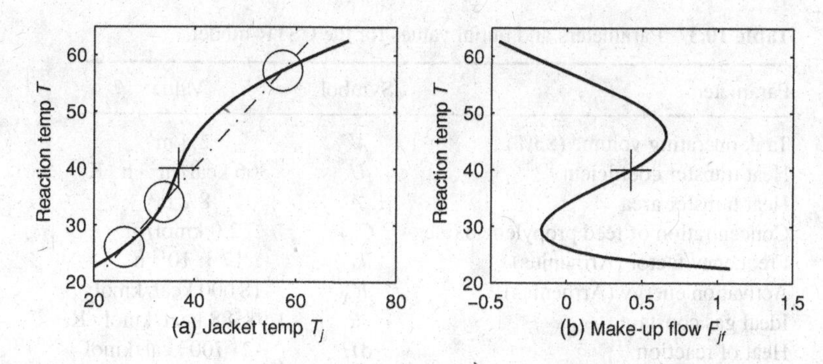

Figure 10.6 Static, open-loop reaction temperature T as a function of jacket temperature (a). The diagonal line (dash-dotted) corresponds to $y = x$, that is, the line where the two temperatures are equal. There are three equilibrium points (circled), and the desired operating point (+) is in an unstable region. According to the equation for the steady-state temperature (b), there are several possible temperature solutions for a fixed make-up flow, and their dynamic gains have opposite sign. (figequi.m)

10.2.1 CSTR Model

The effective tank volume is 2.4 m^3 of fluid, and the residence time of the fluid in the tank is approximately 13 min with an inflow of reactor feed mix at 11 m^3/h, or, equivalently, 183 l/min which is in the order of a car wash. There are three major time-dependent processes in the model:

- the time development of the reaction temperature T;
- the time development of the reaction concentration C_A; and
- the time development of the jacket temperature T_j.

The reaction rate is the major nonlinearity, modelled as an exponential function (*Arrhenius equation*). The differential equations that describe the dynamics can be found in the book by Bequette (2003).

Figure 10.6 shows that the uncontrolled reaction has three equilibria, of which the one in the middle region is unstable. The graph indicates that there is positive feedback in the middle region. A small positive disturbance, which increases the temperature of the jacket, causes the reaction temperature to increase more than the jacket temperature. Heat flows to the jacket, and its temperature increases further. The situation persists until the reaction temperature equals that of the jacket – at the higher equilibrium point. At this point, a small increase in jacket temperature causes a heat flow to the reaction, which will decrease the jacket temperature again.

The controller

The main controlled output is the *reaction temperature T* (Figure 10.5). During the reaction, the temperature tends to rise, but it is controlled by the jacket flow with the *jacket temperature* T_j. Controller TC1 sends a setpoint to controller TC2, which in turn controls the *make-up flow* F_{jf} at the make-up supply, in order to reach its setpoint. The configuration is a *cascade controller*.

Table 10.3 Parameters and initial values for the CSTR model.

Parameter	Symbol	Value
Tank operating volume (85%)	V	2.4 m^3
Heat transfer coefficient	U	366 $kcal/m^2 \cdot h \cdot K$
Heat transfer area	A	8.2 m^2
Concentration of feed propylene oxide	C_{Af}	2.0 $kmol/m^3$
Frequency factor (Arrhenius)	k_0	17×10^{12} h^{-1}
Activation energy (Arrhenius)	E_A	18 000 kcal/kmol
Ideal gas constant	R	1.98588 $kcal/kmol \cdot K$
Heat of reaction	dH	−21 700 kcal/kmol
Setpoint reactor temperature	T_r	40.0°C
Reactor feed temperature	T_f	15.6°C
Flow rate of reactor feed	F	11 m^3/h
Volumetric heat capacity	ρc_p	853 $kcal/m^3K$
Max capacity of make-up flow	$\sup F_j$	2.5 m^3/h
Initial jacket temperature	T_{j0}	15.0°C
Jacket volume	V_j	$\frac{1}{4}V$
Jacket make-up temperature, cold	T_{jf1}	12.0°C
Jacket make-up temperature, hot	T_{jf2}	55.0°C
Volumetric heat capacity, jacket	$(\rho c_p)_j$	$1.04 \times \rho c_p$

The Simulink model includes a split-range valve, which enables not only coolant into the jacket, but also hot fluid for starting the reaction. When the control signal is positive, a valve opens for a hot fluid, which is at a high fixed temperature T_{jf2}, set in a parameter file (Table 10.3). When the control signal is negative, a valve opens for the cold fluid, which is at a low fixed temperature T_{jf1}, set in the parameter file.

A second measurement, the *reactant concentration* C_A, is available in the simulation model. It may be used for the controller, if preferred. When the CSTR operates in steady state, half of the feed is converted, that is, the concentration C_A is 50% of the feed concentration C_{Af}.

The feed flow through the reactor is assumed constant, so there is no change in the level of the reactant. The reactant volume and its area of contact for heat exchange with the jacket, also remain constant. The model neglects heat developed by the stirrer.

The controller should aim to perform as well as possible, that is, minimize the magnitude and the duration of *dips* (downward excursions) and *flares* (upward excursions) resulting from disturbances. The model includes two disturbances: the *feed temperature* temporarily rises 2.5 K, then it comes down again, and the *concentration of feed* temporarily rises 0.1 $kmol/m^3$, before coming down again. Each disturbance kicks in at a specific time, and lasts for ten hours.

Requirements

The following are primary requirements, which should be met, if possible:

- The start-up response should preferably be without overshoot, before pulling in to the desired setpoint of the reaction, which is 40°C.

- Oscillatory responses should be well damped, and without overshoot.
- In steady state, the controller should achieve the desired setpoint within a band of ± 0.3°C.

A secondary requirement is to minimize the time to reach the setpoint during the start-up phase.

10.2.2 Results and Discussion

Above a certain temperature ($T \approx 30$°C), the reactant temperature rises rapidly, and cooling must be applied immediately in order to attenuate the acceleration. The feed mix is itself fairly cold ($T_f = 15.6$°C), and that contributes to the cooling, but the jacket flow must do the rest in order to keep the temperature steady at the setpoint ($T_r = 40$°C). It is useful to split the control problem into two phases: the start-up phase, and the regulation phase. During start-up, the temperature variation is relatively large, while in the regulation phase, the excursions from the setpoint are relatively small.

Figure 10.7 shows the result of controlling the process with a crisp PI controller and a PI+FPD controller. For the start-up phase ($t < 5$ h), the two controllers reach $T = 38.6$°C in 1 h 11 min, but then there is some oscillation before the temperature settles after about 5 h. The fuzzy controller exhibits relatively fast oscillation periods (15 min) before pulling in to the setpoint, while the PI controller oscillates with a lower frequency, but a larger amplitude. The performance of both controllers is comparable, and the overshoot is negligible, but the fuzzy controller commands more variation in the jacket temperature (to TC2).

For the regulation phase, the fuzzy controller is a little slower than the PI controller, but it reduces dips and flares significantly, and the response to the second disturbance is less oscillatory. The first disturbance is a first-order type (on at $t = 10$, off at $t = 20$) and the second disturbance is a second-order type (on at $t = 30$, off at $t = 40$).

The control strategy is to let the crisp PI controller take charge of the coarse control, that is, to control the large variation during the start-up phase and to provide the integrator signal for the steady state. Figure 10.8 shows the configuration, which consists of an inner loop (TC2) for the jacket temperature and an outer loop (TC1) for the reaction temperature. An FPD controller performs fine control around the setpoint, by delivering a correction, which is

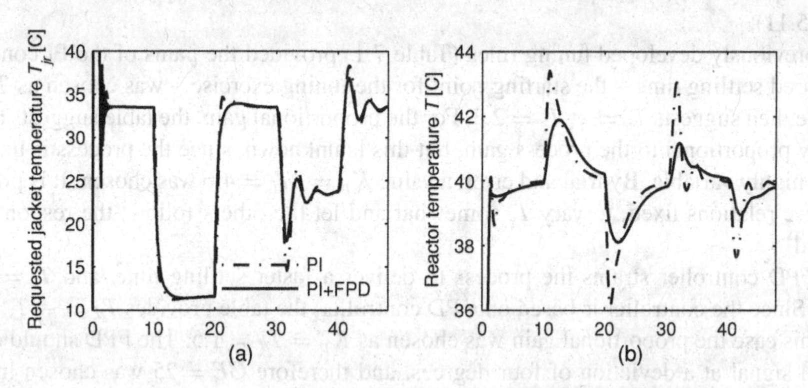

Figure 10.7 Requested jacket temperature from TC1 (a) and reactor temperature T (b). The fuzzy system is compared with the PI controlled system. The vertical axes have been truncated. (figcstr.m)

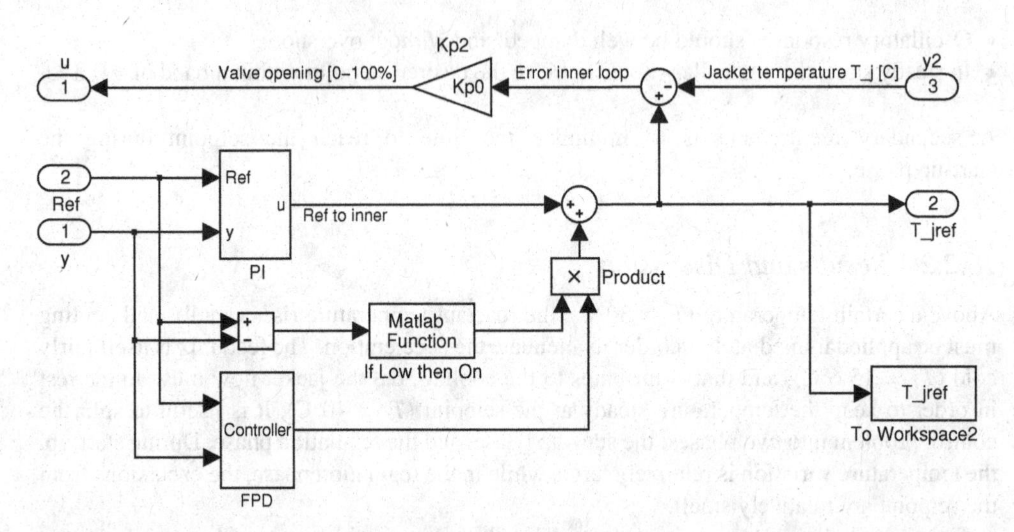

Figure 10.8 The TC1 and TC2 controllers. TC1 consists of a PI controller in collaboration with an FPD controller. A gain schedule limits its additive contribution to be near the setpoint. (cstr.mdl)

added to the PI control signal. A gain schedule sets the FPD into operation only when near the setpoint.

Derivative action was ruled out in the coarse controller, because it causes oscillations owing to the rather sudden changes in reaction temperature. However, derivative action is useful in the case of small variations around the setpoint, because it attenuates dips and flares. An FPD controller with a saturation surface works well in this situation, because the process is open-loop unstable, and tight control is necessary. The FPD is only necessary in the regulation phase, because the PI controller works well for the start-up phase.

A soft z-shaped membership function performs the gain scheduling. It has breakpoints in 4 and 8 which shows that it is specifically designed for a fuzzy band around the setpoint, that is ± 4 K with a fuzzy transition to ± 8 K. The control surface is of the saturation type, but with squared membership functions in order to make it even steeper around the setpoint (Figure 5.11).

The previously developed tuning rules (Table 7.1) provided the gains of the PI controller. The desired settling time – the starting point for the tuning exercise – was chosen as $T_s = 9$. The table then suggests $T_i = \frac{2}{7.8} T_s = 2.3$. For the proportional gain, the table suggests a value inversely proportional to the process gain, but this is unknown, since the process is unstable, and it is highly variable. By trial and error, a value $K_p = 2T_i = 4.6$ was chosen. It is possible, with these relations fixed, to vary T_s somewhat and let the others follow; the responses are still good.

The FPD controller strains the process to deliver a faster settling time, and $T_s = 6$ was chosen. Since the controller is based on a PD controller, the table provides $T_d = \frac{2}{7.8} T_s = 1.5$, and in this case the proportional gain was chosen as $K_p = T_d = 1.5$. The FPD should deliver maximal signal at a deviation of four degrees, and therefore $GE = 25$ was chosen in order to exploit the standard universe $[-100, 100]$ to its full extent. The previously developed gain relationships (Table 4.1) then provide $GCE = T_d \times GE$ and $GU = K_p / GE$.

With respect to the requirements specifications, it was possible to tune the PI+FPD controllers to reach the setpoint without overshoot during the start-up phase. The reactor temperature reaches $T = 39.5°C$ in 1 h 17 min, and it stays within a band of 0.5°C from the setpoint, although there is some oscillation before pulling in to the setpoint. The controllers regulate the disturbances without overshoot, and the excursions from the setpoint are within a narrow band (±1.9 K).

10.3 Idle Speed Control of a Car Engine

If a car driver switches the window heater on to defrost the rear window, the engine must deliver the energy. In control terms, this constitutes a load on the engine and the battery. While idling, the speed of the engine may drop audibly, when such a load is switched on. Air conditioning is one of the heaviest loads, but also head lamps, power steering, electric window heating, and electric window drives are significant loads. The objective of this example is to design a controller, such that the idle speed is kept as steady as possible, despite variations in the load. The evenness of the idle speed is a sales parameter, and manufacturers try to ensure that engines idle at a constant speed (usually 600–1200 rpm).

10.3.1 Engine Model

Figure 10.9 is a simulation model of an engine. The model can be manipulated through the air intake as well as the timing of the ignition of the spark plug. Base values provide inputs that keep the model in trim, corresponding to the operating point, which is 1000 rpm. The controlled output is the engine speed of rotation n. Also, the manifold air pressure is available for a controller, if preferred.

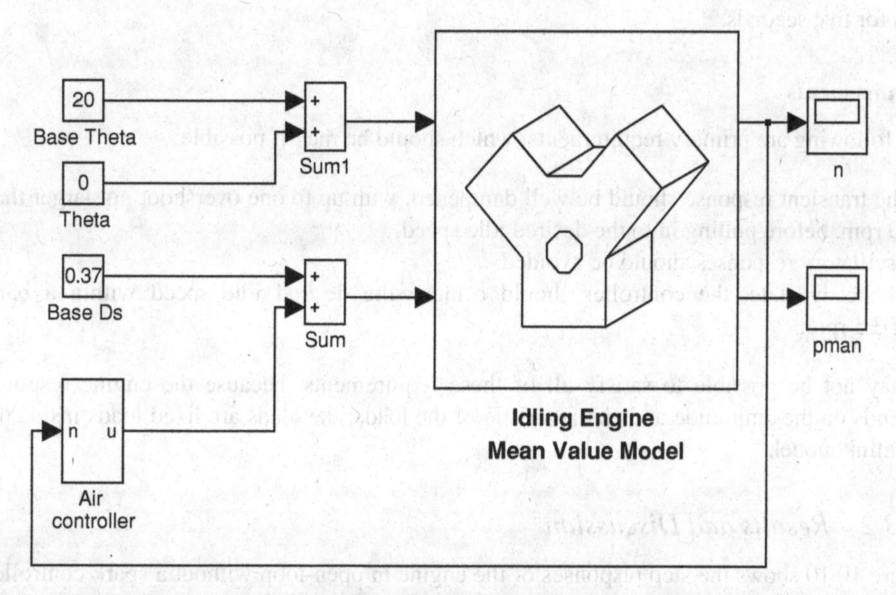

Figure 10.9 The model has two inputs and two outputs. In the figure, the air controller controls the air bypass valve to keep the idle speed steady (engistd.mdl).

The engine is a four-cylinder British Leyland, 1275 ccm GT A series, spark ignition engine. There are three major physical processes in the engine:

- the intake of air and expulsion of the exhaust gases;
- the injection of fuel; and
- the spark ignition of the air/fuel mixture.

There are delays, nonlinearities, and discrete operations which influence the way the engine behaves. The model is due to Elbert Hendricks and his co-workers (Hendricks and Sorenson 1990).

Idle speed control

The main control input is the air flow through the idle speed valve (*air bypass valve*). When the engine idles, the throttle plate is closed, and the air flow into the engine is controlled by the air bypass valve in terms of its degree of opening, which is a fraction between zero and one. A second control input is the spark ignition timing in terms of the angle θ, which is the advance before the piston reaches its upper position (*top dead centre*). It is faster acting, but less effective. The controlled output is the engine rotational speed n, measured in revolutions per minute (rpm), but divided by 1000 to scale the numbers. A second output, the manifold pressure, is available in the simulation model (it is usually not directly available in practice, only in the laboratory). When idling, manifold pressure is low, say, half an atmosphere.

The controller should aim to perform as well as possible, that is, minimize the magnitude and the duration of dips (downward excursions) and flares (upward excursions) resulting from putting a load on the system and taking it off, respectively. The model includes two loads: electric windows on/off, and air conditioning on/off. Each load kicks in at a specific time, and lasts for five seconds.

Requirements

The following are primary requirements, which should be met, if possible:

- The transient response should be well dampened, with up to one overshoot not larger than 20 rpm, before pulling in to the desired idle speed.
- Oscillatory responses should be avoided.
- At steady state, the controller should achieve the desired idle speed within a band of ±5 rpm.

It may not be possible to satisfy all of these requirements, because the engine response depends on the amplitude and characteristic of the loads; the loads are fixed loads inside the Simulink model.

10.3.2 Results and Discussion

Figure 10.10 shows the step responses of the engine in open-loop without a spark controller. The input step is added to the base value on the air valve. The static gain from the air valve to the rotational speed n is 4.7 times. The response looks like a linear first-order response.

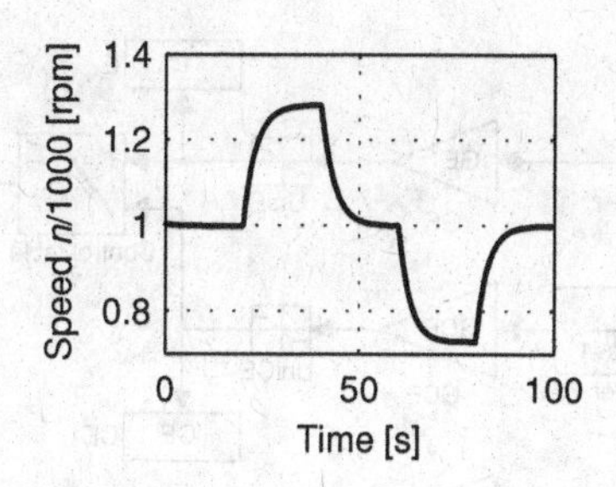

Figure 10.10 Step responses. The air valve is stepped up ($t = 20$), back to zero ($t = 40$), down ($t = 60$) and back to zero ($t = 80$). (figstepengine.m)

However, the crank shaft is a double integrator and the manifold pressure is a third state variable. The amplitude of the response is the same for positive and negative steps, and the settling time is 12.6 s.

The control problem is a regulation type of problem (disturbance rejection), which implies that integral action is necessary. When a load comes on, the integrator must supply the necessary control signal to balance the load in steady-state. The integral action pulls the engine back to the setpoint, but some deviation is allowed – according to the specifications – since that will barely be audible.

Controllers

Figure 10.11 shows simulation runs with three different controllers from three consecutive design attempts. The first is a linear FPID controller based on a crisp PID controller. The first dip, which occurs when the first load switches on, amounts to about 100 rpm, and the engine does not quite make it to the setpoint before the load is switched off again. The flare resulting

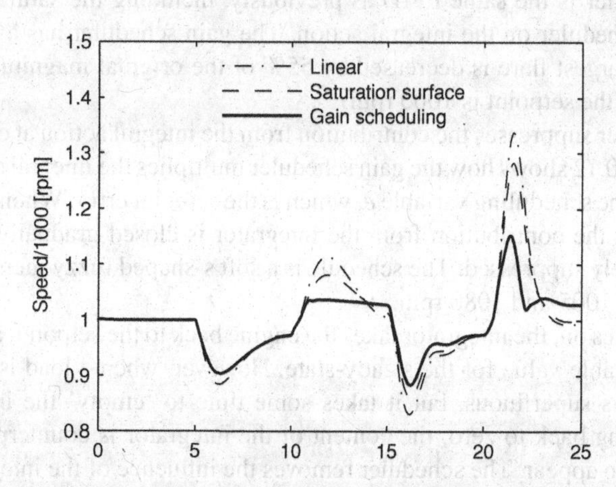

Figure 10.11 Three runs with three controllers. Electric windows are turned on ($t = 5$) and off ($t = 10$), and air conditioning is turned on ($t = 15$) and off ($t = 20$). (figengine.m)

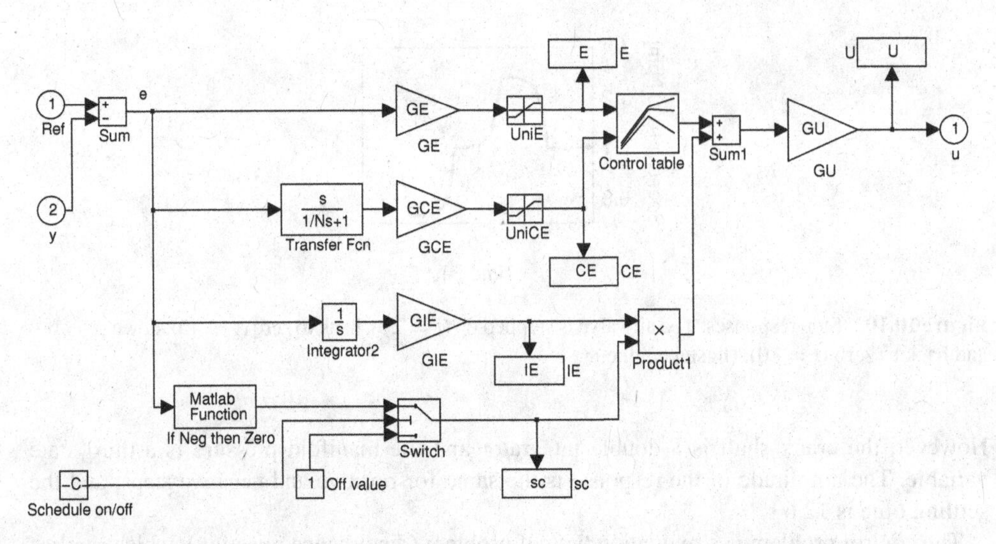

Figure 10.12 Fuzzy PID controller with gain scheduling on the integrator. (engistd.mdl)

from removing the load is slightly larger in magnitude than the dip. The second load is more severe, and the engine dips to 835 rpm. The return from the dip is overdamped. When the second load is taken off again, a large flare up to 1330 rpm occurs. This is the most difficult situation to handle.

The second controller is the same FPID, but the control surface is now of the saturation type (Figure 5.6). This choice was made, because tight control is necessary to try to keep the dips and flares down in magnitude. We are looking for more damping, but we keep the original gain settings in order to make a comparison. The response improved, and the peaks decreased by 17–29%.

The third controller is the same FPID as previously, including the saturation surface, but now with a gain scheduler on the integral action. The gain scheduler has a significant effect on the flares. The largest flare is decreased by 55% of the original magnitude, and it is now down to 1150 rpm (the setpoint is 1000 rpm).

The gain scheduler suppresses the contribution from the integral action at certain times. The diagram in Figure 10.12 shows how the gain scheduler multiplies the integral action by a factor, which depends on the scheduling variable e, which is the setpoint error. When the engine speed is above 1005 rpm, the contribution from the integrator is closed gradually until 1080 rpm, where it is completely suppressed. The schedule is a soft s-shaped fuzzy membership function with breakpoints at 1005 and 1080 rpm.

When a load comes on, the integrator takes the engine back to the setpoint, and the integrator then contains a suitable value for the steady-state. However, when a load is suddenly turned off, the integrator is superfluous, but it takes some time to 'empty' the integrator. During the period of coming back to zero, the content of the integrator is counterproductive, and it allows large flares to appear. The scheduler removes the influence of the integrator only in the case of flares, not dips. The schedule is smooth, but it still has a rather abrupt effect on the speed signal.

It is rather interesting that the model seems linear in the operating range. It is thus the integrator that causes the large flare ($t = 20$), not the model. The problem is similar to the integrator windup problem discussed earlier (Section 4.9.3).

The timing of the spark has a positive effect on the magnitude of the dips and flares. As a control input, it acts immediately, which is convenient, since the dips and flares grow fast, and it is an advantage to start the control action as early as possible. The spark input attenuates the dips and flares. The model operates with a constant base angle value (20°) for idling, and the controller adds a differential contribution. However, the model limits internally the maximum advance angle (to 45°). A crisp proportional controller is sufficient for the spark angle.

Tuning

The design is based on a crisp controller, and we apply the previously developed tuning rules for a PID controller (Table 7.1). The starting point is to select a desired settling time T_s. The open-loop tests showed a settling time of 12.6 s, and for the closed-loop system we selected $T_s = 2.25$ s. This value was selected after some experimentation, though. The settling time is related to the 2% criterion, and the desired time constant is thus $\tau = -T_s/\ln 0.02 = 2.25/3.9 = 0.58$. The time constant refers to the envelope of the response.

The table determines T_d from τ ($T_d = \tau/2 = 0.29$). In practice, it is a little easier to work with the settling time, rather than the time constant, since the former is easier to check on a plot. In theoretical derivations, however, it is easier to use the time constant.

According to the design specifications, only little overshoot is acceptable. With a damping ratio $\zeta = 1$, the theoretical overshoot is 13.5% (free response of a second-order system). Having chosen ζ, the table provides the integral gain from the derivative gain ($T_i = 4\zeta T_d = 4 \times 0.29 = 1.2$). The damping ratio could be increased in an attempt to attenuate the peaks more, but that also slows the response.

Finally, the table determines the proportional gain from the derivative gain, the integral gain, and the static gain of the process. The open-loop responses showed that the latter gain is approximately $K = 4.7$. The table then proposes a proportional gain $\left(K_p = \gamma T_i T_d / K = 10 \times 1.2 \times 0.29/4.7 = 0.70\right)$. It is a proposal, in the sense that the proportional gain should be as high as possible, in order to achieve the desired settling time (or time constant). However, the engine is unable to meet the requirement of a desired settling time, which is too fast. The size of the gain is limited by the stability margin of the closed-loop system; if the gain is too high, the response will oscillate.

In our case, we stayed with the proposals of the table. Nevertheless, the tuning procedure is iterative, and by going back to the beginning, the desired settling time was adjusted to achieve the best response. The actual settling time achieved varies in the different load responses, but the largest flare, corresponding to switching the air conditioning off, decays more or less with the desired settling time.

The spark controller attenuates the peaks, and its proportional gain was turned up as high as possible, that is, until the response starts to oscillate and then de-tuned to about 2/3 of that value ($K_{p\theta} = 800$).

The FPID gains follow from the crisp PID gains. The universe of the error e was chosen as ± 200 rpm, which corresponds to ± 0.2 in the model, because the speed is normalized by dividing by 1000. The gain on the error was chosen as $GE = 500$ in order to fit $E = GE * e$

to the standard universe. The previously developed gain relationships (Table 4.1) give us $GCE = T_d \times GE$, $GU = K_p/GE$, and $GIE = GE/T_i$. Again, all settings follow from the initial choice of desired settling time T_s, which is very convenient in practice.

The performance satisfies the first design specification, since there is hardly any overshoot after a dip or a flare. Regarding the second specification, the responses are slightly oscillatory, but the amplitude is negligible compared to the magnitude of the dips and the flares. Regarding the third specification, however, the response does not reach the steady state within ±5 rpm. This is a trade-off that was made in order to attenuate the peaks, since a slower response has smaller peaks. The smaller peaks were preferred, assuming the peaks are more audible than a steady-state error.

10.4 Balancing a Ball on a Cart

The simulation study in this example concerns a controller for balancing a steel ball on a cart. The ball-balancer, or *cart-ball system*, is a demonstration of the basic concepts of a nonlinear and multi-variable control problem. For certain initial conditions, the system is even a demonstration of a non-minimum phase system (the step-response starts in the opposite direction to the reference). It is a variant of the inverted pendulum problem.

In the example, a linear controller stabilizes the ball balancer. With certain design choices, a fuzzy controller is equivalent to a summation, and thus it can replace the linear controller.

Figure 10.13 shows the cart-ball system. The cart moves on a pair of tracks horizontally mounted on a heavy support, and a steel ball rolls on an arc made of two curved, parallel pipes. The control objective is to balance the ball on the top of the arc and at the same time place the cart in a desired position.

The controller design is based on a linearized model of the system,

$$\dot{\mathbf{x}} = \boldsymbol{A}\mathbf{x} + \boldsymbol{B}\mathbf{u} \tag{10.1}$$

$$\mathbf{y} = \boldsymbol{C}\mathbf{x} \tag{10.2}$$

The model is a linear state-space model, where $\mathbf{x}$ is a vector of state-variables, $\dot{\mathbf{x}}$ is a vector of time derivatives $d\mathbf{x}/dt$, $\mathbf{u}$ is a vector of inputs to the system, and $\mathbf{y}$ is a vector of output variables. The matrices $\boldsymbol{A}$, $\boldsymbol{B}$, and $\boldsymbol{C}$ are matrices of appropriate dimensions containing real numbers. A mathematical model from *first principles* – the basic laws of physics – describe the behaviour of the system. The model is nonlinear, but it can be linearized.

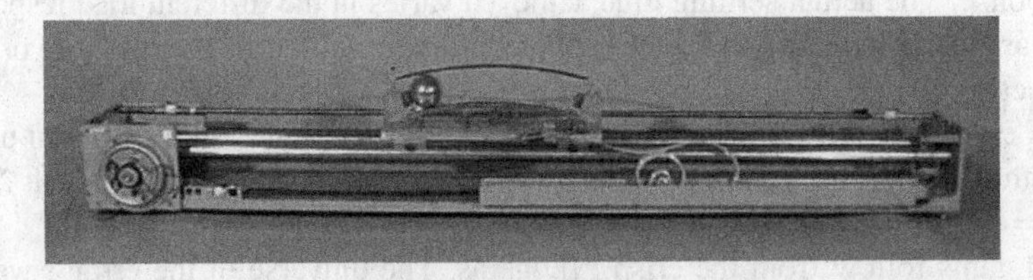

Figure 10.13 Laboratory rig. The cart moves back and forth, commanded by a controller, in order to balance the ball. The length of the rig is 1.5 m (4.9 ft).

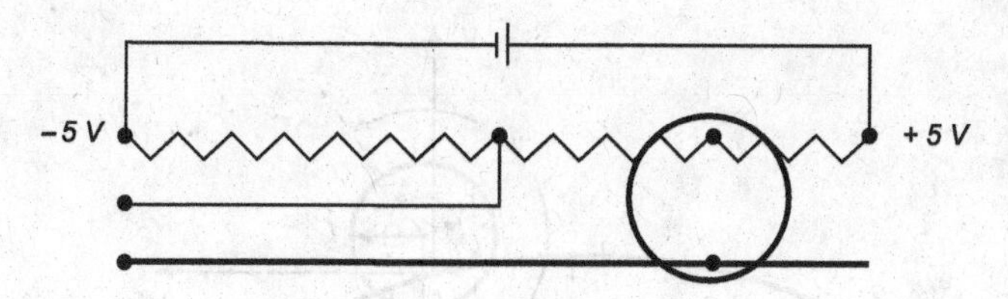

Figure 10.14 Ball position measurement. The steel ball acts as a voltage divider.

The laboratory rig is equipped with power supply and equipment for both analogue and digital control. By pushing the cart left and right manually, it is possible to get the ball near the top of the arc, but it is impossible for a normal human being to position the cart at a particular position at the same time. However, an automatic control-system is capable of achieving this balance. The position of the cart and angle of the ball from the vertical are measured variables, and the manipulated variable is the horizontal force acting on the cart.

The ball rolls on two curved pipes: one is made of aluminium, while the other is a coil of resistance wire. The angle of the ball from the vertical is determined by measuring its position on the pipes. The ball, being made of steel, connects the pipes electrically, and it acts as a voltage divider, producing a voltage proportional to the position (Figure 10.14).

The position of the cart is measured in a similar manner using a carbon wheel contact, which rolls on a coil alongside the rails.

The rails are cylindrical bars mounted on the support, and the wheels of the cart are small, low-friction ball-bearings that roll on the bars. A wire pulls the cart, passing over a pulley at one end and a wire drum at the other end, both attached to the support. The wire drum is driven by a current-driven direct current (DC) print-motor. Although the motor is current-driven, we assume that the armature voltage is proportional to the current and in turn that the torque is proportional to the armature voltage. This is an approximation, but it is a relatively fast DC motor with small electrical and mechanical time constants.

10.4.1 Mathematical Model

The current to the motor is a function of the variables y, $\dot{y}$, φ, and $\dot{\varphi}$, where y is the position of the cart and φ is the angular deviation from the vertical of the ball. The velocity signals are not directly measured, but obtained by differentiation in operational amplifiers.

All directions are assumed positive towards the right. We apply the basic physical equations related to the vertical reaction force V and the horizontal reaction force H. Friction forces are neglected. Figure 10.15 defines the symbols.

By Newton's second law, the horizontal movement of the ball is,

$$m\frac{\mathrm{d}^2}{\mathrm{d}t^2}\left[y + (R + r)\sin\varphi\right] = H \tag{10.3}$$

the vertical movement of the ball is

$$m\frac{\mathrm{d}^2}{\mathrm{d}t^2}\left[(R + r)\cos\varphi\right] = V - mg \tag{10.4}$$

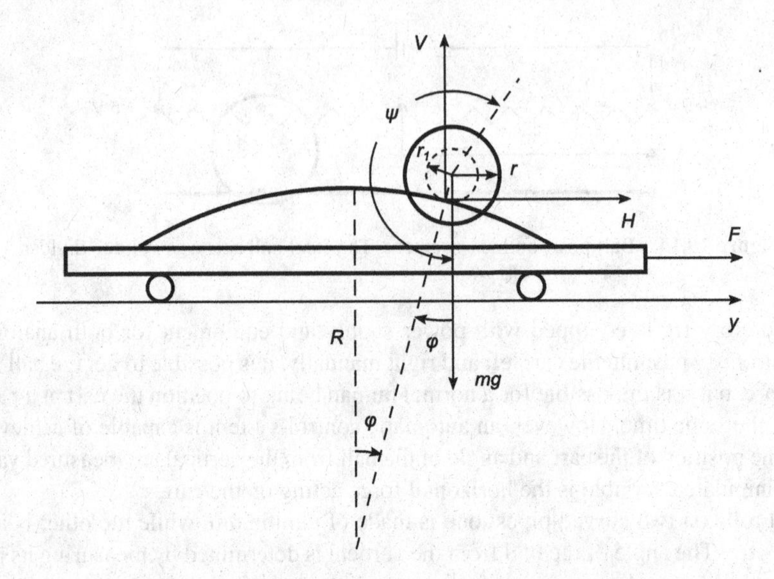

Figure 10.15 Definition of symbols and directions.

the rotational movement of the ball is

$$I\ddot{\psi} = r\,(V\sin\varphi - H\cos\varphi) \tag{10.5}$$

and the horizontal movement of the cart is

$$M\ddot{y} = F - H \tag{10.6}$$

The geometrical relationship between φ and ψ is

$$\varphi R = r\,(\psi - \varphi) \Leftrightarrow \psi = \frac{R+r}{r}\varphi \tag{10.7}$$

The variables ψ, V, and H can be eliminated from Equations (10.3)–(10.7), yielding two second-order differential equations in φ and y:

$$(M+m)\,\ddot{y} = -m\,(R+r)\left(\ddot{\varphi}\cos\varphi - \dot{\varphi}^2\sin\varphi\right) + F \tag{10.8}$$

$$I\frac{R+r}{r}\ddot{\varphi} = mr\,(R+r)\left(-\ddot{\varphi}\sin^2\varphi \;- \dot{\varphi}^2\cos\varphi\sin\varphi\right)$$
$$+\,mgr\sin\varphi + Mr\ddot{y}\cos\varphi - Fr\cos\varphi \tag{10.9}$$

They are nonlinear because they contain trigonometric functions and products of variables. They are coupled such that $\ddot{y}$ appears on the left side of Equation (10.8) and on the right side of Equation (10.9), and the converse is true of $\ddot{\varphi}$. According to the first equation, the force F directly affects the acceleration of the cart $\ddot{y}$, which is also affected by the ball. According to the second equation, the force causes a torque $Fr\cos\varphi$ on the ball, affecting the angular acceleration of the ball directly. The acceleration of the cart $\ddot{y}$ also affects the angular acceleration of the ball.

We linearize by introducing the following small-angle approximations:

$$\cos\varphi \approx 1, \sin\varphi \approx \varphi, \cos^2\varphi \approx 1, \sin^2\varphi \approx 0, \dot{\varphi}^2 \approx 0 \tag{10.10}$$

with $|\varphi| \leq 0.22$ rad. Consequently, we achieve a reduced system of equations:

$$(M+m)\ddot{y} = -m(R+r)\ddot{\varphi} + F$$
$$I\frac{R+r}{r}\ddot{\varphi} = mgr\varphi + Mr\ddot{y} - Fr$$

After rearranging,

$$\ddot{y} = -\frac{m^2r^2g}{MI+mI+mr^2M}\varphi + \frac{mr^2+I}{MI+mI+mr^2M}F$$
$$\ddot{\varphi} = \frac{mr^2g(M+m)}{(R+r)\left(MI+mI+mr^2M\right)}\varphi - \frac{mr^2}{(R+r)\left(MI+mI+mr^2M\right)}F$$

and introducing the substitution variables,

$$a = -\frac{m^2r^2g}{MI+mI+mr^2M}$$
$$b = \frac{mr^2+I}{MI+mI+mr^2M}$$
$$c = \frac{mr^2g(M+m)}{(R+r)\left(MI+mI+mr^2M\right)}$$
$$d = -\frac{mr^2}{(R+r)\left(MI+mI+mr^2M\right)}$$

as well as the state-variables $x_1 = y$, $x_2 = \dot{y}$, $x_3 = \varphi$, and $x_4 = \dot{\varphi}$, a simple linear state-space model emerges:

$$\dot{x}_1 = \dot{y}$$
$$\dot{x}_2 = a\varphi + bF$$
$$\dot{x}_3 = \dot{\varphi}$$
$$\dot{x}_4 = c\varphi + dF$$

Table 10.4 Physical data for the cart-ball rig.

Component	Measure	Symbol	Value
Cart	Length		0.35 m
	Width		0.12 m
	Radius of the arc	R	0.50 m
	Mass (including equivalent mass of motor and transmission)	M	3.1 kg
Ball	Maximum angle	φ	±0.22 rad
	Radius	r_1	0.0275 m
	Rolling radius	r	0.025 m
	Mass	m	0.675 kg
	Moment of inertia $\frac{2}{5}mr_1^2$	I	0.204×10^{-3} kg m^2
Support	Bar length		1.4 m
	Bar diameter		0.025 m
	Motor power		21 W
	Motor voltage	U	13 V
	Motor transmission ratio	$U : F$	1:1
	Motor speed		3700 rpm
	Gravitational acceleration	g	9.81 m s^{-2}

In matrix form, the equations are

$$\dot{\mathbf{x}} = \begin{bmatrix} 0 & 1 & 0 & 0 \\ 0 & 0 & a & 0 \\ 0 & 0 & 0 & 1 \\ 0 & 0 & c & 0 \end{bmatrix} \mathbf{x} + \begin{bmatrix} 0 \\ b \\ 0 \\ d \end{bmatrix} u \tag{10.11}$$

The input u comprises just one input, the driving force F, which can be substituted by the voltage U to the motor, as their relationship is 1 V to 1 N; this is how the motor and the gear were designed.

With the data in Table 10.4 the actual values of the constants are

$$a = -1.34, b = 0.301, c = 14.3, d = -0.386$$

A *signal flow graph* provides an overview. Given a state-space model with matrices containing zeros and non-zero elements, the flow of the signals can be mapped into a directed graph, or *digraph*; the digraph portrays the couplings in the model (Figure 10.16).

The node set is given by one input node and four state nodes. The arc set is given by the non-zero entries in the matrices, that is, if $a_{ij} \neq 0$, then there exists an arc from the jth state node to the ith state node, and if $b_{ij} \neq 0$, then there exists an arc from the jth input node to ith state node. The numbers a_{ij} and b_{ij} are assigned to the arcs as *weights*. Each weight a_{ij} and b_{ij} is implicitly associated with an integration as well.

Insertion of a feedback connection from the node φ to the input node F will create a loop $\varphi - F - \dot{\varphi} - \varphi$. Alternatively, a feedback connection from the output node y to the input node F creates a larger loop, $y - F - \dot{\varphi} - \varphi - \dot{y} - y$, as well as another loop, $y - F - \dot{y} - y$. In

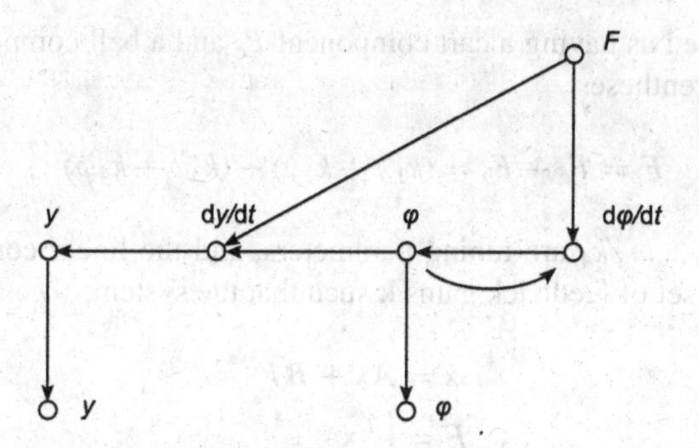

Figure 10.16 Signal flow in the state-space model.

the first case, there is no feedback from the cart $\{y, \dot{y}\}$ to the ball $\{\varphi, \dot{\varphi}\}$, while in the second case, there is feedback through the ball into the cart. Therefore, a ball controller can be designed and tuned independent of how the cart behaves, while a cart controller will be influenced by the behaviour of the ball.

Since state-space models are not unique – a given physical plant may be modelled by several state-space models – the digraph reflects the flow of signals in the model, and not necessarily the physical plant itself. However, in this case, the state-variables have a physical interpretation.

The control task is to design a controller that balances the ball and places the cart in the middle of the track, given the following initial values (Table 10.4)

$$\mathbf{x}_0 = [0.525 \quad 0 \quad -0.22 \quad 0]^T$$

The cart is thus at the far right end of the track: the bar length is 1.40 m, the cart length is 0.35 m, and the position y is half the bar length minus half the cart length or 0.525 m. The ball is on the left, leaning against its end-stop: the angle φ is negative, and the magnitude of the maximum angle is 0.22 rad. Therefore, the controller must initially pull the cart left to get the ball up on top.

10.4.2 Step 1: Design a Crisp PD Controller

To find a set of feedback gains, we shall apply state feedback control, before designing a PD controller. A state feedback controller generates a control signal from the values of the state variables,

$$F = \mathbf{k} \cdot \mathbf{x}$$

or,

$$F = k_1 y + k_2 \dot{y} + k_3 \varphi + k_4 \dot{\varphi} \tag{10.12}$$

The force can be viewed as having a cart component F_c and a ball component F_b, simply by placing two sets of parentheses,

$$F = F_c + F_b = (k_1 y + k_2 \dot{y}) + (k_3 \varphi + k_4 \dot{\varphi})$$

The feedback gains $k_1, \ldots, k_4$ are tuning parameters, and the linear control problem is the following: to design a set of feedback gains $\mathbf{k}$ such that the system,

$$\dot{\mathbf{x}} = \boldsymbol{A}\mathbf{x} + \boldsymbol{B}F \tag{10.13}$$

$$F = \mathbf{k}^T \mathbf{x} \tag{10.14}$$

is stable. Note that at this point we approximate the real system with a linear model, and especially that the limitation of 13 V on the motor voltage is missing from the equation.

By inserting Equation (10.14) in Equation (10.13), the closed-loop system equations are obtained:

$$\dot{\mathbf{x}} = \boldsymbol{A}\mathbf{x} + \boldsymbol{B}\mathbf{k}^T\mathbf{x} = \left(\boldsymbol{A} + \boldsymbol{B}\mathbf{k}^T\right)\mathbf{x} \tag{10.15}$$

Stability of the closed-loop system is guaranteed if all eigenvalues of the closed-loop system matrix $\boldsymbol{A} + \boldsymbol{B}\mathbf{k}^T$ are in the left half of the complex plane. A closer investigation will show that all ks must be strictly positive (apply, for example, Routh's stability criterion). Consequently, all four state-variables must be available to the controller. In the laboratory rig, only the positions are directly measurable, but the velocities are computed electronically, and therefore indirectly observable.

It is important to realize that the cart has positive feedback. Since k_1 must be positive, a positive (negative) position y will generate a positive (negative) contribution F_c to the control signal. For example, as all directions are assumed positive towards the right, the position term of F_c will try to push the cart right when it is on the right side of the centre of the track, that is, *away* from the centre. Whenever F_c pushes the cart away, the ball will roll to the opposite side, and F_b will provide a signal of opposite sign, demanding the cart to pull back towards the centre.

It is difficult, if at all possible, to find four stabilizing gains by hand-tuning; that is a reason for taking this detour around state-space modelling. One possibility is to optimize the frequency response with respect to the sensitivity circle defined earlier (Section 7.6) using the optimizer function `fminsearch` in MATLAB®. A controller with the sensitivity $M_s = 2$ results from the following values

$$\mathbf{k}^T = \begin{bmatrix} 21 & 21 & 194 & 46 \end{bmatrix} \tag{10.16}$$

Figure 10.17 shows the response of both the cart and the ball. The two control signal components F_c and F_b are more or less in counter-phase. Other settings are possible, depending on the choice of sensitivity, and other design methods are also possible, for example, pole placement or optimal control (e.g. the `lqr` function in MATLAB®'s Control Toolbox).

We will assume

$$F_b = 194\varphi + 46\dot{\varphi},$$

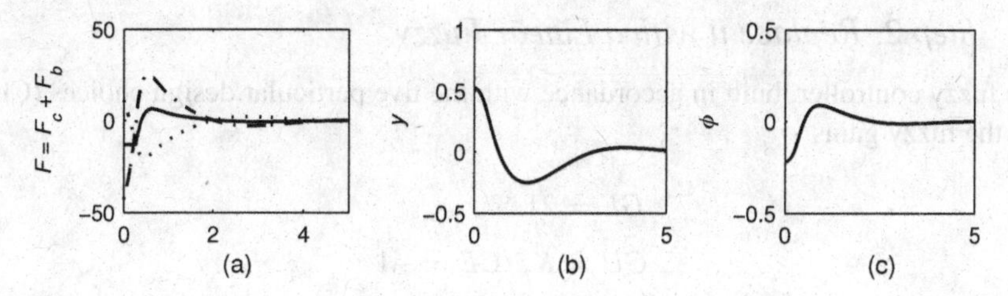

Figure 10.17 Response with $\mathbf{k} = \begin{bmatrix} 21 & 21 & 194 & 46 \end{bmatrix}^T$ and no limitation on the control signal. The control signal (a) is the sum of a component F_c from the cart and a component F_b from the ball. The cart starts at 0.525 m and settles after about 4 seconds (b). The ball starts at −0.22 (−12.6°) and settles a little earlier (c). (figpend.m)

using the gains k_3 and k_4 as above, and design a PD controller for F_c on the basis of the gains k_1 and k_2 as above. Using Equation (10.13), we have

$$\begin{aligned}\dot{\mathbf{x}} &= \mathbf{A}\mathbf{x} + \mathbf{B}F \\ &= \mathbf{A}\mathbf{x} + \mathbf{B}(F_c + F_b) \\ &= \mathbf{A}\mathbf{x} + \mathbf{B}(194\varphi + 46\dot{\varphi}) + \mathbf{B}F_c\end{aligned}$$

For F_c, we use

$$F_c = K_p(e + T_d\dot{e})$$

which we can substitute with a fuzzy controller later. Define

$$e = Ref - y$$

With $Ref = 0$, corresponding to the centre of the track, substitution yields

$$\begin{aligned}F_c &= K_p(-y + T_d(-\dot{y})) \\ &= -K_p y - K_p T_d \dot{y}\end{aligned}$$

We see that $-K_p$ corresponds to k_1 and $-K_p T_d$ corresponds to k_2, therefore

$$\begin{aligned}K_p &= -21 \\ T_d &= 1 \\ k_3 &= 194 \\ k_4 &= 46\end{aligned}$$

Thus we have achieved a set of tuning parameters to start with.

10.4.3 Step 2: Replace it with a Linear Fuzzy

A linear fuzzy controller, built in accordance with the five particular design choices (Chapter 3), with the fuzzy gains

$$GE = 21$$

$$GU = K_p/GE = -1$$

$$GCE = T_d * GE = 21$$

provides the same response as the PD controller. The magnitude of the position y is at most 0.525 m, and therefore the magnitude of the error $E = GE * e$ is at most $21 * 0.525 = 11$, which is inside the 100-limit of the universe. For the change in error $CE = GCE * \dot{e}$, the magnitude of $\dot{e}$ is at most 1.22 m/s (from simulation), and CE is at most $21 * 1.22 = 26$, which is also within the limit of the universe. Thus, there is no saturation in the universe, and the controller is equivalent to the PD controller. We could apply α-scaling to exploit the full universe, but it is not necessary when the controller is linear; the main point is to avoid saturation in the premise universes.

10.4.4 Step 3: Make it Nonlinear

The linear control surface can now be replaced by the nonlinear standard control surfaces successively.

The Nyquist plot in Figure 10.18 shows the relative robustness of the four surfaces. The linear frequency response touches the sensitivity circle, as desired. The saturation surface is less robust, and the two remaining surfaces cut through the sensitivity circle. The gain settings are as stated earlier.

We can rule out the quantizer surface and the deadzone surface, since the system is unstable in open loop, and their Nyquist curves cut into the sensitivity circle inconveniently. The linear surface and the saturation surface behave similar to each other, but near the sensitivity circle the saturation surface is slightly closer to the point of instability. We can expect that a saturation type of surface will provide a slightly faster response, but also with a lesser stability margin.

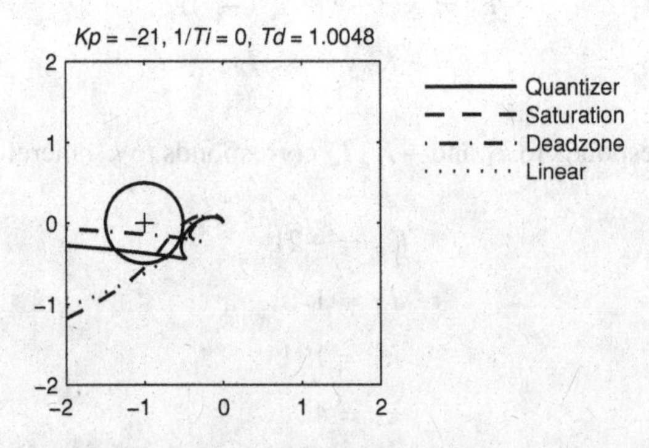

Figure 10.18 Nyquist plot. Cart-ball system controlled by the four standard control surfaces with gains $GE = 21$, $GCE = 21$, and $GU = -1$. (figdf.m)

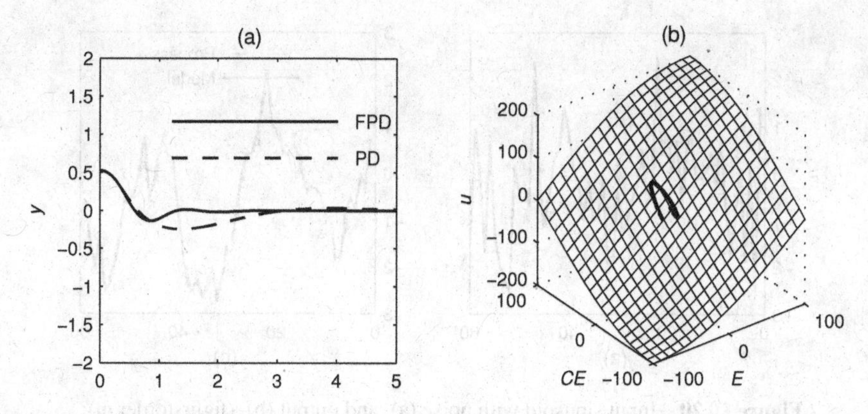

Figure 10.19 Response with settings $GE = 21$, $GCE = 21$, $GU = -1$. The FPD controlled system and the equivalent PD controlled system are shown together (a). The control surface is of the switching type (b). (figpend.m)

10.4.5 Step 4: Fine-Tune it

We now go on to study the step response in the time domain, with the control action limited by the amplifier to ±13 V or, equivalently, ±13 N. The average squared error (ASE) was used as a performance index, in order to compare different test runs.

Figure 10.19 shows the response resulting from fixed gains, but with a control surface of the switching type (Section 5.11). The standard saturation surface was also tested, but the switching surface provided a better performance index. In comparison with the equivalent crisp PD controller, the FPD has a faster response, less overshoot, and a faster settling time. For comparison, the crisp PD achieved $ASE = 0.0303$ while the FPD achieved $ASE = 0.0172$.

It is very difficult to adjust the tuning gains further, as the system quickly becomes unstable. The Nyquist plot (Figure 10.18) supports this observation, as it shows a relatively small stability margin. An increase of GU by 1% drives the performance index to $ASE = 0.0171$. Higher values of GU destabilize the system, while lower values ($0.6 \leq GU < 1$) result in slower responses, but also is more robust. The ratio $GCE/GE = 1$ corresponds to the derivative gain T_d in the equivalent crisp controller, and T_d corresponds to a desired time constant (Table 7.1). The response apparently does have a dominating time constant approximately equal to one. The Nyquist plots of the standard saturation surface and the linear controller are rather close to each other (Figure 10.18), which suggests that similar results might be achieved by a linear controller, but with other gain settings.

10.5 Dynamic Model of a First-Order Process with a Nonlinearity

This example identifies a model of a first-order, discrete-time process. The process is linear, but the system is nonlinear, as the control signal is limited by an actuator with saturation. The model is trained on sampled input-output data, and its performance is compared with the training data.

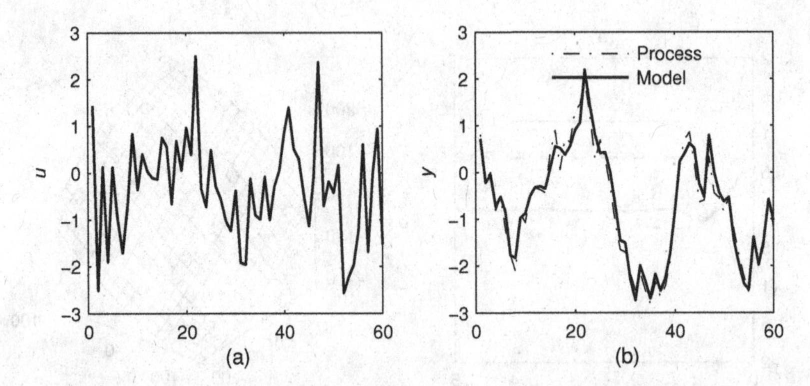

Figure 10.20 Input sinusoid with noise (a), and output (b). (figfirstorder.m)

The process is:

$$y_{k+1} = 0.8y_k + v(u)_k + n$$

Here y_{k+1} is the scalar output at sampling instant $k + 1$ and v_k is the input to the linear process at sampling instant k. The noise n is normally distributed, with zero mean value and a standard deviation of one. The process is stable, since the coefficient 0.8 is less than one. The actuator is a limited function $-0.8 \leq v(u) \leq 0.8$ of the control signal u; that is, its characteristic is a saturator (Figure 10.23 (a)).

In order to generate training data, the system was excited with a sinusoidal input, which generated a sequence of 60 input–output data pairs (y_k, u_k). The input signal was $u = -\sin(a)$ with $0 \leq a \leq 5\pi$.

10.5.1 Supervised Model

The model is of the gain scheduling type. Since the actuator limits the external signal u, the model uses u as the scheduling variable. When the amplitude of u is within the limits of the actuator – the proportional band – the system is linear. The rule base consists of two local models for the two saturated regions, and when the system operates in the proportional band of the actuator, the rule base interpolates between the two local models outside of the linear region.

Two-parameter local models

Figure 10.20 shows the input signal, and the figure compares the output from the process and the model. The model follows the process well, although it misses a few of the very sharp spikes. The process and the model use the same input data, and the model is trained on those data.

The model is based on the following structure of the rule base:

$$\begin{aligned} &\text{If } u_k \text{ is Neg then } \hat{y}_{k+1} = w_{11}y_k + w_{12}u_k \\ &\text{If } u_k \text{ is Pos then } \hat{y}_{k+1} = w_{21}y_k + w_{22}u_k \end{aligned} \tag{10.17}$$

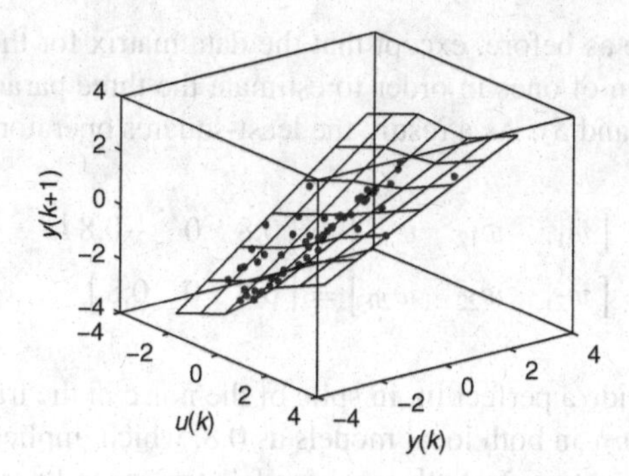

Figure 10.21 Surface fit to data. The surface is drawn as transparent, and there are points on both sides of the surface. (figfirstorder.m)

The universe for u was chosen by inspection as $[-3, 3]$, and the same was used for the universe of the process output y. The set Neg was defined as a trapezoidal membership function with a linear part that matches the saturator; that is, the function is one below $u = -0.8$ and zero above $u = +0.8$. The set Pos is just its negation, that is, Pos = 1 − Neg.

Assuming that the limits of the actuator are known – that is, we are allowed to use the information – we formed two sets of data: S_1 is the set of all triplets (y_{k+1}, y_k, u_k) for which $u_k \leq -0.8$, and S_2 is the set of all triplets for which $u_k \geq 0.8$. The set S_1 was used to find the weights of the first rule, and S_2 was used for the second rule. MATLAB®'s \-operator provided a least-squares solution (Equation (9.11) with the following results:

$$\begin{aligned} \begin{bmatrix} w_{11} & w_{12} \end{bmatrix} &= \begin{bmatrix} 0.94 & 0.37 \end{bmatrix} \\ \begin{bmatrix} w_{21} & w_{22} \end{bmatrix} &= \begin{bmatrix} 0.62 & 0.45 \end{bmatrix} \end{aligned} \tag{10.18}$$

Once the weights are determined, the model in Equation (10.17) can simulate the system. The relationship $y_{k+1} = f(y_k, u_k)$ can be viewed as a three-dimensional surface, and the model can be seen as a surface fit. Figure 10.21 shows the surface and the data points. The surface contains two planes that are connected by a middle section. Due to symmetry, the two planes should be parallel, but they are not, because the set of training data is relatively small.

It was easy to use the least-squares optimizer for the design of the gain scheduling model with just two rules. However, the method requires that the provided supervision – the decision to use ±0.8 as the limits for the data sets – is admissible in practice.

Three-parameter local models

Offsets can easily be included in the model structure. It increases the number of adjustable parameters, which is a disadvantage, but it provides an extra degree of freedom and thus a better fit.

The new rule base has the following extended structure:

$$\begin{aligned} &\text{If } u_k \text{ is Neg then } \hat{y}_{k+1} = w_{11} y_k + w_{12} u_k + w_{10} \\ &\text{If } u_k \text{ is Pos then } \hat{y}_{k+1} = w_{21} y_k + w_{22} u_k + w_{20} \end{aligned} \tag{10.19}$$

Everything is the same as before, except that the data matrix for the least-squares routine is extended with a column of ones in order to estimate the three parameters, instead of two, based on the data sets S_1 and S_2. As a result, the least-squares operator returns the weights

$$\begin{bmatrix} w_{11} & w_{12} & w_{10} \end{bmatrix} = \begin{bmatrix} 0.8 & 0 & -0.8 \end{bmatrix}$$
$$\begin{bmatrix} w_{21} & w_{22} & w_{20} \end{bmatrix} = \begin{bmatrix} 0.8 & 0 & 0.8 \end{bmatrix}$$

These parameters provide a perfect fit, in spite of the noise in the training data. The model correctly identifies w_{11}, w_{21} in both local models as 0.8, which implies that the global model will have the true time constant. In both local models the second parameters (w_{12}, w_{22}) are zero, which apparently implies that the models neglect the input u. However, the models concern the ranges where the actuator saturates, and its value is constant (± 0.8) in these regions. The constant value is captured by the offset parameters (w_{10}, w_{20}). Within the region of the actuator's proportional band, the rule base interpolates between the two local offsets (w_{10}, w_{20}), and the interpolation is sufficient to model the input. Unseen future perturbations of the input will affect the scheduling variable – thereby the model output also – and the model thus takes the input signal into account.

10.5.2 Semi-Automatic Identification by a Modified HCM

The previous two models required crucial knowledge about the breakpoints of the saturator (± 0.8). In a practical system, these could be hidden from observation. We tried hard c-means clustering (HCM) in order to detect the breakpoints semi-automatically. The HCM algorithm must be initialized with a set of cluster centres or the membership matrix M (Section 9.4.2). Knowing that the nonlinearity contains three linear regions, we set the desired number of clusters to three.

It was possible to identify three clusters, and build two local models from two of the clusters. The performance relative to the training data was not bad, but the clusters were inaccurate. Unfortunately, the algorithm finds clusters that are unrelated to the characteristic of the nonlinearity. The problem is mostly the Euclidean distance measure, which clusters points that are close in the three-dimensional space (Figure 10.21), even across the boundaries of the linear regions of the nonlinearity. There are variants of the FCM algorithm that search, for instance, elliptical clusters using the Mahalanobis distance (Bezdek in Babuška 1998). In our case, there is a simpler solution.

Assume that we know from process knowledge that the input u separates the linear regions via the actuator – this is, after all, the reason for choosing u as the scheduling variable. We would thus prefer one cluster in each region, corresponding to a partitioning of the universe of u, without too much overlap. We therefore redefined the original distance measure of the HCM algorithm.

The distance from a point P_1 with the coordinates (u_k, y_k, y_{k+1}) to the cluster centre P_0 with the coordinates $\left(u_k^0, y_k^0, y_{k+1}^0\right)$ is now defined as

$$\left\| \overrightarrow{P_0 P_1} \right\| = | \; u_k - u_k^0 \; |$$

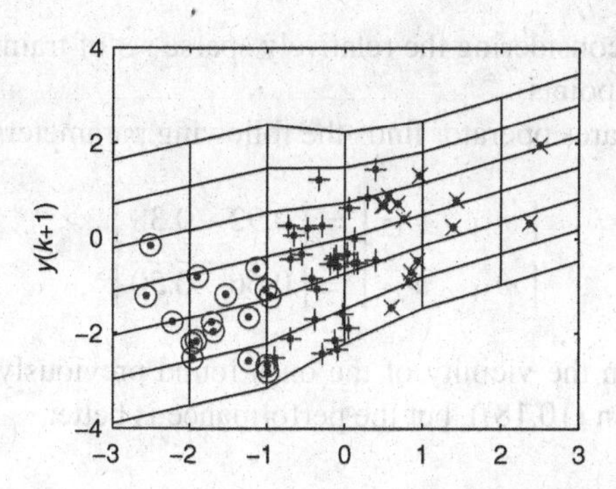

Figure 10.22 Projection of the approximated surface, and the clusters, on the (u_k, y_{k+1})-plane. Each cluster is marked with a different symbol (x, o, +). The clusters are separated by vertical lines perpendicular to the u-axis. (figfirstorder.m)

It is simply the difference between the u-coordinates of the two points. With this definition, points that belong to cluster c_i will be separated by a plane ($u =$ constant) from points that belong to neighbouring clusters (c_{i-1}, c_{i+1}). HCM should then find three clusters, in three rectangular regions, separated by planes perpendicular to the u-axis. The location of a separating plane is halfway between the u-coordinates of two neighbouring centres.

Two-parameter local models

HCM finds three neighbouring clusters that are located one after the other as desired. Figure 10.22 shows the separation into three clusters of 17, 29, and 14 data points counted from negative values of u to positive values. The training set S_1 is set to the leftmost cluster, and S_2 is set to the rightmost cluster. The separating lines are at $u_1 = -0.91$ and $u_2 = 0.43$.

Figure 10.23 compares the membership functions with the characteristic of the actuator. The shoulder-points of the membership functions are located at u_1 and u_2. The overlapping region of the membership functions is not quite the same as the actuator's proportional band,

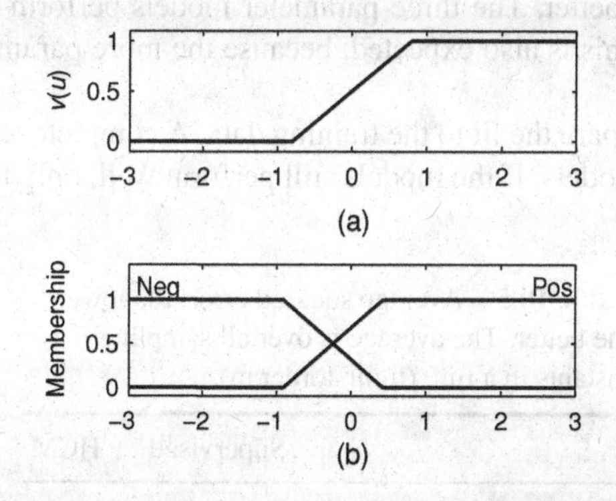

Figure 10.23 The actuator's characteristic (a). The break-points of the identified membership functions (b) are slightly misaligned, relative to the break-points of the actuator. (figfirstorder.m)

but that is to be expected considering the relatively sparse set of training data (S_1 contains 17 points and S_2 contains 14 points).

MATLAB®'s least-squares operator finds the following parameters from S_1 and S_2:

$$\begin{bmatrix} w_{11} & w_{12} \end{bmatrix} = \begin{bmatrix} 0.92 & 0.38 \end{bmatrix}$$
$$\begin{bmatrix} w_{21} & w_{22} \end{bmatrix} = \begin{bmatrix} 0.66 & 0.50 \end{bmatrix}$$

These parameters are in the vicinity of the ones found previously in the supervised two-parameter model (Equation (10.18)), but the performance is better.

Three-parameter local models

Finally, we included offsets in the model structure, as in Equation (10.19). The data matrix prepared for the least-squares routine included a column of ones in order to estimate three parameters, instead of two, based on the data sets S_1 and S_2. As a result, MATLAB®'s \-operator returned the following weights:

$$\begin{bmatrix} w_{11} & w_{12} & w_{10} \end{bmatrix} = \begin{bmatrix} 0.8 & 0 & -0.8 \end{bmatrix}$$
$$\begin{bmatrix} w_{21} & w_{22} & w_{20} \end{bmatrix} = \begin{bmatrix} 0.76 & 0.13 & 0.55 \end{bmatrix}$$

The first set of parameters is equal to the set of parameters of the perfect model, while the second set of parameters is different from those of the perfect model. The result of running this model is very good, although not perfect.

Performance measures

Table 10.5 collects the average of the squared errors for every run of a model. This gives an idea of the relative accuracy of the four models. The supervised models perform well, but that is to be expected since they take advantage of process knowledge in the design. Nevertheless, the HCM model with two parameters can compete with the supervised version, and its performance measure is even slightly better. The three-parameter models perform (much) better than the two-parameter models. This is also expected, because the more parameters in the model, the more flexibility.

The numbers only compare the fit to the training data. A complete test must include a set of test data, unseen by the models. If the models still perform well, only then have they captured the underlying structure.

Table 10.5 Average squared error; the lower the better. The average is over all sampling instants in a run. (figfirstorder.m)

	Supervised	HCM
Two parameter models	0.0564	0.0546
Three parameter models	≈ 0	0.0097

10.6 Summary

The examples demonstrate that there is much to gain from the linear control theory. The recommendation is to extract as much process knowledge as possible, especially when a mathematical model is available. For example, knowing that all four feedback gains must be strictly positive in the cart-ball example, is valuable information.

On the other hand, although the design procedure prescribes a linear start, the examples demonstrate that the final nonlinear design can be quite far from a textbook linear design. This demonstrates that the nonlinear domain has room for skills and imagination.

Regarding the question of performance, the examples show that the nonlinear design improved the linear design. There is no guarantee, however, that the nonlinear design always performs better, because we have not tried to find an optimal design or optimal settings. However, the fuzzy controller performs at least as well as its initial PID controller, and then it gives more options for adjustments.

Whether a fuzzy controller is more suitable in practice depends also on other parameters, such as cost of implementation and maintenance as well as available skills, tools, and the flexibility of the design. Fuzzy controllers have been successful in industry and consumer products owing to their almost human interface, which is another matter than performance.

10.7 Further State-Space Analysis of the Cart-Ball System*

The linear state-space model provides further insight into the controllability and stability of the system.

Open loop

Using Laplace notation $s = \mathrm{d}/\mathrm{d}t$, the general state-space model is

$$s\mathbf{x} = \boldsymbol{A}\mathbf{x} + \boldsymbol{B}\mathbf{u}$$

$$\mathbf{y} = \boldsymbol{C}\mathbf{x}$$

Rearranging the state equations,

$$(s\boldsymbol{I} - \boldsymbol{A})\mathbf{x} = \boldsymbol{B}\mathbf{u}$$

where $\boldsymbol{I}$ is the identity matrix, and solving for $\mathbf{x}$, we get

$$\mathbf{x} = (s\boldsymbol{I} - \boldsymbol{A})^{-1}\boldsymbol{B}\mathbf{u}$$

By substituting this value into the output equations, we obtain

$$\mathbf{y} = \boldsymbol{C}(s\boldsymbol{I} - \boldsymbol{A})^{-1}\boldsymbol{B}\mathbf{u}$$

Thus, the transfer functions from the inputs to the outputs are

$$\frac{\mathbf{y}}{\mathbf{u}} = \boldsymbol{C}\,(s\boldsymbol{I} - \boldsymbol{A})^{-1}\,\boldsymbol{B} \tag{10.20}$$

$$= \boldsymbol{C}\frac{\text{adj}\,(s\boldsymbol{I} - \boldsymbol{A})}{\det\,(s\boldsymbol{I} - \boldsymbol{A})}\boldsymbol{B} \tag{10.21}$$

where adj is the adjoint of the matrix and det is the determinant. It is a matrix expression representing several transfer functions, one from each input to each output in general. The denominator governs the stability of the system. Therefore, we are interested in the characteristic equation

$$\det\,(s\boldsymbol{I} - \boldsymbol{A}) = 0 \tag{10.22}$$

It is a polynomial whose roots are the poles of the system, and its solution is an eigenvalue problem. An eigenvalue λ of $\boldsymbol{A}$ satisfies

$$\boldsymbol{A}\mathbf{v} = \lambda\mathbf{v}$$

where $\mathbf{v}$ is the eigenvector corresponding to λ. Rearranging, we get

$$(\boldsymbol{A} - \lambda\boldsymbol{I})\mathbf{v} = 0$$

For any proper solution to exist, the matrix $(\boldsymbol{A} - \lambda\boldsymbol{I})$ must be singular, or

$$\det(\boldsymbol{A} - \lambda\boldsymbol{I}) = 0$$

which is equivalent to Equation (10.22). Thus, the eigenvalues of $\boldsymbol{A}$ are the solutions to the characteristic equation, and therefore the eigenvalues of $\boldsymbol{A}$ are also the poles of the system.

In our case, the characteristic polynomial is

$$\det\,(s\boldsymbol{I} - \boldsymbol{A}) = \det\left(\begin{pmatrix} s & 0 & 0 & 0 \\ 0 & s & 0 & 0 \\ 0 & 0 & s & 0 \\ 0 & 0 & 0 & s \end{pmatrix} - \begin{pmatrix} 0 & 1 & 0 & 0 \\ 0 & 0 & a & 0 \\ 0 & 0 & 0 & 1 \\ 0 & 0 & c & 0 \end{pmatrix}\right)$$

$$= \det\begin{pmatrix} \boxed{s} & -1 & 0 & 0 \\ 0 & \boxed{s} & -a & 0 \\ 0 & 0 & s & -1 \\ 0 & 0 & -c & s \end{pmatrix}$$

$$= s^2(s^2 - c)$$

The matrix is block triangular, with three blocks in the diagonal, indicated by boxes, and the determinant of the whole is just the product of the determinants of each diagonal block. The roots are $\lambda_i = \{0, 0, \sqrt{c}, -\sqrt{c}\}$, and we observe that the constant a is absent. Thus, in the presence of uncertainty, an inaccurate a does not affect the stability of the model, while an inaccurate c does.

We can also directly relate the eigenvalues to the physics. The double eigenvalue in zero results from the movement of the cart – compare Newton's second law Equation (10.6) – and the second-degree polynomial $(s^2 - c)$ results from the internal loop concerning the ball. We are thus dealing with a double integrator, which we have studied earlier, combined with a second-order system.

The adjoint is

$$\operatorname{adj}(s\boldsymbol{I} - \boldsymbol{A}) = \operatorname{adj}\begin{pmatrix} \mathbf{s} & -1 & 0 & 0 \\ 0 & \mathbf{s} & -a & 0 \\ 0 & 0 & \mathbf{s} & \mathbf{-1} \\ 0 & 0 & \mathbf{-c} & \mathbf{s} \end{pmatrix}$$
$$= \begin{pmatrix} -cs + s^3 & -c + s^2 & as & a \\ 0 & -cs + s^3 & as^2 & as \\ 0 & 0 & s^3 & s^2 \\ 0 & 0 & cs^2 & s^3 \end{pmatrix}$$

The adjoint of any n-by-n matrix $\boldsymbol{M}$ is the transpose of the matrix of cofactors of the elements m_{ij}. A cofactor is the signed minor of the element m_{ij} and the sign is $(-1)^{i+j}$. The minor is the determinant of the $(n-1)$-square submatrix obtained by deleting row i and column j. Because of the structure of $\boldsymbol{B}$,

$$\boldsymbol{B} = \begin{pmatrix} 0 \\ b \\ 0 \\ d \end{pmatrix}$$

we are only interested in the second and fourth column of the adjoint, and furthermore just the first row, in order to find the transfer function from the input to the cart position y, the first state-variable. Thus,

$$\frac{y}{F} = \boldsymbol{C}\frac{\operatorname{adj}(s\boldsymbol{I} - \boldsymbol{A})}{\det(s\boldsymbol{I} - \boldsymbol{A})}\boldsymbol{B}$$
$$= \begin{pmatrix} 1 & 0 & 0 & 0 \end{pmatrix} \frac{\begin{pmatrix} -cs + s^3 & -c + s^2 & as & a \\ 0 & -cs + s^3 & as^2 & as \\ 0 & 0 & s^3 & s^2 \\ 0 & 0 & cs^2 & s^3 \end{pmatrix}}{s^2(s^2 - c)} \begin{pmatrix} 0 \\ b \\ 0 \\ d \end{pmatrix}$$
$$= \frac{(-c + s^2)\,b + ad}{s^2(s^2 - c)}$$

The numerator is of the second degree, and the roots are

$$\pm\sqrt{\frac{-ad}{b}+c} = \pm 3.549$$

Thus, we have a zero in the right half-plane, and the system is a non-minimum phase system, which means that its step response will initially set out in the 'wrong' direction, away from the final value.

Closed loop

With a control law

$$\mathbf{u} = \mathbf{k}^T\mathbf{x}$$

the closed-loop system equations are

$$\begin{aligned} s\mathbf{x} &= \boldsymbol{A}\mathbf{x} + \boldsymbol{B}\mathbf{u} \\ &= \boldsymbol{A}\mathbf{x} + \boldsymbol{B}\mathbf{k}^T\mathbf{x} \\ &= \left(\boldsymbol{A} + \boldsymbol{B}\mathbf{k}^T\right)\mathbf{x} \end{aligned}$$

The control problem is to construct the vector $\mathbf{k}$ such that the closed-loop system matrix has all its eigenvalues in the left half-plane. If, and only if, the pair $(\boldsymbol{A}, \boldsymbol{B})$ is *controllable*, the eigenvalues can be assigned to any values in the complex plane arbitrarily by means of static state feedback. The pair is controllable if

$$\operatorname{rank}\left[\boldsymbol{A}^0\boldsymbol{B}, \boldsymbol{A}^1\boldsymbol{B}, \ldots, \boldsymbol{A}^{n-1}\boldsymbol{B}\right] = n$$

where n is the order of the system and the comma notation means that the matrices are concatenated ('glued') with each other. In our case,

$$\boldsymbol{A}^0\boldsymbol{B} = \boldsymbol{I}\boldsymbol{B} = \begin{pmatrix} 0 \\ b \\ 0 \\ d \end{pmatrix}$$

$$\boldsymbol{A}^1\boldsymbol{B} = \begin{pmatrix} 0 & 1 & 0 & 0 \\ 0 & 0 & a & 0 \\ 0 & 0 & 0 & 1 \\ 0 & 0 & c & 0 \end{pmatrix}\begin{pmatrix} 0 \\ b \\ 0 \\ d \end{pmatrix} = \begin{pmatrix} b \\ 0 \\ d \\ 0 \end{pmatrix}$$

$$\boldsymbol{A}^2\boldsymbol{B} = \begin{pmatrix} 0 & 0 & a & 0 \\ 0 & 0 & 0 & a \\ 0 & 0 & c & 0 \\ 0 & 0 & 0 & c \end{pmatrix}\begin{pmatrix} 0 \\ b \\ 0 \\ d \end{pmatrix} = \begin{pmatrix} 0 \\ ad \\ 0 \\ cd \end{pmatrix}$$

$$A^3 B = \begin{pmatrix} 0 & 0 & 0 & a \\ 0 & 0 & ac & 0 \\ 0 & 0 & 0 & c \\ 0 & 0 & c^2 & 0 \end{pmatrix} \begin{pmatrix} 0 \\ b \\ 0 \\ d \end{pmatrix} = \begin{pmatrix} ad \\ 0 \\ cd \\ 0 \end{pmatrix}$$

Thus, we test whether

$$\text{rank} \begin{pmatrix} 0 & b & 0 & ad \\ b & 0 & ad & 0 \\ 0 & d & 0 & cd \\ d & 0 & cd & 0 \end{pmatrix}$$

is equal to the system order. By inserting numbers, it is seen that the rank is indeed 4, and the system is therefore pole assignable. The constant d appears in many locations, and if d were close to zero, the determinant would be close to zero; this would occur if the mass of the cart was much larger than the mass of the ball.

Knowing that we can place the poles arbitrarily, we would like to place a set of poles in a 'good' location and work backwards to find a corresponding feedback vector **k**. The closed loop characteristic polynomial is

$$P = \det\left(s\boldsymbol{I} - \boldsymbol{A} - \boldsymbol{B}\mathbf{k}^T\right) \tag{10.23}$$

$$= \det\left(\begin{pmatrix} s & 0 & 0 & 0 \\ 0 & s & 0 & 0 \\ 0 & 0 & s & 0 \\ 0 & 0 & 0 & s \end{pmatrix} - \begin{pmatrix} 0 & 1 & 0 & 0 \\ 0 & 0 & a & 0 \\ 0 & 0 & 0 & 1 \\ 0 & 0 & c & 0 \end{pmatrix} - \begin{pmatrix} 0 \\ b \\ 0 \\ d \end{pmatrix} \begin{pmatrix} k_1 & k_2 & k_3 & k_4 \end{pmatrix}\right) \tag{10.24}$$

$$= \det \begin{pmatrix} s & -1 & 0 & 0 \\ -bk_1 & s - bk_2 & -a - bk_3 & -bk_4 \\ 0 & 0 & s & -1 \\ -dk_1 & -dk_2 & -c - dk_3 & s - dk_4 \end{pmatrix} \tag{10.25}$$

$$= s^4 + s^3(-bk_2 - dk_4) + s^2(-c - bk_1 - dk_3) \tag{10.26}$$

$$+ s(bck_2 - adk_2) + bck_1 - adk_1 \tag{10.27}$$

The characteristic polynomial can also be written in terms of the desired poles:

$$P = (s - \lambda_1)(s - \lambda_2)(s - \lambda_3)(s - \lambda_4) \tag{10.28}$$

$$= s^4 \tag{10.29}$$

$$+ s^3(-\lambda_1 - \lambda_2 - \lambda_3 - \lambda_4) \tag{10.30}$$

$$+ s^2(\lambda_1\lambda_2 + \lambda_1\lambda_3 + \lambda_1\lambda_4 + \lambda_2\lambda_3 + \lambda_2\lambda_4 + \lambda_3\lambda_4) \tag{10.31}$$

$$+ s(-\lambda_1\lambda_2\lambda_3 - \lambda_1\lambda_2\lambda_4 - \lambda_1\lambda_3\lambda_4 - \lambda_2\lambda_3\lambda_4) + \lambda_1\lambda_2\lambda_3\lambda_4 \tag{10.32}$$

Matching the coefficients in Equations (10.26) and (10.28), we have four linear equations in the unknowns $k_1, \ldots, k_4$:

$$
\begin{aligned}
-bk_2 - dk_4 &= -\lambda_1 - \lambda_2 - \lambda_3 - \lambda_4 \\
-c - bk_1 - dk_3 &= \lambda_1\lambda_2 + \lambda_1\lambda_3 + \lambda_1\lambda_4 + \lambda_2\lambda_3 + \lambda_2\lambda_4 + \lambda_3\lambda_4 \\
bck_2 - adk_2 &= -\lambda_1\lambda_2\lambda_3 - \lambda_1\lambda_2\lambda_4 - \lambda_1\lambda_3\lambda_4 - \lambda_2\lambda_3\lambda_4 \\
bck_1 - adk_1 &= \lambda_1\lambda_2\lambda_3\lambda_4
\end{aligned}
$$

or, in matrix form,

$$
\begin{pmatrix} 0 & -b & 0 & -d \\ -b & 0 & -d & 0 \\ 0 & bc-ad & 0 & 0 \\ bc-ad & 0 & 0 & 0 \end{pmatrix}
\begin{pmatrix} k_1 \\ k_2 \\ k_3 \\ k_4 \end{pmatrix}
= \begin{pmatrix} -\lambda_1 - \lambda_2 - \lambda_3 - \lambda_4 \\ \lambda_1\lambda_2 + \lambda_1\lambda_3 + \lambda_1\lambda_4 + \lambda_2\lambda_3 + \lambda_2\lambda_4 + \lambda_3\lambda_4 \\ -\lambda_1\lambda_2\lambda_3 - \lambda_1\lambda_2\lambda_4 - \lambda_1\lambda_3\lambda_4 - \lambda_2\lambda_3\lambda_4 \\ \lambda_1\lambda_2\lambda_3\lambda_4 \end{pmatrix}
+ \begin{pmatrix} 0 \\ c \\ 0 \\ 0 \end{pmatrix}
$$

The system of equations has a solution since the determinant $d^2\,(bc - ad)^2$ is non-zero (the coefficient matrix is upper triangular, and thus the determinant is the product of the diagonal elements). Given a set of desired poles $\lambda_1, \ldots, \lambda_4$, we can thus find a unique solution $k_1, \ldots, k_4$.

How do we then decide what 'good' pole locations are? As a rule of thumb, the farther away the pole is from the origin, the larger is the magnitude of the control signal required. If a pole λ is located at $\sigma + j\omega$, then, roughly speaking, ω affects the frequency of the response and σ affects the damping. Thus, poles on the real axis cause purely exponential terms in the impulse response, and complex poles may cause sinusoidal components. For a stable system, the closed-loop poles must be in the left half of the complex plane. Poles at the origin are allowed, but should be restricted in number; a single pole at the origin generates a constant term in the impulse response, two poles generate a ramp and a constant term, and so on. Poles on the imaginary axis are not desirable, since they generate undamped sinusoidal components.

A simple way to select the poles is to use prototype response poles (Franklin *et al.* 1991, p. 388, Table 6.1b),

$$
\begin{aligned}
\lambda_1 &= -0.6573 - j0.8302 \\
\lambda_2 &= -0.6573 + j0.8302 \\
\lambda_3 &= -0.9047 - j0.2711 \\
\lambda_4 &= -0.9047 + j0.2711
\end{aligned}
$$

These refer to a prototype response of a fourth-order system with little overshoot. The four poles can be multiplied by a constant w in order to obtain some variation in the speed of the response. An initial set of gains is

$$k = \begin{bmatrix} 0.26 & 0.84 & 48.6 & 8.74 \end{bmatrix}^T \qquad (w = 1)$$

These are rather low values, compared to the ones we used previously in Equation (10.16), and consequently the control action will be rather low in magnitude. A faster response can be obtained by scaling the prototype eigenvalues to $\{w\lambda_1, w\lambda_2, w\lambda_3, w\lambda_4\}$ $(w > 1)$. In practice the obtainable speed of the response is limited by the limits of the amplifier (± 13 V), and w should be in the range $1 < w < 5$.

10.7.1 Nonlinear Equations

The motor has friction between the brushes and commutator as well as in the bearings. There is also friction in the drum grooves, where the transmission wire is seated. Both static and dynamic friction must be expected, and sustained limit cycles in closed loop are a clear indication. The linear model ignores friction.

As a first approximation we may model friction as a force proportional to the velocity of the cart in the direction opposite to the driving force. In the equations we may then apply, instead of the driving force F, a reduced force

$$F - fy \tag{10.33}$$

Here, f is a physical constant that models the magnitude of the frictional effect. The constant f can be measured at constant velocity ($F = 0$). We are, however, just interested in discovering how friction affects the system matrix $\boldsymbol{A}$.

Substitution of (10.33) for F in (10.13) results in two new non-zero terms in $\boldsymbol{A}$,

$$\boldsymbol{A}(2, 2) = -bf$$

$$\boldsymbol{A}(4, 2) = -df$$

As a result, one eigenvalue will remain zero, whereas the other three will be non-zero.

Another question is the magnitude of the error introduced by the linearization of Equation (10.11). If we instead refrain from using linearization, and rearranging Equations (10.8) and (10.9), we obtain the nonlinear state-space equations:

$$\dot{x}_1 = x_2 \tag{10.34}$$

$$\dot{x}_2 = -\frac{m^2r^2g\cos x_3 \sin x_3 - \left(RmI + mrI + m^2r^3 + Rm^2r^2\right)x_4^2 \sin x_3}{MI + mI + mr^2M + m^2r^2\sin^2 x_3} \tag{10.35}$$

$$+ \frac{mr^2 + I}{MI + mI + mr^2M + m^2r^2\sin^2 x_3}F \tag{10.36}$$

$$\dot{x}_3 = x_4 \tag{10.37}$$

$$\dot{x}_4 = \frac{mr^2g\,(M + m)\sin x_3 - m^2r^2\,(R + r)\,x_4^2\cos x_3 \sin x_3}{(R + r)\left(MI + mI + mr^2M + r^2m^2\sin^2 x_3\right)} \tag{10.38}$$

$$- \frac{mr^2\cos x_3}{(R + r)\left(MI + mI + mr^2M + r^2m^2\sin^2 x_3\right)}F \tag{10.39}$$

A simulation using the nonlinear state Equations (10.36) rather than the linear Equations (10.11) showed that the difference in the peak values of a response is less than 3%.

10.8 Notes and References*

The CSTR model is implemented based on the equations in the recommendable book by Bequette (2003), which contains a wealth of illustrative practical problems and solutions from the process industry.

The background for the idle speed example is the Ford Advanced Project on Internal Engine Idle Speed Control (Maclay and Dorey 1995). Different control techniques – including classical, fuzzy, H-infinity, predictive, and variable structure methods – were implemented by control specialists from British universities – and one Danish – on a Ford Mondeo engine. The controllers were assessed for their transient and steady-state performance, sensitivity to load disturbances, etc. There is a publication from this project related to the sliding mode design (Bhatti, Spurgeon, Dorey, and Edwards 1999). For an overview of approaches to idle speed control, see the article by Ye (2007)

The ball-balancer was built for teaching students of electrical engineering about automatic control, originally with a focus on state-space control theory. It is educational, because the laboratory rig is sufficiently slow for visual inspection of different control strategies, and the mathematical model is sufficiently complex to be challenging. The system was built during an MSc project, and later the mathematical model was published in an educational journal (Jørgensen 1974). A simulator of the same system was built in MATLAB® much later for a course on the Internet (Jantzen 2003).

The example concerning the model of a dynamic system was inspired by an example in the book by Babuška (1998). His book uses a more advanced clustering algorithm (the Gustafson and Kessel algorithm), and it is interesting to compare the approaches.

There are further control design examples, which could be candidates for future demonstration examples, in the collection of Simulink control design examples from MathWorks.

References

Aracil J and Gordillo F 2004 Describing function method for stability analysis of PD and PI fuzzy controllers. *Fuzzy Sets and Systems* **143**, 233–249.

Assilian S and Mamdani E 1974a Learning control algorithms in real dynamic systems. *Proc. Fourth Int. Conf. On Digital Computer Applications to Process Control, Zürich*, pp. 13–20 IFAC/IFIP. Springer.

Assilian S and Mamdani EH 1974b An experiment in linguistic synthesis with a fuzzy logic controller. *Int. J. Man Machine Studies* **7**(1), 1–13.

Åström KJ and Hägglund T 2006 *Advanced PID Control*. Instrumentation, Systems, and Automation Society (ISA), ISA, 67 Alexander Drive, PO Box 12277, Research Triangle Park, North Carolina 27709, USA.

Åström KJ and Wittenmark B 1984 *Computer controlled systems – theory and design*. Prentice-Hall.

Åström KJ and Wittenmark B 1995 *Adaptive Control* 2nd edn. Addison-Wesley.

Atherton D 2011 *An Introduction to Nonlinearity in Control Systems*. Ventus Publishing ApS, Available from: bookboon.com [cited 17 Feb 2013].

Atherton DP 1975 *Nonlinear Control Engineering* unabridged edn. Van Nostrand Reinhold Company.

Atherton DP 1982 *Nonlinear Control Engineering* student edn. Van Nostrand Reinhold Company.

Babuška R 1998 *Fuzzy Modeling For Control*. Kluwer Academic Publishers.

Babuška R 1999 An overview of fuzzy modelling and model-based fuzzy control In *Fuzzy Logic Control: Advances in Applications* (ed. Verbruggen HB and Babuška R) Robotics and Intelligent Systems – Vol 23 World Scientific pp. 3–35. ISBN 981-02-3825-8.

Bennett S 1993 Development of the PID controller. *IEEE Control Systems* **13**(5), 58–65.

Bequette B 2003 *Process Control: Modeling, Design, and Simulation* PTR. Prentice Hall.

Bezdek J and Pal SK 1992 *Fuzzy models for pattern recognition*. IEEE Press, New York. (Selected reprints).

Bhatti A, Spurgeon, SK DR, and Edwards C 1999 Sliding mode configurations for automotive engine control. *International Journal of Adaptive Control and Signal Processing* **13**, 49–69.

Braae M and Rutherford D 1979a Selection of parameters for a fuzzy logic controller. *Fuzzy Sets and Systems* **2**, 185–199.

Braae M and Rutherford D 1979b Theoretical and linguistic aspects of the fuzzy logic controller. *Automatica* **15**, 553–577.

Cohen G and Coon G 1953 Theoretical consideration of retarded control. *Trans. of American society of Mechanical Engineers, ASME* **75**, 827–834.

Cominos P and Munro N 2002 PID controllers: recent tuning methods and design to specification. *IEE Proceedings – Control Theory And Applications* **149**(1), 46–53.

Cuesta F, Gordillo F, Aracil J, and Ollero A 1999 Stability analysis of nonlinear multivariable Takagi-Sugeno fuzzy control systems. *IEEE Transactions on Fuzzy Systems* **7**(5), 508–520.

DiStefano J, Stubberud A, and Williams I 1995 *Schaum's Outline of Theory and Problems of Feedback and Control Systems* Schaum's Outline Series 2nd edn. McGraw-Hill.

Driankov D, Hellendoorn H, and Reinfrank M 1996 *An introduction to fuzzy control* 2nd edn. Springer-Verlag.

Duda RO, Hart PE, and Stork DG 2001 *Pattern Classification* 2. edn. Wiley-Interscience.

Foundations of Fuzzy Control: A Practical Approach, Second Edition. Jan Jantzen.

Edwards C and Spurgeon SK 1998 *Sliding Mode Control: Theory and Applications* Systems and Control Book Series. Taylor & Francis.

Farinwata SS, Filev D, and Langari R (eds.) 2000 *Fuzzy Control: Synthesis And Analysis*. Wiley.

Franklin GF, Powell JD, and Emami-Naeini A 1991 *Feedback Control of Dynamic Systems* Electrical and Computer Engineering: Control Engineering 2nd edn. Addison-Wesley.

Franksen OI 1979 Group representation of finite polyvalent logic In *Proceedings 7th IFAC Triennial World Congress, Helsinki* (ed. Niemi A) International Federation of Automatic Control, IFAC. Pergamon Press.

Fukami S, Mizumoto M, and Tanaka K 1980 Some considerations of fuzzy conditional inference. *Fuzzy Sets and Systems* **4**, 243–273.

Gelb A and Vander Velde WE 1968 *Multiple-Input Describing Functions and Nonlinear System Design*. McGraw-Hill, Also available from: http://ocw.mit.edu [cited 17 Feb 2013].

Gordillo F, Aracil J, and Álamo T 1997 Determining limit cycles in fuzzy control systems *Proceedings of 6th International Fuzzy Systems Conference*, vol. 1, pp. 193–198. IEEE.

Gupta MM and Sinha NK (eds.) 1996 *Intelligent Control Systems: Theory and practice*. IEEE Press.

Guzman J, Åström K, Dormido S, Hägglund T, Berenguel M, and Piguet Y 2008 Interactive learning modules for PID control. *IEEE Control Systems* **28**(5), 118–134.

Gyöngy IJ and Clarke DW 2006 On the automatic tuning and adaptation of PID controllers. *Control Engineering Practice* **14**, 149–163.

Hájek P 1998 *Metamathematics of Fuzzy Logic* Trends in Logic. Kluwer.

Hametner C and Jakubek S 2013 Local model network identification for online engine modelling. *Information Sciences* **220**, 210–225.

Haykin S 2009 *Neural Networks and Learning Machines* 3rd edn. Pearson Education.

Hendricks E and Sorenson S 1990 Mean value modelling of spark ignition engines. *SAE technical paper 900616*.

Holmblad LP and Østergaard JJ 1982 Control of a cement kiln by fuzzy logic In *Fuzzy Information and Decision Processes* (ed. Gupta and Sanchez) North-Holland Amsterdam pp. 389–399. (Reprint in: FLS Review No 67, FLS Automation A/S, Høffdingsvej 77, DK-2500 Valby, Copenhagen, Denmark).

Holmblad LP and Østergaard JJ 1995 The FLS application of fuzzy logic. *Fuzzy Sets and Systems* **70**, 135–146.

IEC 2000 Programmable controllers – part 7: Fuzzy control programming. Technical Report IEC 61131, International Electrotechnical Ccommission (IEC). Draft available from http://www.fuzzytech.com/binaries/ieccd1.pdf [cited on 17 Feb 2013].

Isaksson A and Hägglund T (eds.) 2002 *Special section on PID Control* vol. 149.

Jakubek S, Hametner C, and Keuth N 2008 Total least squares in fuzzy system identification: An application to an industrial engine. *Engineering Applications of Artificial Intelligence* **21**, 1277–1288.

Jang JSR and Sun CT 1995 Neuro-fuzzy modeling and control In *Proceedings of the IEEE* (ed. 3), **83**, 378–406.

Jang JSR, Sun CT, and Mizutani E 1997 *Neuro-Fuzzy and Soft Computing* MATLAB Curriculum Series. Prentice Hall.

Jantzen J 1995 Array approach to fuzzy logic. *Fuzzy Sets and Systems* **70**, 359–370.

Jantzen J 2003 Internet learning in control: A fuzzy control course In *Prepr. ACE 2003, The 6th IFAC Symposium on Advances in Control Education* (ed. Lindfors J), pp. 27–33. IFAC.

Jantzen J, Verbruggen H, and Østergaard JJ 1999 Fuzzy control in the process industry: Common practice and challenging perspectives In *Practical Applications of Fuzzy Technologies* (ed. Zimmermann HJ) Dubois and Prade (Eds), The Handbooks of Fuzzy Sets Series Kluwer chapter 1, pp. 3–56.

Jespersen T 1981 Self-organizing fuzzy logic control of a pH-neutralisation process. Technical Report 8102, Electric Power Eng. Dept., Technical University of Denmark.

Jørgensen V 1974 A ball-balancing system for demonstration of basic concepts in the state-space control theory. *Int. J. Elect. Enging Educ.* **11**, 367–376.

Kickert W and Mamdani E 1978 Analysis of a fuzzy logic controller. *Fuzzy Sets and Systems* **1**, 29–44.

Kickert WJM and Van Nauta Lemke HR 1976 Application of a fuzzy controller in a warm water plant. *Automatica* **12**(4), 301–308.

Kiszka JB, Kochanska ME, and Sliwinska DS 1985 The influence of some fuzzy implication operators on the accuracy of a fuzzy model. *Fuzzy Sets and Systems* **15**, (Part1) 111–128; (Part 2) 223–240.

Kosko B 1992 *Neural Networks and Fuzzy Systems. A Dynamical Systems Approach to Machine Intelligence*. Prentice-Hall.

Larsen PM 1981 Industrial applications of fuzzy logic control In *Fuzzy Reasoning and its Applications* (ed. Mamdani EH and Gaines BR) Academic Press London pp. 335–342.

Lee CC 1990 Fuzzy logic in control systems: Fuzzy logic controller. *IEEE Trans. Systems, Man & Cybernetics* **20**(2), 404–435.

Lewis R 1990 *Practical Digital Image Processing* Ellis Horwood Series in Digital and Signal Processing. Ellis Horwood Ltd, New York, etc.

Li HX and Gatland HB 1995 A new methodology for designing a fuzzy logic controller. *IEEE Trans. Systems, Man & Cybernetics* **25**(3), 505–512.

Lin CT and Lee CSG 1996 *Neural Fuzzy Systems: A Neuro-Fuzzy Synergism to Intelligent Systems*. Prentice Hall PTR.

Luenberger DG 1969 *Optimization by Vector Space Methods* Series in Decision and Control. Wiley.

Maclay D and Dorey R 1995 A controller and design implementation environment for the idle speed control of an internal combustion engine. *IEE Colloquium Digest* **14**, 1–3.

Mamdani E and Baaklini N 1975 Prescriptive method for deriving control policy in a fuzzy-logic controller. *Electronics Letters* **11**(25/26), 625–626.

Mamdani EH 1977 Application of fuzzy logic to approximate reasoning using linguistic synthesis. *IEEE Transactions on Computers* **C-26**(12), 1182–1191.

MathWorks 2012 *Fuzzy Logic Toolbox for Use with Matlab: User's Guide*. online edn The MathWorks Inc. Available from www.mathworks.se [cited 22 Jul 2012].

Michels K, Klawonn F, Kruse R, and Nürnberger A 2006 *Fuzzy Control: Fundamentals, Stability and Design of Fuzzy Controllers*. Springer.

Mizumoto M 1992 Realization of PID controls by fuzzy control methods *First Int. Conf. on Fuzzy Systems*, pp. 709–715 The Institute of Electrical and Electronics Engineers, Inc, San Diego.

Mizumoto M 1995 Realization of PID controls by fuzzy control methods. *Fuzzy Sets and Systems* **70**, 171–182.

Mizumoto M, Fukami S, and Tanaka K 1979 Some methods of fuzzy reasoning In *Advances in Fuzzy Set Theory Applications* (ed. Gupta, Ragade, and Yager) North-Holland, New York.

Møller G 1986 A logic programming tool for qualitative system design. *APL Quote Quad (APL86 conference proceedings)* **16**(4), 266–271.

Møller GL 1998 *On the Technology of Array-Based Logic* PhD thesis Technical University of Denmark, Electric Power Engineering Dept., DK-2800 Lyngby, Denmark (2nd ed.).

Murakami S, Takemoto F, Fulimura H, and Ide E 1989 Weld-line tracking control of arc welding robot using fuzzy logic controller. *Fuzzy Sets and Systems* **32**(2), 221–237.

Nauck D, Klawonn F, and Kruse R 1997 *Foundations of Neuro-Fuzzy Systems*. John Wiley and Sons.

Nelles O 2001 *Nonlinear System Identification*. Springer-Verlag.

Nguyen HT and Walker EA 2000 *A first course in fuzzy logic* 2nd edn. Chapman & Hall, New York.

Nise N 1995 *Control Systems Engineering* 2nd edn. Benjamin/Cummings.

Østergaard JJ 1977 Fuzzy logic control of a heat exchanger system In *Fuzzy Automata and Decision Processes* (ed. Gupta MM, Saridis GN, and Gaines BR) North-Holland Amsterdam pp. 285–320.

Østergaard JJ 1990 Fuzzy II: The new generation of high level kiln control. *Zement Kalk Gips (Cement-Lime-Gypsum)* **43**(11), 539–541.

Østergaard JJ 1996 High level control of industrial processes In *Proc. TOOLMET '96* (ed. Yliniemi L and Juuso E), pp. 1–12. University of Oulu, Control Engineering Laboratory, Linnanmaa, FIN-90570 Oulu, Finland.

Palm R, Driankov D, and Hellendoorn H 1997 *Model Based Fuzzy Control*. Springer.

Passino KM and Yurkovich S 1998 *Fuzzy Control*. Addison Wesley Longman, Inc.

Pedrycz W 1993 *Fuzzy control and fuzzy systems* 2nd edn. Wiley and Sons.

Precup RE and Hellendoorn H 2011 A survey on industrial applications of fuzzy control. *Computers in Industry* **62**, 213–226.

Procyk TJ and Mamdani EH 1979 A linguistic self-organizing process controller. *Automatica* **15**, 15–30.

Qiao W and Mizumoto M 1996 PID type fuzzy controller and parameters adaptive method. *Fuzzy Sets and Systems* **78**, 23–35.

Ross T 2010 *Fuzzy Logic with Engineering Applications* 3rd edn. Wiley.

Rugh W and Shamma J 2000 Research on gain scheduling. *Automatica* **36**, 1401–1425.

Rundqwist L 1991 Anti-reset windup for PID controllers In *Proc. 11th triennial world congress of the International Federation of Automatic Control, IFAC* (ed. Jaakso and Utkin), pp. 453–458. Pergamon Press.

Sala A, Guerra TM, and Babuška R 2005 Perspectives of fuzzy systems and control. *Fuzzy Sets and Systems* **156**, 432–444.

Self K 1990 Designing with fuzzy logic. *IEEE Spectrum* **27**(11), 42–44 + 105.

Siler W and Ying H 1989 Fuzzy control theory: The linear case. *Fuzzy Sets and Systems* **33**, 275–290.

Šiljak D 1968 *Nonlinear Systems: The Parameter Analysis and Design*. John Wiley & Sons.

Slotine JJE and Li W 1991 *Applied Nonlinear Control*. Prentice Hall.

Smith LC 1979 Fundamentals of control theory. *Chemical Engineering* **86**(22), 11–39. (Deskbook issue).

Stoll RR 1979 *Set Theory and Logic* dover edn. Dover Publications, New York. (org 1963).

Sugeno M (ed.) 1985 *Industrial applications of fuzzy control*. North-Holland.

Sugeno M, Murofushi T, Mori T, Tatematsu T, and Tanaka J 1989 Fuzzy algorithmic control of a model car by oral instructions. *Fuzzy Sets and Systems* **32**(2), 207–219.

Takagi T and Sugeno M 1985 Fuzzy identification of systems and its applications to modeling and control. *IEEE Trans. Systems, Man & Cybernetics* **15**(1), 116–132.

Tanaka K, Sano M, and Suzuki K 1991 A new tuning method of fuzzy controllers *Proc. IFSA91*, pp. 207–210. IFSA.

Tso SK and Fung YH 1997 Methodological development of fuzzy-logic controllers from multivariable linear control. *IEEE Trans. Systems, Man & Cybernetics* **27**(3), 566–572.

von Altrock C 1995 *Fuzzy Logic and Neurofuzzy Applications Explained*. Prentice Hall.

von Altrock C 1996 *Fuzzy Logic and Neurofuzzy Applications In Business And Finance*. Prentice Hall PTR.

Wang LX 1997 *A Course in Fuzzy Systems and Control* international edn. Prentice Hall PTR.

Wenstøp F 1980 Quantitative analysis with linguistic values. *Fuzzy Sets and Systems* **4**(2), 99–115.

Yamakawa T and Miki T 1986 The current mode fuzzy logic integrated circuits fabricated by the standard CMOS process. *IEEE Trans. Computers* **35**(2), 161–167.

Yamazaki T 1982 *An improved algorithm for a self-organising controller and its experimental analysis* PhD thesis Queen Mary College, London Dept. of Electrical and Electronic Engineering.

Yamazaki T and Mamdani EH 1982 On the performance of a rule-based self-organizing controller *Proc. IEEE Conf on Applications of Adaptive and Multivariable Control*, Hull.

Yasunobu S, Miyamoto S, and Ihara H 1983 Fuzzy control for automatic train operation system *Proc. Int. Congress on Control in Transportation Systems* IFAC/IFIP/IFORS, Baden-Baden.

Yazdi H 1997 *Control and Supervision of Event-Driven Systems* PhD thesis Technical University of Denmark Dept.

Ye Z 2007 Modeling, identification, design, and implementation of nonlinear automotive idle speed control systems: An overview. *IEEE Transactions on Systems, Man, and Cybernetics, Part C: Applications and Reviews*.

Zadeh L 1994 Soft computing and fuzzy logic. *IEEE Software* **11**(6), 48–56.

Zadeh LA 1965 Fuzzy sets. *Inf. and Control* **8**, 338–353.

Zadeh LA 1973 Outline of a new approach to the analysis of complex systems and decision processes. *IEEE Trans. Systems, Man & Cybernetics* **1**, 28–44.

Zadeh LA 1975 The concept of a linguistic variable and its application to approximate reasoning. *Information Sciences* **8**, 43–80.

Zadeh LA 1984 Making computers think like people. *IEEE Spectrum* **21**, 26–32.

Zadeh LA 1988 Fuzzy logic. *IEEE Computer* **21**(4), 83–93.

Ziegler J and Nichols N 1942 Optimum settings for automatic controllers. *Transactions of the American Society of Mechanical Engineers (ASME)* **64**, 759–768.

Ziegler J and Nichols N 1943 Process lags in automatic-control circuits. *Transactions of the American Society of Mechanical Engineers (ASME)* **65**, 433–444.

Zimmermann HJ 1993 *Fuzzy set theory – and its applications* 2nd edn. Kluwer, Boston.

Zimmermann HJ (ed.) 1999 *Practical Applications of Fuzzy Technologies* The Handbooks of Fuzzy Sets. Kluwer.

Index

Foundations of Fuzzy Control: A Practical Approach, Second Edition. Jan Jantzen.